Catalan Numbers

Selected Chapters of Number Theory: Special Numbers

Print ISSN: 2810-9716
Online ISSN: 2810-9724

Series Editor: Elena Deza (*Moscow Pedagogical State University, Russia*)

Published

Vol. 4 *Catalan Numbers*
    by Elena Deza

Vol. 3 *Stirling Numbers*
    by Elena Deza

Vol. 2 *Perfect and Amicable Numbers*
    by Elena Deza

Vol. 1 *Mersenne Numbers and Fermat Numbers*
    by Elena Deza

Selected Chapters of Number Theory: Special Numbers – Volume 4

Catalan Numbers

Elena Deza
Moscow Pedagogical State University, Russia

NEW JERSEY · LONDON · SINGAPORE · BEIJING · SHANGHAI · HONG KONG · TAIPEI · CHENNAI · TOKYO

Published by

World Scientific Publishing Co. Pte. Ltd.
5 Toh Tuck Link, Singapore 596224
USA office: 27 Warren Street, Suite 401-402, Hackensack, NJ 07601
UK office: 57 Shelton Street, Covent Garden, London WC2H 9HE

Library of Congress Cataloging-in-Publication Data
Names: Deza, Elena, author.
Title: Catalan numbers / Elena Deza.
Description: New Jersey : World Scientific, [2025] | Series: Selected chapters of number theory :
　　special numbers ; 2810-9716 ; volume 4 | Includes bibliographical references and index.
Identifiers: LCCN 2024024468 | ISBN 9789811293221 (hardcover) |
　　ISBN 9789811293238 (ebook for institutions) | ISBN 9789811293245 (ebook for individuals)
Subjects: LCSH: Catalan numbers (Mathematics)
Classification: LCC QA164 .D495 2025 | DDC 511.3/5--dc23/eng/20241029
LC record available at https://lccn.loc.gov/2024024468

British Library Cataloguing-in-Publication Data
A catalogue record for this book is available from the British Library.

Copyright © 2025 by World Scientific Publishing Co. Pte. Ltd.

All rights reserved. This book, or parts thereof, may not be reproduced in any form or by any means, electronic or mechanical, including photocopying, recording or any information storage and retrieval system now known or to be invented, without written permission from the publisher.

For photocopying of material in this volume, please pay a copying fee through the Copyright Clearance Center, Inc., 222 Rosewood Drive, Danvers, MA 01923, USA. In this case permission to photocopy is not required from the publisher.

For any available supplementary material, please visit
https://www.worldscientific.com/worldscibooks/10.1142/13840#t=suppl

Desk Editors: Balasubramanian Shanmugam/Rosie Williamson/Shi Ying Koe

Typeset by Stallion Press
Email: enquiries@stallionpress.com

Mathematics is the queen of the sciences and number theory is the queen of mathematics.

— Gauss

God invented the integers; all else is the work of man.

— Kronecker

I know numbers are beautiful. If they aren't beautiful, nothing is.

— Erdős

Preface

Special numbers, among integers, are an important part of number theory, general mathematics, and several applied areas (such as cryptography).

While many of the "names" of these numbers (Fermat numbers, Mersenne numbers, Fibonacci numbers, Catalan numbers, Bernoulli numbers, Stirling numbers, etc.) are known to every mathematician, detailed information about them is often scattered throughout the special numbers literature. There are only a few books which systematically present their history and properties while giving the main proofs.

Moreover, even though such books exist (see, for example, [Bond93], [CoGu96], [Hogg69], [Line86], [LiNi96], [KLS01], [SlPl95], [Uspe76], [Voro61]), many new results have appeared during the several decades since their publication. So, this material needs to be updated and reworked.

We should also mention several classical monographs on number theory ([Apos86], [Dick05], [HaWr79], [MSC96], [Ribe96], [Sier64], etc.) and combinatorics ([Comt74], [FlSe09], [GKP94] [GGL95Û] [Hall70], [Knut97], [LiWi92], [Mazu10], [Rior80], [Stan97], etc.) These are basic books of fundamental importance. But they treat number theory and combinatorics as a whole, and special numbers receive only limited attention.

Our first book on special numbers, the monograph *Figurate Numbers* [DeDe12] identifies all such problems for polygonal numbers and their different generalizations, giving a full retrospective of the history of the most commonly known class of special integers.

In the series *Selected Chapters of Number Theory: Special Numbers*, we consider, following a similar structure, the history and properties of several other famous classes of special numbers. The books offered in the series are: *Mersenne Numbers and Fermat Numbers*, *Perfect and Amicable Numbers*, *Stirling Numbers*, *Catalan Numbers*, *Bernoulli Numbers*, and *Euler Numbers*.

The main purpose of the series is to give a complete presentation of the theory of several classes of classical special numbers, considering many of their properties, facts, and theorems with full proofs.

In particular, we expect to:

- find and organize much of the scattered material;
- present updated material in a clear, fully detailed, and unified way;
- consider the complete range of well-known and hidden connections of a given set number with different mathematical problems;
- draw up a system of multilevel tasks.

The first book of the series, *Mersenne Numbers and Fermat Numbers*, was published in 2021 [Deza21].

Mersenne numbers, i.e., positive integers of the form $\mathcal{M}_n = 2^n - 1$, $n \in \mathbb{N}$, were considered by Euclid (4th century BC) and later by many Arabic and European mathematicians as an attempt to find all perfect numbers. The numbers of the form $2^n - 1$ were named after Marin Mersenne (1588–1648), a French monk and mathematician who discussed them in his work *Cogita Physico Mathematics* (1644) and stated some conjectures about the numbers' occurrence.

A *Fermat number* F_n is a positive integer of the form $F_n = 2^{2^n} + 1$, $n \in \mathbb{Z}$, $n \geq 0$.

Fermat numbers were studied by the 17th-century French lawyer and mathematician Pierre de Fermat (1607–1665), who conjectured (1650) that these numbers would always give a prime for $n = 0, 1, 2, \ldots$. In fact, the first five Fermat numbers are primes:

$F_0 = 3$, $F_1 = 5$, $F_2 = 17$, $F_3 = 257$, and $F_4 = 65537$. But in 1732, Leonhard Euler (1707–1783), a Swiss-Russian mathematician, physicist, astronomer, geographer, logician, and engineer, found that F_5 is a composite number. So, Fermat's conjecture that F_n is always prime is clearly false. Moreover, no further Fermat primes beyond F_4 have been found.

Nowadays, the theory of Mersenne and Fermat numbers contains many interesting mathematical facts and theorems, as well as significant applications. The mathematical part of this theory is closely connected with classical arithmetic and number theory. It contains much information about the divisibility properties of Mersenne and Fermat numbers, their connections with other classes of special numbers, etc. Moreover, Mersenne and Fermat numbers are used in the construction of new large primes and have numerous applications in contemporary cryptography.

The second book of the series, published in 2023 [Deza23], was devoted to perfect and amicable numbers, as well as their numerous analogs and generalizations.

A *perfect number* is a positive integer n which is equal to the sum of its positive divisors: $n = \sum_{d|n, d<n} d$. The first four perfect numbers are 6, 28, 496, and 8128.

Perfect numbers were studied by Pythagoras and his followers, more for their mystical properties than for their number-theoretic properties. The first recorded mathematical result concerning perfect numbers occurs in Euclid's *Elements*, written around 300 BC. Euclid proved that all even perfect numbers have the form $2^{k-1}(2^k - 1)$, where $2^k - 1$ is a prime. Two millennia later, L. Euler proved the converse of Euclid's result by proving that every even perfect number has to be of this form. So, the *Euclid–Euler theorem* states that *an even positive integer is a perfect number if and only if it can be represented as a product, $2^{k-1}(2^k - 1)$, where $2^k - 1$ is a prime.* In fact, it is a Mersenne prime; therefore, there exists a one-to-one relationship between even perfect numbers and Mersenne primes; each Mersenne prime generates one even perfect number, and vice versa.

As for odd perfect numbers, it is unknown whether any exist, though various results have been obtained.

Amicable numbers are two different natural numbers (m, n) related in such a way that the sum of the proper divisors of each is equal to the other number: $n = \sum_{d|m, d<m} d$, and $m = \sum_{d|n, d<n} d$. The smallest pair of amicable numbers, $(220, 284)$, was known to the Pythagoreans, who credited them with many mystical properties. A general formula by which some of these numbers could be derived was invented circa 850 by the Iraqi mathematician Thābit ibn Qurra. *Thābit's rule* states that *if $p = 3 \cdot 2^{n-1} - 1$, $q = 3 \cdot 2^n - 1$, and $r = 9 \cdot 2^{2n-1} - 1$ are primes, then $A = 2^n pq$ and $B = 2^n r$ are a pair of amicable numbers*. This formula gives the pairs $(220, 284)$ for $n = 2$, $(17296, 18416)$ for $n = 4$, and $(9363584, 9437056)$ for $n = 7$, but no other such pairs are known.

The third book of the series, published in 2024 [Deza24], was devoted to Stirling numbers as well as their analogs and generalizations. There are two classes of Stirling numbers. *Stirling numbers of the second kind* $S(n, k)$ count the number of partitions of a given n-set into k non-empty subsets, for $n = 1, 2, \ldots$, $k = 1, 2, \ldots, n$. *Stirling numbers of the first kind* $s(n, k)$ are defined by the equality $s(n, k) = (-1)^{n+k}|s(n, k)|$, where $|s(n, k)|$ is the number of ways of decomposing a given n-set into k cycles, for $n = 1, 2, \ldots, k = 1, 2, \ldots, n$. Using Stirling numbers of the second kind, we can express the ordinary powers x^n of x through the falling powers $x^{\underline{k}}$ of x: $x^n = \sum_{k=0}^n S(n, k) x^{\underline{k}}$. Conversely, using Stirling numbers of the first kind, we can express the falling factorials (falling powers) $x^{\underline{n}}$ of x through the ordinary powers x^k of x: $x^{\underline{n}} = \sum_{k=0}^n s(n, k) x^k$. These two sets of special numbers have a combinatorial nature. The construction of Stirling numbers, many of their properties, and their main applications are closely connected with classical problems in combinatorics.

This present book, the fourth in the series, is devoted to Catalan numbers, their relatives, and their generalizations. This number set is not as famous as the classes of special numbers outlined above, but it has enormous combinatorial applications, many interesting properties, and different connections with other well-known sets of positive integers.

The book deals with all these aspects in detail.

In Chapter 1, we consider some important questions which will be used later. In this chapter of the book, we collect the main combinatorial configurations and the formulas for their calculations, the binomial and multinomial theorems, classical combinatorial identities, the basic notions of graph theory, the main facts of the theory of set partitions, as well as the basic definitions and statements of the theory of permutations, the main facts of the theory of recurrent equations and generating functions, and the basic notations related to asymptotic formulas. We present all the necessary definitions, give many examples, and collect the main properties of the considered mathematical objects, mainly without proofs but with a large list of possible references.

Chapter 2 collects the main facts about *Pascal's triangle*. Starting with the history of the topic, we consider the recurrent construction of this famous number triangle and collect its elementary properties; we then reveal the connection of the constructed triangle with the numbers C_n^k of k-combinations of an n-set, as well as with binomial coefficients. Moreover, we give the closed formula for the elements of Pascal's triangle, construct the generating functions of certain sequences containing binomial coefficients, and present a large collection of properties of the triangle, including the question of divisibility of its elements. We also discuss various connections between Pascal's triangle and other classes of special numbers, including Catalan numbers.

Chapter 3 contains a rich review of the mathematical theory of *Catalan numbers*, with all the necessary proofs and many examples. Starting with the history of the subject, we consider the recurrent definition of these numbers and collect the most important counting problems related to Catalan numbers. Moreover, we give the closed formula for the nth Catalan number, construct the generating function of the sequence of Catalan numbers, list many properties of this number set, and present a large collection of possible connections of Catalan numbers with other classes of special numbers, including the elements of Pascal's triangle, Stirling numbers of the second kind, the factorial numbers, and Mersenne numbers.

Chapter 4 contains some important questions concerning the well-known relatives of Catalan numbers. We present here a detailed review of the theory of Motzkin numbers, the closest relatives of Catalan numbers. We list and prove the main properties of Schröder numbers, Delannoy numbers, and Narayana numbers, which are also closely connected with Catalan numbers. For the convenience of the reader, we give a short abstract of the properties of Stirling numbers, Bell numbers, and factorial numbers, which are discussed in detail in [Deza24].

In a small Chapter 5, we construct the base of the theory of Catalan numbers, as well as Motzkin numbers, Schröder numbers, Delannoy numbers, Narayana numbers, and binomial coefficients, starting with a common approach: any class of these special numbers is defined as the number of certain paths on the integer lattice $\mathbb{Z}^2$.

Chapter 6 collects some remarkable individual numbers related to the theory of Catalan numbers, as well as to the general theory of combinatorial special numbers.

In Chapter 7, a "Mini Dictionary" lists all the classes of special numbers mentioned in the text.

Chapter 8, "Exercises," contains an exclusive collection of problems connected with the theory of Catalan numbers, either with full solutions or, at least, with short hints of proofs.

The target audience of the book consists of university professors and students (especially at the graduate level) interested in number theory, combinatorics, general algebra, and related fields, as well as school teachers and a general audience interested in mathematics at an amateur level.

Specifically, the recurrent and combinatorial constructions of all considered number sets are accessible to undergraduate students, while some properties are slightly more difficult as they require some (still basic) mathematical and computational background.

The book is organized in such a way that it can be used as a source material for the individual scientific work of university students and even, partially, school students. In fact, the material in the book has already been used for the preparation of bachelor's and master's theses by many students of Moscow Pedagogical State University.

About the Author

Elena (Ivanovna) Deza is a full professor at Moscow Pedagogical State University, where she has been since 1988. Her research focuses on analytic number theory, discrete mathematics, the theory of discrete metric spaces, and problems of mathematical education. She has written over 50 research papers, several mathematical textbooks and books on teaching mathematics in Russian, and several books/monographs in English, including *Figurate Numbers* (2012), *Mersenne Numbers and Fermat Numbers* (2021), *Perfect and Amicable Numbers* (2023), and *Stirling Numbers* (2024). Her mathematical teaching publications are mainly published in the following leading Russian journals: *Mathematics in School*, *Kvant*, and the *Herald of the University of Russian Academy of Education*. One of her special courses at the Moscow Pedagogical State and Moscow Independent Universities is "Topics in Special Numbers".

Contents

Preface	vii
About the Author	xv
Notations	xix

1 Preliminaries — 1
- 1.1 Combinatorial Configurations 1
- 1.2 Graphs 14
- 1.3 Partitions 21
- 1.4 Permutations 31
- 1.5 Recurrent Relations 39
- 1.6 Generating Functions 44
- 1.7 Asymptotic Formulas 52

2 Pascal's Triangle — 61
- 2.1 Construction of Pascal's Triangle 61
- 2.2 Pascal's Triangle and Binomial Coefficients 72
- 2.3 Closed Formula for Elements of Pascal's Triangle . . . 75
- 2.4 Generating Functions of the Sequences of Elements of Pascal's Triangle 78
- 2.5 Properties of Pascal's Triangle 80
- 2.6 Polynomials Related to Pascal's Triangle 101
- 2.7 Pascal's Triangle in the Family of Special Numbers . . 107

3 Catalan Numbers — 133

3.1 Construction of Catalan Numbers 133
3.2 Catalan Numbers and Counting Problems 136
3.3 Closed Formula for Catalan Numbers 166
3.4 Generating Function of the Sequence of Catalan Numbers 171
3.5 Properties of Catalan Numbers 175
3.6 Polynomials Related to Catalan Numbers 193
3.7 Catalan Numbers in the Family of Special Numbers 199

4 Relatives of Catalan Numbers — 215

4.1 Motzkin Numbers 215
4.2 Schröder Numbers 250
4.3 Delannoy Numbers 276
4.4 Narayana Numbers 298
4.5 Stirling Numbers 325
4.6 Bell Numbers 338
4.7 Factorial Numbers 348

5 Catalan Numbers and Their Relatives on Integer Lattice — 361

5.1 Integer Lattice: Basic Notions 361
5.2 Catalan Numbers on Integer Lattice 367
5.3 Motzkin Numbers on Integer Lattice 373
5.4 Schröder Numbers on Integer Lattice 379
5.5 Narayana Numbers on Integer Lattice 386
5.6 Delannoy Numbers on Integer Lattice 392
5.7 Binomial Coefficients on Integer Lattice 397

6 Zoo of Numbers — 405

7 Mini Dictionary — 413

8 Exercises — 421

Bibliography — 447

Index — 459

Notations

- X (Y, Z, A, B, C, S, etc.) — a set; $X = \{x_1, \ldots, x_n\}$ — a *finite set*; $|X| = n$ — the number n of elements in the finite set $X = \{x_1, \ldots, x_n\}$; 2^X — the *power set* of X, i.e., the set of all its subsets; $|2^X| = 2^n$.
- $X \times Y = \{(x, y) | x \in X, y \in Y\}$ — the *Cartesian product* of the sets X and Y, i.e., the set of ordered pairs (x, y) of the elements $x \in X$ and $y \in Y$.
- $\mathbb{N} = \{1, 2, 3, \ldots\}$ — the set of *positive integers* (or *natural numbers*); $V_n = \{1, 2, 3, \ldots, n\}$ — the set of the first n positive integers.
- $\mathbb{Z} = \{\ldots, -3, -2, -1, 0, 1, 2, 3, \ldots\}$ — the set of *integers*.
- $\mathbb{Q}$ — the set of *rational numbers*.
- $\mathbb{R}$ — the set of *real numbers*.
- $\mathbb{C} = \{a + b \cdot i \,|\, a, b \in \mathbb{R}, i^2 = -1\}$ — the set of *complex numbers*.
- $b|a$ — a non-zero integer b divides an integer a: $a = bc$, where $c \in \mathbb{Z}$.
- $\gcd(a_1, \ldots, a_n)$ — *greatest common divisor* of integers $a_1, \ldots, a_n$, at least one of which is non-equal to 0, i.e., the greatest integer dividing $a_1, \ldots, a_n$. If $\gcd(a_1, \ldots, a_n) = 1$, the numbers $a_1, \ldots, a_n$ are called *relatively prime*, or *coprime*; if $\gcd(a_i, a_j) = 1$ for any distinct $i, j \in \{1, \ldots, n\}$, the numbers $a_1, \ldots, a_n$ are called *pairwise relatively prime*.

- $rest(a, b)$ — the *reminder* of an integer a after its *integer division* by a positive integer b: $a = b \cdot q + rest(a, b)$, where $q, rest(a, b) \in \mathbb{Z}$ and $0 \leq rest(a, b) < b$.
- $P = \{2, 3, 5, 7, 11, 13, 17, 19, \ldots\}$ — the set of *prime numbers*, i.e., positive integers with exactly two positive integer divisors; $p, q, p_1, p_2, \ldots, p_k, \ldots, q_1, q_2, \ldots, q_s, \ldots$ — prime numbers.
- $n = p_1^{\alpha_1} \cdot \ldots \cdot p_k^{\alpha_k}$, $p_1 < \cdots < p_k \in P$, $\alpha_1, \ldots, \alpha_k \in \mathbb{N}$ — the *prime factorization* of a positive integer $n > 1$, i.e., its representation as a product of positive integer powers of different prime numbers $p_1, \ldots, p_k$.
- $S = \{4, 6, 8, 9, 10, 12, 14, 15, 16, 18, \ldots\}$ — the set of *composite numbers*, i.e., the positive integers with more than two positive integer divisors.
- $n = (c_s c_{s-1} \cdots c_1 c_0)_g = c_s g^s + c_{s-1} g^{s-1} + \cdots + c_1 g + c_0$, $g \in \mathbb{N} \backslash \{1\}$, $0 \leq c_i \leq g-1$, $c_s \neq 0$ — representation of a positive integer n in base g.
- $\lfloor x \rfloor$, $x \in \mathbb{R}$ — *floor function*: the largest integer less than or equal to x.
- $\lceil x \rceil$, $x \in \mathbb{R}$ — *ceiling function*: the smallest integer greater than or equal to x.
- $[x]$, $x \in \mathbb{R}$ — *nearest integer function*: the nearest integer to x.
- $a \equiv b \pmod{n}$ — an integer a is *congruent* to an integer b modulo n, $n \in \mathbb{N}$, i.e., $n | (a-b)$.
- $[a]_n = \mathbf{a}_n = \{x \in \mathbb{Z} \mid x \equiv a \pmod{n}\} = \{\ldots, a-2n, a-n, a, a+n, a+2n, a+3n, \ldots\}$ — *residue class (of a) modulo n*: the set of all integers congruent to a modulo n. Any representative r_a of $[a]_n$ is called a *residue of a modulo n*. The smallest non-negative representative of $[a]_n$ is called the *smallest non-negative residue of a modulo n*; it is the *reminder* $rest(a, n)$ of a after its integer division by n. The smallest absolute value representative of $[a]_n$ is called the *minimal residue of a modulo n*.
- $(\mathbb{Z}/n\mathbb{Z}, +, \cdot)$, where $\mathbb{Z}/n\mathbb{Z} = \{[0]_n, [1]_n, [2]_n, \ldots, [n-1]_n\}$ — the *factor ring* (or *quotient ring*) modulo n; for a prime number p, the ring $(Z/pZ, +, \cdot)$ is a (finite) field.
- π (or α, β, τ, ...); $\pi = (\pi_1, \pi_2, \ldots, \pi_n) = \pi_1 \pi_2 \cdots \pi_n$ — *permutation* of the set $\{1, 2, \ldots, n\}$, i.e., a bijection π of $\{1, 2, \ldots, n\}$ into

itself, $\pi(k) = \pi_k$, $k = 1, 2, \ldots, n$; S_n — *symmetric group* of all permutations on $\{1, 2, \ldots, n\}$.
- $\{a_n\}_{n=0}^{\infty} = \{a_n\}_n$ — sequence $a_0, a_1, a_2, \ldots, a_n, \ldots$; $\sum_{n=0}^{\infty} a_n = a_0 + a_1 + a_2 + \cdots + a_n + \cdots$ — the sum of the elements of the sequence $\{a_n\}_n$; $\prod_{n=0}^{\infty} a_n = a_0 \cdot a_1 \cdot a_2 \cdots a_n \cdots$ — the product of the elements of the sequence $\{a_n\}_n$.
- $f(x) = a_0 + a_1 x + a_2 x^2 + \cdots + a_n x^n + \cdots$, $|x| < r$ — *generating function* of the sequence $a_0, a_1, a_2, \ldots, a_n, \ldots$.
- $g(x) = a_0 + a_1 \frac{x}{1!} + a_2 \frac{x^2}{2!} + \cdots + a_n \frac{x^n}{n!} + \cdots$, $|x| < r$ — *exponential generating function* of the sequence $a_0, a_1, a_2, \ldots, a_n, \ldots$.
- $b_0 c_{n+k} + b_1 c_{n+k-1} + \cdots + b_n c_k = 0$, $b_0, \ldots, b_n \in \mathbb{R}$, $b_0 \neq 0$ — *linear recurrent equation of the nth order* for the sequence $c_0, c_1, c_2, \ldots, c_{n-1}$, $c_n = -\frac{b_1}{b_0} c_{n-1} - \cdots - \frac{b_n}{b_0} c_0$, $c_{n+1} = -\frac{b_1}{b_0} c_n - \cdots - \frac{b_n}{b_0} c_1$, $\ldots$, $c_{n+k} = -\frac{b_1}{b_0} c_{n+k-1} - \cdots - \frac{b_n}{b_0} c_k$, *cdots*, with the *initial values* $c_0, c_1, \ldots, c_{n-1}$.
- $A = ((a_{ij}))$, $1 \leq i, j \leq n$ — *square $n \times n$ matrix* with real entries a_{ij}. $\det A$ or $|A|$ — *determinant* of a given $n \times n$ matrix $A = ((a_{ij}))$.
- $\log_a x$, $a, x \in \mathbb{R}$, $a > 0, a \neq 1$, $x > 0$ — *logarithm x with base a*, i.e., a number $y \in \mathbb{R}$ such that $a^x = y$; $\log x = \ln x$ — *natural logarithm* $\log_e x$; $\lg x$ — *decimal logarithm* $\log_{10} x$.
- $n!$ — *factorial* of a non-negative integer n: $n! = 1 \cdot 2 \cdots n$, $n \in \mathbb{N}$; $0! = 1$.
- $n\sharp$ — *primorial* of a positive integer n: the product of all primes less than or equal to n.
- $n_\alpha! = n(n-\alpha)(n-2\alpha)\cdots$, $\alpha \in \mathbb{N}$ — *multifactorial* of a positive integer n; $n!!$ — *double factorial* of a positive integer n: $n!! = n \cdot (n-2) \cdot (n-4) \cdots$.
- $x^{\underline{n}}$, $x \in \mathbb{R}$, $n \in \mathbb{N}$ — *falling factorial* of x: $x^{\underline{n}} = x \cdot (x-1) \cdots (x-n+1)$; $x^{\underline{0}} = 1$.
- $x^{\overline{n}}$, $x \in \mathbb{R}$, $n \in \mathbb{N}$ — *rising factorial* of x: $x^{\overline{n}} = x \cdot (x+1) \cdots (x+n-1)$; $x^{\overline{0}} = 1$.
- $\binom{n}{m}$ — *binomial coefficients*; C_n^m — the number of *combinations*, i.e., the ways to choose a non-ordered subset of m distinct elements from a fixed set of n distinct elements; $\binom{n}{m} = C_n^m = \frac{n!}{m!(n-m)!}$, $0 \leq m \leq n$. They form *Pascal's triangle*.

- $\binom{n}{n_1,\ n_2,\ \ldots,\ n_k}$, where $n_1, n_2, \ldots, n_k$ are non-negative integers, such that $n_1 + n_2 + \cdots + n_k = n$ — *multinomial coefficients*; $P_{(n_1,n_2,\ldots,n_k)}$ — the number of *permutations with repetitions, where the first element is used n_1 times, ..., the kth element is used n_k times*, i.e., the ways to construct an ordered multiset of $n = n_1 + n_2 + \cdots + n_k$ elements from a fixed non-ordered multiset of n elements, where the first element is used n_1 times, ..., the kth element is used n_k times; $\binom{n}{n_1,\ n_2,\ \ldots,\ n_k} = P_{(n_1,n_2,\ldots,n_k)} = \frac{n!}{n_1! n_2! \cdots n_k!}$; $\binom{n}{k,l}$ — *trinomial coefficients*.
- P_n — the number of *permutations of n elements*, i.e., the ways to construct an ordered set of n distinct elements from a fixed non-ordered set of n distinct elements: $P_n = n!$, $n \geq 0$.
- A_n^k (or P_k^n, nP_k, $^n P_k$, $P_{n,k}$, $P(n,k)$) — the number of *partial permutations from n elements by k elements*, i.e., the ways to choose an ordered subset of k distinct elements from a fixed set of n distinct elements: $A_n^k = \frac{n!}{(n-k)!}$, $n, k \geq 0$, $n \geq k$.
- $\overline{A}_n^k$ — the number of *partial permutations from n elements by k elements with repetitions*, i.e., the ways to choose an ordered multiset of k (not necessarily distinct) elements from a fixed set of n distinct elements: $\overline{A}_n^k = n^k$, $n, k \geq 0$.
- $\overline{C}_n^k$ (or $\left(\!\binom{n}{k}\!\right)$) — the number of *combinations from n elements by k elements with repetitions*, i.e., the ways to choose a non-ordered multiset of k (not necessarily distinct) elements from a fixed set of n distinct elements: $\overline{C}_n^k = C_{n+k-1}^k = \frac{(n+k-1)!}{k!(n-1)!}$, $n, k \geq 0$.
- $f(x) = O(g(x))$ — *big-O notation*: for a complex-valued function f and a real-valued function g, $g(x) > 0$, for all large enough values of x, there exists a positive real number C and a real number x_0 such that $|f(x)| \leq C \cdot g(x)$ for all $x \geq x_0$.
- $f(x) = o(g(x))$ — *little-o notation*: for a complex-valued function f and a real-valued function g, $g(x) > 0$, for all large enough values of x, $\lim_{x \to \infty} \frac{f(x)}{g(x)} = 0$.
- $f(x) \sim g(x)$: for a complex-valued function f and a real-valued function g, $g(x) > 0$ for all large enough values of x, $\lim_{x \to \infty} \frac{f(x)}{g(x)} = 1$.

- $e = 2.718281828\ldots$ — *Euler's number*: $e = \lim_{n\to\infty}(1+\frac{1}{n})^n$.
- $\gamma = 0.5772156649\ldots$ — the *Euler–Mascheroni constant*: $\gamma = \lim_{n\to\infty}\left(\sum_{k=1}^{n}\frac{1}{k} - \log n\right)$.
- $\Gamma(z)$ — *gamma function*: $\Gamma(z) = \int\limits_0^\infty t^{z-1}e^{-t}\,dt$, $z = x + iy \in \mathbb{C}$, $x > 0$; $\Gamma(n+1) = n!$ for $n \in \mathbb{N}$.
- δ_{nm} — *Kronecker delta function*: $\delta_{nm} = 1$ for $n = m$ and $\delta_{nm} = 0$ for $n \neq m$.

Chapter 1
Preliminaries

1.1. Combinatorial Configurations
Basic definitions

1.1.1. To eliminate the influence of a particular type of object, we use the general language of set theory: we talk about sets, their subsets, and the operations of union, intersection, difference, symmetric difference, Cartesian product, etc., of two or more sets. Usually, we speak about finite sets. For a set A consisting of elements $a_1, a_2, \ldots, a_n$, we write $A = \{a_1, a_2, \ldots, a_n\}$. We call the number n of the elements of the set A the *cardinality* of the set A and denote it by $|A| : |A| = n$. In this case, we use for a set A the name n-*set*.

Sets that take order into account are called *ordered sets*, or *tuples*. An ordered set consisting of elements $a_1, a_2, \ldots, a_n$, is written as $(a_1, a_2, \ldots, a_n)$. So, using a 3-set $A = \{1, 2, 3\}$, we get six ordered sets: $(1, 2, 3), (1, 3, 2), (2, 1, 3), (2, 3, 1), (3.1, 2)$, and $(3, 2, 1)$, which are called the *permutations* of the elements of the set A.

1.1.2. A *multiset* (or *bag*, *mset*) is a modification of the concept of a set that, unlike a set, allows multiple instances of each of its elements. The number of instances given for each element is called the *multiplicity* of that element in the multiset.

As with sets, and in contrast to tuples, order does not matter in discriminating multisets; therefore, $\{1,1,2,3\}$ and $\{1,2,1,3\}$ denote the same multiset.

Multisets that take order into account are called *ordered multsets* (or *strings*).

For example, using the (non-ordered) multiset $A = \{1,1,2\}$, we get three ordered multisets: $(1,1,2)$, $(1,2,1)$, and $(2,1,1)$, which are called *permutations with repetitions*.

The *cardinality* of a multiset is the sum of the multiplicities of all its elements. For example, in the multiset $\{1,1,2,2,2,3\}$, the multiplicities of the members 1, 2, and 3 are respectively 2, 3, and 1; therefore, the cardinality of this multiset is 6.

1.1.3. *Permutations* of m elements (or *m-permutations*) are all kinds of ordered sets that can be obtained from m different elements of a given m-set. The number of permutations of m elements is indicated by the symbol P_m.

For example, $P_3 = 6$ since, by using an (non-ordered) 3-set $A = \{1,2,3\}$, we can get exactly six permutations (ordered sets): $(1,2,3), (1,3,2), (2,1,3), (2,3,1), (3,1,2)$, and $(3,2,1)$.

Let us now try to get all subsets of the set $A = \{1,2,3\}$, each containing two elements. These subsets are $\{1,2\}, \{1,3\}$, and $\{2,3\}$. They are called *combinations from 3 elements by 2 elements* (or *2-combinations of a 3-set*).

In general, *combinations* from m elements by n elements (or *n-combinations of an m-set*) are all (non-ordered) sets, consisting of n different elements selected from m different elements of a given m-set. The number of combinations from m elements by n elements is indicated by the symbol C_m^n.

For example, $C_3^2 = 3$, as there are exactly three 2-combinations of the 3-set: $\{1,2,3\}$: $\{1,2\}$ and $\{1,3\}, \{2,3\}$.

If we want to obtain from a given 3-set (say, from the set $A = \{1,2,3\}$) all ordered 2-sets, then we get the tuples $(1,2), (2,1), (1,3)$, and $(3,1), (2,3), (3,2)$. They are called *partial permutations from 3 elements by 2 elements* (or *partial 2-permutations of a given 3-set*).

Partial permutations from m elements by n elements (or *partial n-permutations of a given m-set*) are all ordered sets, consisting of n

different elements selected from m different elements of a given m-set. The number of partial n-permutations of a given m-set is indicated by the symbol A_m^n.

So, $A_3^2 = 3$, as there are exactly six partial 2-permutations of the elements: $\{1, 2, 3\}$: $(1, 2)$, $(2, 1)$, $(1, 3)$, $(3, 1)$, $(2, 3)$, and $(3, 2)$.

1.1.4. Similarly, we can define permutations, combinations, and partial permutations with repetitions.

Permutations with repetitions, in which the first element is repeated α_1 times, the second element is repeated α_2 times, ..., the kth element is repeated α_k times (or $(\alpha_1, \ldots, \alpha_k)$-*permutations*), are all ordered multisets, which can be obtained from a non-ordered multiset, in which the first element is repeated α_1 times, the second element is repeated α_2 times, ..., the kth element is repeated α_k times. The number of such permutations with repetitions is indicated by the symbol $P_{(\alpha_1, \ldots, \alpha_k)}$.

For example, permutations with repetitions, in which element 1 repeats two times, and elements 2 and 3 are used each once, i.e., $(2, 1, 1)$-permutations of the multiset $\{1, 1, 2, 3\}$, are $(1, 1, 2, 3)$, $(1, 1, 3, 2)$, $(1, 2, 1, 3)$, $(1, 2, 3, 1)$, $(1, 3, 1, 2)$, $(1, 3, 2, 1)$, $(2, 1, 1, 3)$, $(2, 1, 3, 1)$, $(2, 3, 1, 1)$, $(3, 1, 1, 2)$, and $(3, 1, 2, 1)$, $(3, 2, 1, 1)$. Thus, the number $P_{(2,1,1)}$ of such permutations with repetitions is 12: $P_{(2,1,1)} = 12$. Obviously, $P_{(1,1,1)} = P_3 = 6$. (Compare this result with $P_{(3)} = 1$.)

Combinations with repetitions from m elements by n elements (or *n-combinations with repetitions of a given m-set*) are all non-ordered multisets, consisting of n, not necessarily distinct, elements selected from m different elements of a given m-set. The number of n-combinations with repetitions of a given m-set is indicated by the symbol $\overline{C}_m^n$.

For example, all 2-combinations with repetitions of the 3-set $\{1, 2, 3\}$ are $\{1, 2\}$, $\{1, 3\}$, $\{2, 3\}$, $\{1, 1\}$, $\{2, 2\}$, and $\{3, 3\}$. Thus, it holds that $\overline{C}_3^2 = 6$.

Partial permutations with repetitions from m elements by n elements (or *partial n-permutations with repetitions of a given m-set*) are all ordered multisets, consisting of n, not necessarily distinct, elements selected from m different elements of a given m-set. The number of partial n-permutations with repetitions of a given m-set is indicated by the symbol $\overline{A}_m^n$.

For example, all partial 2-permutations with repetitions of the set $\{1, 2, 3\}$ are $(1, 2)$, $(2, 1)$, $(1, 3)$, $(3, 1)$, $(2, 3)$, $(3, 2)$, $(1, 1)$, $(2, 2)$, and $(3, 3)$. Thus, $\overline{A}_3^2 = 9$.

Addition and multiplication principles

1.1.5. The main combinatorial configurations are *combinations, partial permutations*, and *permutations*, with or without repetitions.

It is not difficult to show that the number P_m of permutations of m different elements is equal to $m!$, the number A_m^n of partial permutations from m elements by n elements is equal to $\frac{m!}{(m-n)!}$, the number C_m^n of combinations from m elements by n elements is equal to $\frac{m!}{n! \cdot (m-n)!}$, the number $\overline{A}_m^n$ of partial permutations with repetitions from m elements by n elements is equal to m^n, the number $\overline{C}_m^n$ of combinations with repetitions from m elements by n elements is equal to $P_{(m-1,n)} = \frac{(m+n-1)!}{n! \cdot (m-1)!}$, and the number $P_{(\alpha_1, \alpha_2, \ldots, \alpha_k)}$ of permutations with repetitions, in which the ith element is repeated α_i times, $i = 1, \ldots, k$, is equal to $\frac{(\alpha_1 + \alpha_2 + \cdots + \alpha_k)!}{\alpha_1! \cdot \alpha_2! \cdot \ldots \cdot \alpha_k!}$.

1.1.6. In order to prove these formulas, as well as to solve the majority of simple combinatorial problems, one should use the combinatorial *addition and multiplication principles* (or *rules of sum and product*).

These principles, together with the inclusion–exclusion principle, are the basic counting principles, i.e., the fundamental principles of counting. They are often used for different enumerative purposes.

The *rule of sum* is an intuitive principle, stating that if there are A possible outcomes for an event (or ways to do something) and B possible outcomes for another event (or ways to do another thing), and the two events cannot both occur (or the two things can't both be done), then there are $A + B$ total possible outcomes for the events (or total possible ways to do one of the things). Formally, to arrive at this conclusion, we use the following well-known fact of set theory: the sum of the cardinalities of two disjoint sets A and B is equal to the cardinality of their union, $A \cup B$.

The *rule of product* is another intuitive principle, stating that if there are A ways to do something and B ways to do another thing, then there are $A \cdot B$ ways to do both things. Formally, to arrive at this conclusion, we use the following well-known fact of set theory: the product of the cardinalities of two sets A and B is equal to the cardinality of their *Cartesian product* $A \times B = \{(a,b) \, | \, a \in A, b \in B\}$.

1.1.7. Using the multiplication principle, it is easy to prove, for example, the formula

$$A_m^n = \frac{m!}{(m-n)!}.$$

In fact, in order to obtain a partial n-permutation of a given m-set, we should fill n ordered places using m different elements, $a_1, \ldots, a_m$. For the first place to be filled in, we have m possibilities (one can choose any element from $a_1, \ldots, a_m$); to fill the second place, we have $m-1$ possibilities (one can select any element from $a_1, \ldots, a_m$, except the one which was used for the first step); $\ldots$; to fill the nth place, we have $m-n+1$ possibilities. Since we must fill in the first, *and* the second, $\ldots$, *and* the nth places, then, according to the multiplication principle, the total number of possibilities is equal to $m \cdot (m-1) \cdot \ldots \cdot (m-n+1) = \frac{m!}{(n-m)!}$.

The same or similar reasons can be used for proofs of the other combinatorial formulas.

- For the numbers of classical combinatorial configurations, we have the following formulas:

$$P_m = m!; \quad A_m^n = \frac{m!}{(m-n)!}; \quad C_m^n = \frac{m!}{n! \cdot (m-n)!}; \quad \overline{A}_m^n = m^n;$$

$$P_{(\alpha_1, \alpha_2, \ldots, \alpha_k)} = \frac{(\alpha_1 + \alpha_2 + \cdots + \alpha_k)!}{\alpha_1! \cdot \alpha_2! \cdot \ldots \cdot \alpha_k!};$$

$$\overline{C}_m^n = P_{(m-1,n)} = \frac{(m+n-1)!}{n! \cdot (m-1)!}.$$

6   *Catalan Numbers*

Inclusion–exclusion principle

1.1.8. Many combinatorial problems can be solved using the *inclusion–exclusion principle* (or *inclusion–exclusion method*), a counting technique which generalizes the classical method of obtaining the number of elements in the union of two finite sets A and B:

$$|A \cup B| = |A| + |B| - |A \cap B|,$$

where $|S|$ is the cardinality of a finite set S, i.e., the number of elements in S.

The formula expresses the fact that the sum of the cardinalities of two finite sets A and B may be too large since some elements may be counted twice. The double-counted elements are those in the intersection $A \cap B$ of the sets A and B, and the result should be corrected by subtracting the cardinality of the intersection $A \cap B$ of the sets A and B.

A generalization of the results of these examples gives the *inclusion–exclusion principle*.

- *To find the cardinality of the union of n finite sets: include the cardinalities of the sets; exclude the cardinalities of the pairwise intersections; include the cardinalities of the triple-wise intersections; exclude the cardinalities of the quadruple-wise intersections; include the cardinalities of the quintuple-wise intersections; continue until the cardinality of the n-tuple-wise intersection is included (if n is odd) or excluded (if n is even).*

1.1.9. In general, the following statement holds (see, for example, [DeMo10] and [Deza24]).

Theorem (inclusion–exclusion formula). *Consider a set A, consisting of N different elements. Consider k properties $\alpha_1, \alpha_2, \ldots, \alpha_k$, which the elements of the set A can or cannot possess. Let $N(\alpha_i, \alpha'_j)$ be the number of elements of the set A having the property α_i and* not having the property α'_j. Then,

$$N(\alpha'_1, \alpha'_2, \ldots, \alpha'_k)$$
$$= N - N(\alpha_1) - \cdots - N(\alpha_k) + N(\alpha_1, \alpha_2) + \cdots + N(\alpha_{k-1}, \alpha_k)$$
$$- N(\alpha_1, \alpha_2, \alpha_3) - \cdots - N(\alpha_{k-2}, \alpha_{k-1}, \alpha_k) + \cdots$$
$$+ (-1)^k N(\alpha_1, \alpha_2, \ldots, \alpha_k).$$

Binomial theorem and Pascal's triangle

1.1.10. The *binomial theorem* (or *binomial expansion*, *Newton's binom*) describes the algebraic expansion of powers of a binomial (for example, of the binomial $x + y$ or the binomial $1 + x$). According to the theorem, it is possible to expand the polynomial $(x+y)^n$ into a sum that involves the terms of the form $B_n^k x^k y^{n-k}$, where the exponents k and $n-k$ are non-negative integers, and the coefficient B_n^k of each term is a specific positive integer, depending on n and k. For example, for $n = 4$, it holds that

$$(x+y)^4 = x^4 + 4x^3y + 6x^2y^2 + 4xy^3 + y^4.$$

Theorem (binomial theorem). *For any $x, y \in \mathbb{R}$ and for any non-negative integer n, the following equality holds:*

$$(x+y)^n = \binom{n}{0} x^n + \binom{n}{1} x^{n-1} y$$
$$+ \binom{n}{2} x^{n-2} y^2 + \cdots + \binom{n}{n-1} xy^{n-1} + \binom{n}{n} y^n,$$

where the summation is taken over all non-negative integers k, $0 \le k \le n$, and

$$\binom{n}{k} = \frac{n!}{k!(n-k)!}.$$

1.1.11. The coefficients $\binom{n}{0}, \binom{n}{1}, \binom{n}{2}, \ldots, \binom{n}{n}$ of this decomposition are called *binomial coefficients*. Combinatorially, the binomial coefficient $\binom{n}{k}$ can be interpreted as the number of ways to choose k elements from an n-element set, i.e., the number C_n^k of k-compositions of a given n-set.

8   Catalan Numbers

- **Binomial coefficients** $\binom{n}{k}$ *coincide with the numbers* C_n^k *of compositions.*

□ In fact, if we write $(x+y)^n$ as a product,
$$(x+y)(x+y)(x+y)\cdots(x+y),$$
then, according to the distributive law, there will be one term in the expansion for each choice of either x or y from each of the binomials of the product. For example, there will only be one term x^n corresponding to choosing x from each binomial. However, there will be several terms of the form $x^{n-k}y^k$, $k \in \mathbb{N}$, one for each way of choosing exactly k binomials $x+y$ to contribute y. Therefore, after combining like terms, the coefficient of $x^{n-k}y^k$ will be equal to the number C_n^k of ways to choose exactly k elements from an n-set. □

1.1.12. Using summation notation, the *binomial formula* (or *binomial identity*) can be written as

$$(x+y)^n = \sum_{k=0}^n \binom{n}{k} x^{n-k} y^k = \sum_{k=0}^n \binom{n}{k} x^k y^{n-k}.$$

The final expression follows from the previous one by the symmetry of x and y in the first expression and by the symmetry $\binom{n}{k} = \binom{n}{n-k}$ of binomial coefficients.

A simple variant of the binomial formula is obtained by substituting 1 for y so that it involves only a single variable. In this form, the formula is

$$(1+x)^n = \binom{n}{0}x^0 + \binom{n}{1}x^1 + \binom{n}{2}x^2 + \cdots + \binom{n}{n-1}x^{n-1} + \binom{n}{n}x^n,$$

or, equivalently,

$$(1+x)^n = \sum_{k=0}^n \binom{n}{k} x^k.$$

More explicitly, we obtain

$$(1+x)^n = 1 + nx + \frac{n(n-1)}{2!}x^2 + \frac{n(n-1)(n-2)}{3!}x^3 + \cdots + nx^{n-1} + x^n.$$

Here are the first few cases of the binomial theorem:

$$(x+y)^0 = 1,$$
$$(x+y)^1 = x+y,$$
$$(x+y)^2 = x^2 + 2xy + y^2,$$
$$(x+y)^3 = x^3 + 3x^2y + 3xy^2 + y^3,$$
$$(x+y)^4 = x^4 + 4x^3y + 6x^2y^2 + 4xy^3 + y^4,$$
$$(x+y)^5 = x^5 + 5x^4y + 10x^3y^2 + 10x^2y^3 + 5xy^4 + y^5.$$

These examples show that the binomial coefficients for varying n and k can be arranged in the form of a *Pascal's triangle* (see Chapter 2).

Multinomial theorem

1.1.13. The binomial theorem is the most important special case of the *multinomial theorem*, which works with powers of sums of more than two terms.

Theorem (multinomial theorem). *For any $x_1, \ldots, x_m \in \mathbb{R}$ and for any non-negative integer n, the following equality holds:*

$$(x_1 + \cdots + x_m)^n = \sum_{k_1+\cdots+k_m=n, k_i \geq 0} \binom{n}{k_1, k_2, \ldots, k_m} x_1^{k_1} \cdots x_m^{k_m},$$

where the summation is taken over all sequences of non-negative integers $k_1, \ldots, k_m$ such that the sum of all k_i is n, and

$$\binom{n}{k_1, k_2, \ldots, k_m} = \frac{n!}{k_1! \cdot k_2! \cdot \cdots \cdot k_m!}.$$

1.1.14. The coefficients $\binom{n}{k_1,\ldots,k_m}$ are known as *multinomial coefficients*. Combinatorially, the multinomial coefficient $\binom{n}{k_1,\ldots,k_m}$ counts the number of different ways to partition an n-element set into disjoint subsets of sizes $k_1, \ldots, k_m$, i.e., the number $P_{(k_1,\ldots,k_m)}$ of $(k_1, \ldots, k_m)$-partitions.

- The multinomial coefficients $\binom{n}{k_1,\ldots,k_m}$ coincide with the numbers $P_{(k_1,\ldots,k_m)}$ of $(k_1, \ldots, k_m)$-partitions.

□ First, let us open the brackets in the left-hand side of the considered equality:
$$(x_1 + \cdots + x_m)^n = (x_1 + \cdots + x_m) \cdots (x_1 + \cdots + x_m)$$
$$= x_1 \cdots x_1 + \cdots + x_m \cdots x_m.$$

We get a finite number of terms, each of which consists of n factors, with each factor taking one of the values $x_1, \ldots, x_m$. Combining similar factors in each of the obtained terms, we get a finite number of terms of the form $x_1^{k_1} \cdots x_m^{k_m}$, where $k_1, \ldots, k_m$ are non-negative integers, such that their sum is equal to n: $k_1 + \cdots + k_m = n$, $k_1 \geq 0, \ldots, k_m \geq 0$. For a fixed $k_1, \ldots, k_m$, the element $x_1^{k_1} \cdots x_m^{k_m}$ can be obtained from any term, in which x_1 is represented k_1 times, x_2 is represented k_2 times, $\ldots$, x_m is represented k_m times. The numbers of such terms is equal to $P_{(k_1,\ldots,k_m)}$. The statement holds. □

It is easy to see that we get the binomial theorem for $k = 2$, noting that if $k_1 + k_2 = n$, then $k_2 = n - k_1$, and using the equality, $P_{(k_1, n-k_1)} = \frac{n!}{k_1!(n-k_1)!} = \binom{n}{k_1} = \binom{n}{n-k_1}$.

Basic combinatorial identities

There are many useful and beautiful identities connected with combinatorial constructions. For proofs of these identities, a variety of methods are used.

Direct validation

1.1.15. The simplest of these methods is the method of direct algebraic validation. For example, in order to prove the identity
$$C_n^k = C_{n-1}^k + C_{n-1}^{k-1},$$
it is enough to perform the following transformations:
$$C_{n-1}^k + C_{n-1}^{k-1} = \frac{(n-1)!}{k!(n-k-1)!} + \frac{(n-1)!}{(k-1)!(n-k)!}$$
$$= \frac{n!}{k!(n-k)!}\left(\frac{n-k}{n} + \frac{k}{n}\right) = \frac{n!}{k!(n-k)!} = C_n^k.$$

Checking the identity
$$C_n^k = C_n^{n-k}$$
is even easier:
$$C_n^k = \frac{n!}{k!(n-k)!} = \frac{n!}{(n-k)!(n-(n-k))!} = C_n^{n-k}.$$

Combinatorial approach

1.1.16. However, the same identities can be proved using only combinatorial reasons.

For example, to prove the identity $C_n^k = C_{n-1}^k + C_{n-1}^{k-1}$, it is enough to note that, by fixing one element, for example, a_1, from the set of n different elements $a_1, \ldots, a_n$, we can state that any k-combination of elements $a_1, \ldots, a_n$ either contains this fixed element a_1 or does not contain it. The number of k-combinations of the first type is C_{n-1}^k: one should choose all k elements from the $(n-1)$-set $a_2, \ldots, a_n$ that do not contain a_1. The number of k-combinations of the second type is C_{n-1}^{k-1}: using the element a_1 as a member of the considered combination, we should "get" the remaining $k-1$ elements of this combination from the set $a_2, \ldots, a_n$ not containing a_1.

The identity $C_n^k = C_n^{n-k}$ is based on a simple combinatorial observation: choosing k elements from n elements, we left $n-k$ elements (from n elements) untouched, i.e., selecting them, for example, for our colleague.

On the other hand, the identity
$$C_n^0 C_m^k + C_n^1 C_m^{k-1} + \cdots + C_n^{k-1} C_m^1 + C_n^k C_m^0 = C_{n+m}^k$$
can be proven if we consider the following combinatorial problem.

- *How many ways exist to choose a team from n men and m women?*

□ The answer is as follows: there are exactly C_{n+m}^k possibilities to choose k people from $n+m$ people. On the other hand, in a given team, there can be zero men, one man, two men, ..., k men. So, if in the team there are exactly i men (and, hence, $k-i$ women), we have $C_n^i C_m^{k-i}$ possibilities. So, comparing these two representations, we obtain that

$$C_{n+m}^k = C_n^0 C_m^k + C_n^1 C_m^{k-1} + \cdots + C_n^{k-1} C_m^1 + C_n^k C_m^0 = \sum_{i=0}^{k} C_n^i C_m^{k-i}. \square$$

Binomial theorem's approach

1.1.17. Some combinatorial identities can be proved using the binomial theorem (or the multinomial theorem). For example, putting in the binomial formula the values $x = 1, y = 1$, we get the equality $(1+1)^n = C_n^0 + C_n^1 + \cdots + C_n^{n-1} + C_n^n$ or, in other words, the well-known identity

$$C_n^0 + C_n^1 + \cdots + C_n^{n-1} + C_n^n = 2^n.$$

Putting in the binomial formula the values $x = 1, y = -1$, we obtain the identity

$$C_n^0 - C_n^1 + \cdots + (-1)^n C_n^n = 0.$$

Similarly, as

$$(1+a)^n = C_n^0 + C_n^1 a + \cdots + C_n^{n-1} a^{n-1} + C_n^n a^n,$$

we obtain, for example, the identities

$$C_n^0 + 2C_n^1 + 4C_n^2 + \cdots + 2^n C_n^n = 3^n,$$
$$C_n^0 - 3C_n^1 + 9C_n^2 \cdots + (-3)^n C_n^n = (-2)^n.$$

Special combinatorial transforms

1.1.18. In order to calculate some finite combinatorial sums, one should use more special techniques. For example, to find the value of the sum $C_n^1 + 2C_n^2 + \cdots + nC_n^n$, one can note that

$$iC_n^i = i\frac{n!}{i!(n-i)!} = n\frac{(n-1)!}{(i-1)!((n-1)-(i-1))!} = nC_{n-1}^{i-1}.$$

Therefore, it holds that

$$C_n^1 + 2C_n^2 + \cdots + nC_n^n = n(C_{n-1}^0 + C_{n-1}^1 + \cdots + C_{n-1}^{n-1}) = n \cdot 2^{n-1}.$$

Exercises

1. An electrical panel has five switches. How many ways can the switches be positioned up or down if three switches must be up and two must be down?
2. A coat hanger has four knobs, and each knob can be painted any color. If six different colors of paint are available, how many ways can the knobs be painted?
3. How many ways can the letters from the word TREES be ordered such that each "word" starts with a consonant and ends with a vowel?
4. There are nine dots randomly placed on a circle. How many triangles can be formed within the circle?
5. Prove the following identities:
   (a) $C_n^m C_m^k = C_n^k C_{n-k}^{m-k}$;
   (b) $C_n^0 + 2C_n^1 + \cdots + 2^n C_n^n = 3^n$;
   (c) $\displaystyle\sum_{i_1+i_2+\cdots+i_m=n} P_{(i_1,i_2,\ldots,i_m)} = m^n$;
   (d) $C_{n-1}^k + C_{n-2}^{k-1} + \cdots + C_{n-(k+1)}^0 = C_n^k, n > k \geq 0$;
   (e) $C_k^0 C_{n-1}^{k-1} + C_k^1 C_{n-1}^{k-2} + \cdots + C_k^{k-1} C_{n-1}^0 = C_{n+k-1}^{k-1}$;
   (f) $C_n^0 C_m^k + C_n^1 C_m^{k-1} + \cdots + C_n^k C_m^0 = C_{n+m}^k$;
   (g) $(C_n^0)^2 + (C_n^1)^2 + \cdots + (C_n^n)^2 = C_{2n}^n$;
   (h) $C_{n-1}^k + C_{n-2}^{k-1} + C_{n-3}^{k-2} + \cdots + C_{n-k-1}^0 = C_n^k$;
   (i) $C_1^0 + C_2^1 + \cdots + C_{n+1}^n = C_{n+2}^n$;
   (j) $C_2^0 + \cdots + C_{n+2}^n = C_{n+3}^n$;
   (k) $C_{m+2}^2 C_n^n + C_{m+1}^2 C_{n+1}^n + \cdots + C_2^2 C_{n+m}^n = C_{n+m+3}^{n+3}$;
   (l) $P_{(i-1,j,k)} + P_{(i,j-1,k)} + P_{(i,j,k-1)} = P_{(i,j,k)}$.
6. Prove that $\binom{n}{k} = \frac{n}{k}\binom{n-1}{k-1}$.
7. Find the sums:
   (a) $C_n^0 + 2C_n^1 + 3C_n^2 + \cdots + (n+1)C_n^n$;
   (b) $C_n^1 - 2C_n^2 + 3C_n^3 + \cdots + (-1)^{n-1}nC_n^n$;
   (c) $C_n^0 + \frac{1}{2}C_n^1 + \frac{1}{3}C_n^2 + \cdots + \frac{1}{n+1}C_n^n$;

14 Catalan Numbers

(d) $C_n^0 - \frac{1}{2}C_n^1 + \frac{1}{3}C_n^2 + \cdots + (-1)^n \frac{1}{n+1}C_n^n$;

(e) $C_{n+1}^1 + \frac{1}{2}C_{n+1}^2 + \cdots + \frac{1}{2^n}C_{n+1}^{n+1}$;

(f) $C_n^0 + C_n^1 + \cdots + C_n^{[\frac{n}{2}]}$;

(g) $C_n^0 + C_n^1 + \cdots + C_n^{[\frac{n+1}{2}]}$;

(h) $\overline{C}_n^0 + \overline{C}_n^1 + \cdots + \overline{C}_n^m$;

(i) $\overline{A}_n^0 + \overline{A}_n^1 + \overline{A}_n^2 + \cdots + \overline{A}_n^{k-1}$;

(j) $\frac{C_n^1}{C_n^0} + 2\frac{C_n^2}{C_n^1} + 3\frac{C_n^3}{C_n^2} + \cdots + n\frac{C_n^n}{C_n^{n-1}}$;

(k) $\frac{1}{P_1 P_{n-1}} + \frac{1}{P_2 P_{n-2}} + \cdots + \frac{1}{P_n P_0}$;

(l) $C_n^0 \overline{A}_{m-1}^n + C_n^1 \overline{A}_{m-1}^{n-1} + \cdots + C_n^n \overline{A}_{m-1}^0$.

8. Check that the *Vandermond convolution* $C_{n+p}^k = \sum_{m=0}^{M} C_n^m C_p^{k-m}$, where $M = \min\{k, n-1\}$, can be proved using the relationships

$$C_{n+p}^k = C_{n+p-1}^k + C_{n+p-1}^{k-1} = C_{n+p-2}^k + C_{n+p-2}^{k-1}$$
$$+ C_{n+p-2}^{k-1} + C_{n+p-2}^{k-3} = C_{n+p-2}^k + 2C_{n+p-2}^{k-1} + C_{n-2}^{k-3}$$
$$= C_{n-3}^k + 3C_{n-3}^{k-1} + 3C_{n-3}^{k-2} + C_{n-3}^{k-3} = \cdots.$$

1.2. Graphs

Definition and examples

1.2.1. A *graph* (or *undirected graph*, *simple graph*) is a pair $G = (V, E)$, where V is a set whose elements are called *vertices* and E is a set of unordered pairs $\{u, v\}$ of distinct vertices, whose elements are called *edges* (or *links*, *lines*): $E \subseteq \{\{u, v\} \mid u, v \in V, \ u \neq v\}$.

Generally, the vertex set V is taken to be finite (which implies that the edge set E is also finite): $V = \{v_1, v_2, \ldots, v_n\}$. Sometimes, infinite graphs are considered, but they are usually viewed as a special kind of binary relation because most results on finite graphs either do not extend to the infinite case or need a rather different proof.

A *directed graph* (or *digraph*, *directed simple graph*) is a graph whose edges have orientations. Formally, a *directed graph* is a pair

$D = (V, E)$, where V is a set of vertices (or *nodes*, *points*) and E is a set of *arcs* (or *directed edges*, *directed links*, *directed lines*, *arrows*), which are ordered pairs of distinct vertices: $E \subseteq \{(u, v) \mid u, v \in V, u \neq v\}$.

A *multigraph* is a graph which is permitted to have *multiple edges* (or *parallel edges*), that is, edges that have the same end vertices. Thus, two vertices may be connected by more than one edge.

Sometimes, graphs are allowed to contain *loops*, which are edges that join a vertex to itself. To allow loops, the pairs of vertices in E must be allowed to have the same vertex twice. Such generalized graphs are called *graphs with loops* or, simply, graphs when it is clear from the context that loops are allowed.

A *subgraph* of a graph G is another graph formed from a subset of the vertices and edges of G. A *spanning subgraph* is one that includes all vertices of the graph; an *induced subgraph* is one that includes all the edges whose endpoints belong to the vertex subset.

1.2.2. The vertices u and v of an edge $\{u, v\}$ are called the *edge's endpoints*. The edge is said to *join* u and v and to *be incident* on them. When an edge $\{u, v\}$ exists, the vertices u and v are called *adjacent*.

In any graph, the *degree* $d(v)$ of a vertex v is defined as the number of edges that have v as an endpoint. For graphs that are allowed to contain loops, a loop should be counted as contributing two units to the degree of its endpoint.

A vertex v may belong to no edge, in which case it is not joined to any other vertex and is called *isolated*, $d(v) = 0$. A vertex v belonging to one edge is called a *leaf* (or *terminal vertex*, *hanging vertex*), $d(v) = 1$. A vertex v belonging to an even (odd) number of edges is called *even* (*odd*); in this case, $d(v)$ is even (odd).

1.2.3. The *degree sum formula* states that *the sum of the vertex degrees equals twice the number of edges*:

$$\sum_{v \in V} d(v) = 2|E|.$$

□ The proof is very simple: the sum $\sum_{v \in V} d(v)$ counts each edge $\{u, v\}$ twice. □

The *handshaking lemma* states that *the number of odd vertices in every finite undirected graph is even.*

☐ In fact, it holds that

$$2|E| = \sum_{v \in V} d(v) = \sum_{u \in V, d(u)=2k} d(v) + \sum_{v \in V, d(v)=2t+1} d(v).$$

As the values $2|E|$ and $\sum_{u \in V, d(u)=2k} d(v)$ are even, the sum $\sum_{v \in V, d(v)=2t+1} d(v)$ should be even and, hence, should have an even number of summands. ☐

Types of graphs

1.2.4. There are several important types of graphs.

A *complete graph* is a graph in which each pair of vertices is joined by an edge. In other words, a complete graph contains all possible edges. The complete graph on n vertices is denoted by K_n. It has exactly $\frac{n(n-1)}{2}$ edges.

A *bipartite graph* is a simple graph in which the vertex set can be partitioned into two sets, V_1 and V_2, so that no two vertices in V_1 share a common edge and no two vertices in V_2 share a common edge. In a *complete bipartite graph*, every vertex in V_1 is adjacent to every vertex in V_2, but there are no edges within V_1 or V_2. If $|V_1| = n_1$ and $|V_2| = n_2$, the complete bipartite graph is denoted by K_{n_1,n_2}. It contains $n_1 n_2$ edges.

A *path* (or *simple path, path graph, linear graph*) P_n, $n \geq 1$, is a graph in which the vertices can be listed in an order, $v_1, v_2, \ldots, v_n$, such that the edges are $\{v_1, v_2\}, \{v_2, v_3\}, \ldots, \{v_{n-1}, v_n\}$. The number of edges in P_n is $n-1$. If a path graph occurs as a subgraph of another graph, it is called a *path* in that graph.

A *cycle* (or *simple cycle, cycle graph, circular graph*) C_n, $n \geq 3$, is a graph in which the vertices can be listed in an order, $v_1, v_2, \ldots, v_n$, such that the edges are $\{v_1, v_2\}, \{v_2, v_3\}, \ldots, \{v_{n-1}, v_n\}, \{v_n, v_1\}$. The number of edges in C_n is n. If a cycle graph occurs as a subgraph of another graph, it is a *cycle* (or *circuit*) in that graph.

In an undirected graph, an unordered pair of vertices u, v is called connected if a path leads from u to v. Otherwise, the unordered pair is called disconnected.

A *connected graph* is an (undirected) graph in which every two vertices, u and v, are connected by a path leading from u to v. Otherwise, the graph is called a *disconnected graph*.

It is easy to prove that the number m of edges of a connected graph on n vertices is at least $n - 1$. (See, for example, [DeMo10].)

A *planar graph* is a graph that can be embedded in the plane, i.e., drawn on the plane in such a way that its edges intersect only at their endpoints. In other words, it can be drawn in such a way that no edges cross each other. Such a drawing is called a *plane graph* (or *planar embedding of the graph*). A plane graph can be defined as a planar graph with a mapping from every vertex to a point on a plane and from every edge to a plane curve on that plane, such that the extreme points of each curve are the points mapped from its end vertices, and all curves are disjoint except on their extreme points. Every graph that can be drawn on a plane can be drawn on the sphere as well, and vice versa, by means of stereographic projection.

Kuratowski's theorem (see, for example, [DeMo10]) states that:

- *a finite graph is planar if and only if it does not contain a subgraph that is a subdivision of the complete graph K_5 or the complete bipartite graph $K_{3,3}$.*

Trees

1.2.5. A *tree* is an (undirected) graph in which any two vertices are connected by exactly one path; or, equivalently, it is a connected acyclic undirected graph.

A *forest* is an (undirected) graph in which any two vertices are connected by at most one path; or, equivalently, it is an acyclic undirected graph or a disjoint union of trees.

A *polytree* (or *directed tree, oriented tree, singly connected network*) is a directed acyclic graph whose underlying undirected graph is a tree. In other words, if we replace its directed edges with undirected edges, we obtain an undirected graph that is both connected and acyclic.

Let us list several important properties of trees (see [DeMo10] and [Ore80]):

- Every tree on n vertices has $n-1$ edges.
- Every tree is a bipartite graph.
- Every tree with only countably many vertices is a planar graph.
- Every connected graph G admits a *spanning tree*, which is a tree that contains every vertex of G and whose edges are edges of G.
- Every finite tree with n vertices, $n > 1$, has at least two terminal vertices (leaves); this minimal number of leaves is characteristic of path graphs; the maximal number, $n-1$, is attained only by star graphs.
- For any three vertices in a tree, the three paths between them have exactly one vertex in common. More generally, a vertex in a graph that belongs to the three shortest paths among three vertices is called the *median* of these vertices. Because every three vertices in a tree have a unique median, every tree is a *median graph*.

A *rooted tree* is a tree in which one vertex has been designated the *root*.

The edges of a rooted tree can be assigned a natural orientation, either away from or toward the root, in which case the structure becomes a directed rooted tree. When a directed rooted tree has an orientation away from the root, it is called an *arborescence* (or *out-tree*); when it has an orientation toward the root, it is called an *anti-arborescence* (or *in-tree*).

Tree-order is the partial ordering on the vertices of a tree with $u < v$ if and only if the unique path from the root to v passes through u.

In a rooted tree, the *parent* of a vertex v is the vertex connected to v on the path to the root; every vertex has a unique parent, except that the root has no parent. A *child* of a vertex v is a vertex of which

v is the parent. An *ascendant* of a vertex v is any vertex that is either the parent of v or is (recursively) an ascendant of a parent of v. A *descendant* of a vertex v is any vertex that is either a child of v or is (recursively) a descendant of a child of v. A *sibling* of a vertex v is any other vertex on the tree that shares a parent with v. A *leaf* is a vertex with no children. An *internal vertex* is a vertex that is not a leaf.

A *k-ary tree* (for non-negative integers k) is a rooted tree in which each vertex has at most k children. If there are all possible children, we speak about a *full k-ary tree*. 2-ary trees are often called *binary trees*, while 3-ary trees are sometimes called *ternary trees*.

So, a *full binary tree* (or *proper, plane, strict binary tree*) is a binary tree in which every vertex has either zero or two children. In other words, a full binary tree is either a single vertex (a single vertex as the root) or a tree whose root vertex has two subtrees, both of which are full binary trees.

Rooted trees, often with an additional structure such as an ordering of the neighbors at each vertex, are a key data structure in computer science.

Enumeration of trees

1.2.6. A *labeled tree* (in general, a *labeled graph*) is a tree (or graph) in which each vertex is given a unique label.

It is well known (see, for example, [DeMo10]) that:

- there are $2^{\frac{n(n-1)}{2}}$ *graphs on n labeled vertices.*

  Cayley's formula states that:
- there are n^{n-2} *trees on n labeled vertices.*

  A classic proof uses the *Prüfer sequences*, which naturally show a stronger result:
- *the number of trees with vertices $1, 2, \ldots, n$ of degrees $d_1, d_2, \ldots, d_n$, respectively, is the multinomial coefficient*

$$\binom{n-2}{d_1 - 1, d_2 - 1, \ldots, d_n - 1}.$$

A more general problem is to count spanning trees in an undirected graph, which is addressed by the *matrix tree theorem*. (Cayley's formula is the special case of spanning trees in a complete graph.)

Counting the number of unlabeled free trees is a harder problem. No closed formula for the number $t(n)$ of trees with n vertices up to graph isomorphism is known. The first few values of $t(n)$ are $1, 1, 1, 1, 2, 3, 6, 11, 23, 47, \ldots$ (sequence A000055 in the OEIS).

Richard Otter (1948, [Otte48]) proved the asymptotic estimate

$$t(n) \sim C\alpha^n n^{-5/2} \quad \text{as } n \to \infty,$$

with $C = 0.534949606\ldots$ and $\alpha = 2.95576528565\ldots$ (sequence A051491 in the OEIS).

This is a consequence of his asymptotic estimate for the number $r(n)$ of unlabeled rooted trees with n vertices:

$$r(n) \sim D\alpha^n n^{-3/2} \quad \text{as } n \to \infty,$$

with $D = 0.43992401257\ldots$ and the same $\alpha = 2.95576528565\ldots$ (see [Knut97] and [FlSe09]).

The first few values of $r(n)$ are $1, 1, 2, 4, 9, 20, 48, 115, 286, 719, \ldots$ (sequence A000081 in the OEIS).

Exercises

1. Find all trees on n vertices, $n = 1, 2, 3, 4, 5$. Find all binary trees on n vertices, $n = 1, 2, 3, 4, 5$.
2. Prove that any tree is a bipartite graph. Can a tree be a complete bipartite graph?
3. Find all possible graphs on n vertices, $n = 1, 2, 3, 4, 5$. For each constructed graph, find (if possible) several of its spanning trees. For each constructed graph, find its distance matrix $((d_{ij}))$, where d_{ij} is the length of the smallest path connecting the vertices i and j; otherwise, $d_{ij} = \infty$.
4. Prove that a tree with n vertices has exactly $n - 1$ edges.

1.3. Partitions

Definitions and examples

1.3.1. A *partition* of a set is a grouping of its elements into non-empty subsets in such a way that every element is included in exactly one subset, i.e., X is a disjoint union of the subsets.

Formally, a family of sets S is a *partition* of X if the following conditions hold:

- $\emptyset \notin S$, the family S does not contain the empty set;
- $\bigcup_{A \in P} A = X$, the union of the sets in S is equal to X;
- $(\forall A, B \in P)\ A \neq B \implies A \cap B = \emptyset$, the intersection of any two distinct sets in S is empty.

The sets in S are called the *parts* (or *blocks*, *cells*) of the partition.

The number of partitions of a given n-set is the nth *Bell number* $B(n)$. Starting with $n = 1$, we get exactly one partition of a 1-set (say, the set $\{1\}$): $\{\{1\}\}$. For the 2-set $\{1, 2\}$, we have two partitions: $\{\{1, 2\}\}$ and $\{\{1\}, \{2\}\}$. The 3-set $\{1, 2, 3\}$ has five partitions: $\{\{1\}, \{2\}, \{3\}\}$, $\{\{1, 2\}, \{3\}\}$, $\{\{1, 3\}, \{2\}\}$, $\{\{1\}, \{2, 3\}\}$, and $\{\{1, 2, 3\}\}$. So, $B(1) = 1$, $B(2) = 2$, $B(3) = 5$. In fact, taking into account $B(0) = 1$, we obtain that the sequence of Bell numbers starts with the numbers $1, 1, 2, 5, 15, 52, 203, 877, 4140, \ldots$ (sequence A000110 in the OEIS).

1.3.2. A partition of a set X, together with a total order on the parts of the partition, gives a structure, called an *ordered partition* (or a *list of sets*) ([Stan97]). An ordered partition of a finite set may be written as a finite sequence of the sets in the partition: for instance, the three ordered partitions of the set $\{a, b\}$ are $(\{a\}, \{b\})$, $(\{b\}, \{a\})$, and $(\{a, b\})$.

The number of ordered partitions of a given n-set X is the nth *ordered Bell number* (or *Fubini number*) $OB(n)$. The sequence of ordered Bell numbers starts, for $n = 0, 1, 2, \ldots$, with $1, 1, 3, 13, 75, 541, 4683, 47293, 545835, 7087261, \ldots$ (sequence A000670 in the OEIS).

Partitions and binary relations

1.3.3. A *binary relation* over sets X and Y is a new set of ordered pairs (x, y) consisting of elements x in X and y in Y.

Formally, for given sets X and Y, a *binary relation* ω over sets X and Y is a subset of the Cartesian product $X \times Y$ of the sets X and Y, defined as $X \times Y = \{(x, y) \mid x \in X, y \in Y\}$. In other words, a binary relation over sets X and Y is an element of the *power set* $2^{X \times Y}$. The statement $(x, y) \in \omega$ reads "x is ω-related to y" and is denoted by $x\omega y$.

When $X = Y$, a binary relation is called a *homogeneous relation* (or *endorelation*). It is also simply called a (binary) *relation over X*. The set of all homogeneous relations over a set X is the power set $2^{X \times X}$.

A homogeneous relation ω over a set X may be identified with a directed simple graph D, permitting loops, where X is the vertex set of D and ω is the arc set of D. There is an arc (x, y) from a vertex x to a vertex y if and only if $x\omega y$.

Some important properties that a homogeneous relation ω over a set X may have are represented by the following definitions:

- ω is *reflexive* if, for all $x \in X$, $x\omega x$. For example, $\geq$ over $\mathbb{R}$ is a reflexive relation, but $>$ is not.
- ω is *antireflexive* if, for all $x \in X$, not $x\omega x$. For example, $>$ over $\mathbb{R}$ is an antireflexive relation, but $\geq$ is not.
- ω is *symmetric* if, for all $x, y \in X$, $x\omega y$, then $y\omega x$. For example, "is a blood relative of" is a symmetric relation over a set of people.
- ω is *antisymmetric* if, for all $x, y \in X$, $x\omega y$ and $y\omega x$, then $x = y$. For example, $\geq$ and $>$ are antisymmetric relations over $\mathbb{R}$.
- ω is *asymmetric* if, for all $x, y \in X$, $x\omega y$, then not $y\omega x$. For example, $>$ over ω is an asymmetric relation, but $\geq$ is not.
- ω is *transitive* if, for all $x, y, z \in X$, $x\omega y$ and $y\omega z$, then $x\omega z$. For example, "is ancestor of" over a set of people is a transitive relation, while "is parent of" is not.
- ω is *connected* if, for all $x, y \in X$, $x \neq y$, then $x\omega y$ or $y\omega x$. For example, $\geq$ and $>$ are connected relations over $\mathbb{R}$.

- ω is *strongly connected* if, for all $x, y \in X$, $x\omega y$ or $y\omega x$. For example, $\geq$ is a strongly connected relation over ω, while $>$ is not.

1.3.4. A homogeneous relation $\preceq$ over a given set X is called a *preorder* (or *quasiorder, non-strict preorder*), if it is reflexive and transitive. A set equipped with a preorder is called a *preordered set* (or *proset*).

A *partial order* (or *reflexive partial order, weak partial order, non-strict partial order*) is a homogeneous relation $\preceq$ over a set X that is reflexive, antisymmetric, and transitive. In this case, the set X is called a *partially ordered set* (or *poset*).

A *total order* (or *linear order, simple order, connex order, full order*) is a partial order in which any two elements are comparable. That is, a total order is a binary relation $\preceq$ over a given set X that is reflexive, transitive, antisymmetric, and strongly connected. A set equipped with a total order is called a *totally ordered set* (or *linearly ordered set, loset*).

The set of real numbers ordered by the usual "less than or equal to" ($\leq$) or "greater than or equal to" ($\geq$) relations is totally ordered. Hence, each subset of the real numbers is totally ordered, such as the positive integers, integers, and rational numbers.

A binary relation $\sim$ on a set X is called an *equivalence relation* if it is reflexive, symmetric, and transitive.

There are many equivalence relations in mathematics, including the relations "is equal to" on the set of real numbers, "is congruent to modulo n" on the set of positive integers, "is congruent to," and "is similar to" on the set of all triangles.

1.3.5. Every partition S of a set X may be identified with an equivalence relation on X, namely, the relation $\sim_S$, such that, for any $a, b \in X$, we have $a \sim_S b$ if and only if $a \in [b]_S$; equivalently, if and only if $b \in [a]_S$. The notation $\sim_S$ invokes the idea that the equivalence relation may be constructed from the partition.

Conversely, every equivalence relation $\sim$ on a set X may be identified with a partition S on X; its parts are the equivalence classes $X/\sim$. This is why it is sometimes stated informally that "an equivalence relation is the same as a partition." If S is the partition, identified with a given equivalence relation $\sim$, then some authors write $S = X/\sim$. This notation is suggestive of the idea that the partition is the set X divided into cells. The notation also invokes the idea that from the equivalence relation, one may construct the partition.

1.3.6. On the other hand, given a preorder $\preceq$ on X, one may define an equivalence relation $\sim$ on X, such that

$$a \sim b \quad \text{if and only if } a \preceq b, \quad \text{and} \quad b \preceq a.$$

The resulting relation $\sim$ is reflexive since the preorder $\preceq$ is reflexive; transitive, by applying the transitivity of $\preceq$ twice; and symmetric, by definition.

Using this relation, it is possible to construct a partial order $\leq$ on the quotient set $X/\sim$ of the equivalence relation $\sim$, which is the set of all equivalence classes of $\sim$:

$$[x] \leq [y] \quad \text{if and only if } x \preceq y.$$

Conversely, from any partial order on a partition of a set X, it is possible to construct a preorder on X itself. There is a one-to-one correspondence between preorders and pairs (partition, partial order).

1.3.7. A partition of a set X, together with a total order on the parts of the partition, gives an ordered structure, called an *ordered partition*.

Any ordered partition on X gives rise to a strict weak ordering on X, in which two elements $x, y \in X$ are incomparable when they belong to the same part of the partition; otherwise, they inherit the order of the parts that contain them.

In the other direction, for a set X with a strict weak order on it, the equivalence classes of incomparability give a partition of X, in which the sets inherit a total ordering from their elements, giving rise to an ordered partition.

1.3.8. A partition of the set $V_n = \{1, 2, \ldots, n\}$ with the corresponding equivalence relation $\sim$ is called *non-crossing* if it has the following property: *if four elements $a, b, c,$ and d of V_n, with $a < b < c < d$, satisfy $a \sim c$ and $b \sim d$, then $a \sim b \sim c \sim d$.*

The name comes from the following equivalent definition. Imagine the elements $1, 2, \ldots, n$ of V_n drawn as the n vertices of a regular n-gon (in clockwise order). A partition can then be visualized by drawing each block as a polygon (whose vertices are the elements of the block). The partition is non-crossing if these polygons do not intersect.

The number of non-crossing partitions of an n-set is the nth *Catalan number* $C_n = \frac{1}{n+1}\binom{2n}{n}$. The sequence of Catalan numbers, for $n = 0, 1, 2, \ldots$, starts with $1, 1, 2, 5, 14, 42, 132, 429, 1430, 4862, \ldots$ (sequence A000108 in the OEIS).

Partitions of sets and combinatorial configurations

1.3.9. There exist close connections between some kinds of partitions of a given n-set and the classical combinatorial configurations.

In fact, all classical combinatorial configurations are closely connected with different types of decompositions (including ordered and non-ordered partitions) of a given n-set into several parts. As a rule, one speaks about such distributions in terms of different (e.g., marked) balls and different (e.g., marked) boxes, or different (e.g., marked) balls and identical (e.g., plain) boxes.

For example, any partition of a given n-set X into k parts corresponds to the placement of n marked balls in x plain boxes, with no empty boxes allowed. Similarly, any ordered partition of a given n-set X into k parts corresponds to the placement of n marked balls in x marked boxes, with no empty boxes allowed.

In general, we can put some restrictions on the number of balls placed in a given box (for example, consider only placements with no multi-packs allowed; see also $(r_1, r_2, \ldots, r_k)$-partitions in the following). On the other hand, there can be no rules on placement, e.g., empty boxes can be allowed.

1.3.10. Given positive integers $r_1, r_2, \ldots, r_k$, let $(r_1, r_2, \ldots, r_k)$-*partition* of a given n-set X be an ordered set $X_1, X_2, \ldots, X_k$ of subsets of X, such that $X_i \cap X_j = \emptyset$ for $i \neq j$, $X_1 \cup X_2 \cup \cdots \cup X_k = X$, and $|X_1| = r_1, |X_2| = r_2, \ldots, |X_k| = r_k$. It is easy to show that the number $\sigma(r_1, r_2, \ldots, r_k)$ of $(r_1, r_2, \ldots, r_k)$-partitions can be calculated using the combinatorial formula for the number $P_{(r_1, r_2, \ldots, r_k)}$ of corresponding permutations with repetitions.

- For the number $\sigma(r_1, r_2, \ldots, r_k)$ of $(r_1, r_2, \ldots, r_k)$-partitions, the following identity holds:

$$\sigma(r_1, r_2, \ldots, r_k) = P_{(r_1, r_2, \ldots, r_k)} = \frac{(r_1 + r_2 + \cdots + r_k)!}{r_1! r_2! \cdots r_k!}$$

$$= \frac{n!}{r_1! r_2! \cdots r_k!}.$$

□ In fact, the subset $X_1 \subset X$ can be chosen in $C_n^{r_1}$ ways, the subset $X_2 \subset X \setminus X_1$ can be chosen in $C_{n-r_1}^{r_2}$ ways, etc. To choose a subset $X_k \subset X \setminus (X_1 \cup X_2 \cup \cdots \cup X_{k-1})$, we have exactly $C_{n-r_1-r_2-\cdots-r_{k-1}}^{r_k} = C_{r_k}^{r_k} = 1$ ways.

Then, according to the combinatorial rule of product, it holds that

$$\sigma(r_1, r_2, \ldots, r_k) = C_n^{r_1} C_{n-r_1}^{r_2} \cdots C_{n-r_1-r_2-\cdots-r_{k-1}}^{r_k}$$

$$= \frac{n!}{r_1(n-r_1)} \cdot \frac{(n-r_1)!}{r_2!(n-r_1-r_2)!}$$

$$\cdots \cdot \frac{(n-r_1-r_2-\cdots-r_{k-1})!}{r_k!(n-r_1-r_2-\cdots-r_k)!}$$

$$= \frac{n!}{r_1! r_2! \cdots r_k! 0!} = \frac{n!}{r_1! r_2! \cdots r_k!}.$$

However, the same formula can be obtained as a number $P_{(r_1, r_2, \ldots, r_k)}$ of permutations r_1 numbers 1, r_2 numbers 2, ..., r_k numbers k, where i denotes the number of the subset X_i to which an element with this index belongs. □

So, we can say that the number $\sigma(r_1, r_2, \ldots, r_k)$, which can be interpreted as the number of placements of n different balls in k different boxes, such that there should be exactly r_i balls in the ith box, $i = 1, 2, \ldots, k$, is equal to the number $P_{(r_1, r_2, \ldots, r_k)}$ of permutations with repetitions of r_1 numbers 1, r_2 numbers 2, ..., r_k numbers k.

1.3.11. A particular case, $r_1 = r_2 = \ldots = r_k = 1$, gives us $(1, 1, \ldots, 1)$-partition of an n-set X with $n = k$. In this case, the number $\sigma(1, 1, \ldots, 1)$ of $(1, 1, \ldots, 1)$-partitions of a given n-set X, i.e., the ordered partitions, in which each part contains exactly one element, is equal to the number P_n of permutations of n elements:

$$\sigma(1, 1, \ldots, 1) = P_{(1,1,\ldots,1)} = P_n = n!.$$

In terms of boxes, the number P_n of permutations of a given n-set is the number of ways to put n different balls in n different boxes such that there should be exactly one ball in each box. In other words, it is the number of placements of n marked balls in $k = n$ marked boxes with no multi-packs allowed.

1.3.12. From this point of view, one can say that the number A_n^k of k-permutations of a given n-set is the number of ways to place n different balls (not necessarily all balls) in k different boxes such that there should be exactly one ball in each box. In other words, it is the number of placements of n marked balls (not necessarily all balls) in k marked boxes with no multi-packs allowed.

As a rule, we speak about placements in which all balls should be distributed. So, the above example is an exception for our consideration. However, when we speak about partial permutations, there exists an another way to connect them with partitions. In this situation, we can speak about the partition of a string of n different elements into k (or fewer) substrings. The number of such decompositions is

$$P_{(1,\ldots,1,k-1)} = \frac{(n+k-1)!}{(k-1)!} = (n+k-1) \cdot \ldots \cdot (k+1) \cdot k = A_{n+k-1}^{k-1}.$$

1.3.13. Consider now another particular case of $(r_1, r_2, \ldots, r_k)$-partitions. In fact, any k-combination of a given n-set, $n \in \mathbb{N}$, $1 \leq k \leq n - 1$, is a partition of this set into two parts, one of which has k elements and the other has $n - k$ elements, i.e., it is an $(k, n - k)$-partition. So, the number of such partitions is C_n^k.

In terms of boxes, C_n^k is the number of ways to put n different balls in 2 different boxes such that in the first box, there are k balls, and in the second box, there are $n - k$ balls.

In this interpretation, k can be equal to zero (or to n): it means that the first (or second) box can be empty.

If the question is, What is the number of ways to put n different balls in 2 different boxes (without any restriction on the number of balls in each box), the answer is 2^n. It is the number of all subsets of a given n-set. If we choose some subset, we put it in the first box, the remaining balls will be placed in the second box. Note that

$$2^n = C_n^0 + C_n^1 + C_n^2 + \cdots + C_n^n,$$

i.e., we recall a simple fact that in any such distribution, the first box can contain $0, 1, 2, \ldots, n$ objects.

If the considered two boxes are identical, we obtain 2^{n-1} possibilities. And if each box should be non-empty, we obtain $2^{n-1} - 1 = \frac{2^n - 2}{2}$ possibilities.

1.3.14. Going back to the general $(r_1, r_2, \ldots, r_k)$-partitions, we can state that the number $OS(n, k)$ of all ordered partitions of a given n-set into k parts can be represented as

$$OS(n,k) = \sum_{r_1 + \cdots + r_k = n} \sigma(r_1, r_2, \ldots, r_k) = \sum_{r_1 + \cdots + r_k = n} \frac{n!}{r_1! r_2! \cdots r_k!}.$$

In terms of boxes, it is the number of placements of n different balls in k different boxes with no empty boxes allowed (and all balls should be placed).

So, the number of all ordered partitions of a given n-set, which is by definition the nth *ordered Bell number* (or *Fubini number*) $OB(n)$, can now be represented as

$$OB(n) = \sum_{k=1}^{n} \sum_{r_1 + \cdots + r_k = n} \sigma(r_1, r_2, \ldots, r_k)$$

$$= \sum_{k=1}^{n} \sum_{r_1 + \cdots + r_k = n} \frac{n!}{r_1! r_2! \cdots r_k!}.$$

1.3.15. What happens if, by placing n different balls in k different boxes, we allow some of boxes to be empty?

- The number of ways to place n different balls in k different boxes with empty boxes allowed is the number $\overline{A}_k^n = k^n$ of n-permutations with repetitions of a given k-set.

☐ In order to obtain this result, we should simply attach to each ball the number of the corresponding box. ☐

1.3.16. Now, we obtain the number of partitions of an n-set into k non-empty parts such that the order of parts is not important. (By definition, this number is equal to the Stirling number $S(n,k)$ of the second kind.)

- The number $S(n,k)$ of partitions of an n-set into k non-empty, non-ordered parts can be represented as

$$S(n,k) = \frac{1}{k!} \sum_{r_1+\cdots+r_k=n} \sigma(r_1, r_2, \ldots, r_k)$$

$$= \frac{1}{k!} \sum_{r_1+\cdots+r_k=n} \frac{n!}{r_1! r_2! \cdots r_k!}.$$

☐ Each such partition has k non-empty (non-ordered) parts with some cardinalities $r_1, \ldots, r_k$, $r_i \in \mathbb{N}$, $r_1 + \cdots + r_k = n$. Let us choose k positive integers $r_1, \ldots, r_k$ such that $r_1 + r_2 + \cdots + r_k = n$, and consider the various $(r_1, r_2, \ldots, r_k)$-partitions of the set X. The number $\sigma(r_1, r_2, \ldots, r_k)$ of such partitions is determined by the formula above. Since for us, in every such partition, the order of parts (i.e., the order of k different elements) is not important, we get that it is equal to

$$\frac{1}{k!} \sum_{r_1+\cdots+r_k=n} \sigma(r_1, r_2, \ldots, r_k) = \frac{1}{k!} \sum_{r_1+\cdots+r_k=n} \frac{n!}{r_1! r_2! \cdots r_k!}.$$

Note that we use for a proof a simple relation: $OS(n,k) = k! S(n,k)$. ☐

In terms of boxes, it is the number of ways of placing n different balls in k identical boxes with no empty boxes allowed (and all balls should be placed).

Now, we can obtain the number of partitions of an n-set X into several non-empty parts such that the order of parts is not important. (By definition, this number is equal to the nth Bell number $B(n)$.)

30 Catalan Numbers

- The number $B(n)$ of partitions of an n-set into non-empty, non-ordered parts can be represented as

$$\sum_{k=1}^{n} \frac{1}{k!} \sum_{r_1+\cdots+r_k=n} \sigma(r_1, r_2, \ldots, r_k) = \sum_{k=1}^{n} \frac{1}{k!} \sum_{r_1+\cdots+r_k=n} \frac{n!}{r_1! r_2! \cdots r_k!}.$$

Enumeration of partitions

1.3.17. In the following table, we present the numbers of several important ordered partitions of a given n-set X.

$|X|$	Any	Into k parts	$(r_1,\ldots,r_k)$-partitions	$(n, n-k)$-partitions	$(1,\ldots,1)$-partitions
0	1	1	1	1	1
1	1	0, 1	1	1, 1	1
2	3	0, 1, 2	0, 1, 2	1, 2, 1	2
3	13	0, 1, 6, 6	0, 1, 3 + 3, 6	1, 3, 3, 1	6
4	75	0, 1, 14, 36, 24	0, 1, 4 + 6 + 4, 24	1, 4, 6, 4, 1	24
$\vdots$	$\vdots$	$\vdots$	$\vdots$	$\vdots$	$\vdots$
n	$OB(n) = \sum_{k=1}^{n} k! S(n,k)$	$OS(n,k) = k! S(n,k)$	$P_{(r_1,\ldots,r_k)}$	C_n^k	P_n
OEIS	A000670	A019538	A019538	A007318	A000142

The following table lists the numbers of several important non-ordered partitions of a given n-set X.

$|X|$	Any	Into k parts	Non-crossing
0	1	0	1
1	1	0, 1	1
2	2	0, 1, 1	2
3	5	0, 1, 3, 1	5
4	15	0, 1, 7, 6, 1	14
$\vdots$	$\vdots$	$\vdots$	$\vdots$
n	$B(n) = \sum S(n,k)$	$S(n,k)$	C_n
OEIS	A000110	A008277	A000108

Exercises

1. Find all possible partitions of the sets $\emptyset$, $\{1\}$, $\{1,2\}$, $\{1,2,3\}$, $\{1,2,3,4\}$, and $\{1,2,3,4,5\}$. Check the results of these partitions, as given above.
2. Find all possible equivalence relations over the sets $\emptyset$, $\{1\}$, $\{1,2\}$, $\{1,2,3\}$, $\{1,2,3,4\}$, $\{1,2,3,4,5\}$. Compare the obtained results with those concerning the non-ordered partitions of the same sets.

3. Find all possible partial orders over the sets $\emptyset$, $\{1\}$, $\{1,2\}$, $\{1,2,3\}$, $\{1,2,3,4\}$, $\{1,2,3,4,5\}$. Compare the obtained results with those concerning the ordered partitions of the same sets.
4. Find the number of ways n labeled objects can be distributed into k non-empty parcels. Prove that it coincides with the number of special terms in n variables with a maximal degree of k. Check that it is also the number of surjections from an n-element set to a k-element set.

1.4. Permutations

Definitions and examples

1.4.1. A *permutation* of a set is, in rough terms, an arrangement of its members into a sequence or linear order or, if the set is already ordered, a rearrangement of its elements. The number of permutations of n distinct objects is equal to $n!$.

For example, written as tuples, there are six permutations of the set $\{1, 2, 3\}$, namely $(1, 2, 3)$, $(1, 3, 2)$, $(2, 1, 3)$, $(2, 3, 1)$, $(3, 1, 2)$, and $(3, 2, 1)$. These are all the possible orderings of this three-element set.

Formally, a permutation of a set S is defined as a bijection from S to itself.

That is, it is a function from S to S in which every element occurs exactly once as an image value. This is related to the rearrangement of the elements of S in which each element s is replaced by the corresponding $f(s)$. For example, the permutation $(3, 1, 2)$, mentioned above, is described by the function α, defined as $\alpha(1) = 3$, $\alpha(2) = 1$, $\alpha(3) = 2$. In mathematical texts, it is customary to denote permutations using lowercase Greek letters. Commonly, either α and β or σ, τ, and π are used. As the properties of permutations do not depend on the nature of the set elements, they are often the permutations of the set $V_n = \{1, 2, \ldots, n\}$.

1.4.2. All permutations of a set S with n elements form a symmetric group, denoted by S_n, where the group operation is the function composition $\circ$. Thus, for two permutations π and σ in the group S_n,

the following four group axioms hold:
- Closure: if π and σ are in S_n, then so is $\pi \circ \sigma$.
- Associativity: for any three permutations $\pi, \sigma, \tau \in S_n$, $(\pi \circ \sigma) \circ \tau = \pi \circ (\sigma \circ \tau)$.
- Identity: there is an identity permutation, denoted id and defined by $\mathrm{id}(x) = x$ for all $x \in S$; for any $\sigma \in S_n$, $\mathrm{id} \circ \sigma = \sigma \circ \mathrm{id} = \sigma$.
- Invertibility: for every permutation $\pi \in S_n$, there exists an inverse permutation $\pi^{-1} \in S_n$ so that $\pi \circ \pi^{-1} = \pi^{-1} \circ \pi = \mathrm{id}$.

Different representations of permutations

1.4.3. Since writing permutations elementwise, that is, as piecewise functions, is cumbersome, several notations have been invented to represent them more compactly. Cycle notation is a popular choice for many mathematicians due to its compactness and the fact that it makes a permutation's structure transparent, but other notations are still widely used, especially in application areas.

In Cauchy's *two-line notation*, one lists the elements of S in the first row, and for each, its image is listed below it in the second row. For instance, a particular permutation of the set $S = \{1, 2, 3, 4, 5\}$ can be written as $\sigma = \begin{pmatrix} 1 & 2 & 3 & 4 & 5 \\ 2 & 5 & 4 & 3 & 1 \end{pmatrix}$; this means that σ satisfies $\sigma(1) = 2, \sigma(2) = 5, \sigma(3) = 4, \sigma(4) = 3$, and $\sigma(5) = 1$.

If there is a "natural" order for the elements of S, say $x_1, x_2, \ldots, x_n$, then one uses this for the first row of the two-line notation. Under this assumption, one may omit the first row and write the permutation in *one-line notation* as

$$(\sigma(x_1)\ \sigma(x_2)\ \sigma(x_3)\ \cdots\ \sigma(x_n)),$$

that is, as an ordered arrangement of the elements of S. Care must be taken to distinguish the one-line notation from the cycle notation, described in the following. In mathematical literature, a common practice is to omit the parentheses in the one-line notation while using them for the cycle notation. The one-line notation is also called the *word representation* of a permutation. The example above would then be 2 5 4 3 1, since the natural order 1 2 3 4 5 would be assumed for the first row. (It is typical to use commas to separate these entries only if some have two or more digits.) This form is more compact and is common in elementary combinatorics and computer science.

1.4.4. A permutation can be decomposed into one or more disjoint cycles, that is, the *orbits*, which are found by repeatedly tracing the application of the permutation to some elements. For example, the permutation σ with $\sigma(7) = 7$ has a 1-cycle, (7), while the permutation π with $\pi(2) = 3$ and $\pi(3) = 2$ has a 2-cycle (2 3). In general, a cycle of length k, that is, consisting of k elements, is called a *k-cycle*.

The cycle notation describes the effect of repeatedly applying the permutation on the elements of the set. It expresses the permutation as a product of cycles; since distinct cycles are disjoint, this is referred to as *decomposition into disjoint cycles*.

So, the permutation 2 5 4 3 1 (in the one-line notation) could be written as $(125)(34)$ in the cycle notation.

In some contexts, it is useful to fix a certain order for the elements in the cycles and of the (disjoint) cycles themselves. Thus, the *canonical cycle notation* (or *standard representation*, *standard form*) of a permutation satisfies the following conditions: in each cycle, the largest element is listed first; the cycles are sorted in the increasing order of their first elements.

For example, $(312)(54)(8)(976)$ is a permutation in the canonical cycle notation. The canonical cycle notation does not omit 1-cycles.

The cycles (including the fixed points) of a permutation σ of a set with n elements partition this set; the lengths of these cycles form an integer partition of n, which is called the *cycle type* (or *cycle structure*, *cycle shape*) of σ. For example, the cycle type of the permutation $\beta = (1\,2\,5)(3\,4)(6\,8)(7)$ is $(3, 2, 2, 1)$.

This may also be written in a more compact form as $[1^1 2^2 3^1]$. More precisely, the general form is $[1^{\alpha_1} 2^{\alpha_2} \cdots n^{\alpha_n}]$, where $\alpha_1, \ldots, \alpha_n$ are the numbers of cycles of respective length. It is easy to prove the following fact (see, for example, [Deza24]):

- *The number of permutations of a given cycle type is* $\frac{n!}{1^{\alpha_1} 2^{\alpha_2} \cdots n^{\alpha_n} \alpha_1! \alpha_2! \cdots \alpha_n!}$.

By definition, the number of n-permutations with k cycles is equal to the unsigned Stirling number of the first kind $|s(n, k)|$. We consider this class of numbers in Chapter 4.

k-cycles and transpositions

1.4.5. Formally, a *cyclic permutation* (or *cycle*) is a permutation of the elements of some set S which maps the elements of some subset X of S to each other in a cyclic fashion while fixing (that is, mapping to themselves) all other elements of S. If X has k elements, the cycle is called a *k-cycle*. The set X is called the *orbit* of the cycle.

For example, given $S = \{1, 2, 3, 4\}$, the permutation $(1, 3, 2, 4)$ that sends 1 to 3, 3 to 2, 2 to 4, and 4 to 1 (so $X = S$) is a 4-cycle, and the permutation $(1, 3, 2)$ that sends 1 to 3, 3 to 2, 2 to 1, and 4 to 4 (so $X = \{1, 2, 3\}$, and 4 is a fixed element) is a 3-cycle.

- *The number of k-cycles in the symmetric group S_n is given, for $1 \leq k \leq n$, by the following equivalent formulas:*

$$\binom{n}{k}(k-1)! = \frac{n(n-1)\cdots(n-k+1)}{k} = \frac{n!}{(n-k)!k}.$$

☐ In fact, in order to obtain a k-cycle, we should first choose k elements from n elements (in $\binom{n}{k}$ ways) and then arrange the chosen k elements in a cyclic order (in $(k-1)!$ ways). ☐

1.4.6. The only 1-cycle is the identity permutation: all points of this permutation are fixed.

2-cycles are called *transpositions*; such permutations merely exchange two elements, leaving the others fixed.

For example, the permutation $\pi = 1423 = (1)(42)(3)$ that swaps 2 and 4 is a transposition.

The following result is well known (see, for example, [Kost82]):

- *Every permutation of a finite set can be expressed as the product of transpositions.*

Although many such expressions for a given permutation may exist, either they all contain an even number of transpositions or they all contain an odd number of transpositions. Thus, all permutations can be classified as *even* or *odd* depending on this number.

Ascents, descents, excedances, and inversions

1.4.7. In some applications, the elements of the set being permuted will be compared with each other. This requires that the set S have a total order so that any two elements can be compared. The set $V_n = \{1, 2, \ldots, n\}$ is totally ordered by the usual $\leq$ relation, and so it is the most frequently used set in these applications.

An *ascent* of a permutation σ of n elements is any position $i < n$ where the following value is bigger than the current one. That is, if $\sigma = \sigma_1 \sigma_2 \cdots \sigma_n$, then i is an ascent if $\sigma_i < \sigma_{i+1}$.

For example, the permutation $\sigma = 3452167$ has ascents (at positions) 1, 2, 5, and 6.

Similarly, a *descent* is a position $i < n$ with $\sigma_i > \sigma_{i+1}$; if $\sigma = \sigma_1 \sigma_2 \cdots \sigma_n$, then i is a descent if $\sigma_i > \sigma_{i+1}$.

So, every i with $1 \leq i < n$ is either an ascent or a descent of σ.

The number of permutations of n with k ascents is, by definition, the *Eulerian number* (of the first kind) $E(n, k)$; this is also the number of permutations of n with k descents. These numbers satisfy the recurrent relation $E(n, k) = (n-k)E(n-1, k-1) + (k+1)E(n-1, k)$. The triangle of numbers $E(n, k)$, $n = 1, 2, \ldots$, $k = 1, 2, \ldots, n$, starts with the values $1, 1, 1, 1, 4, 1, 1, 11, 11, 1, \ldots$ (sequence A008292 in the OEIS).

An *excedance* of a permutation $\sigma_1 \sigma_2 \cdots \sigma_n$ is an index j such that $\sigma_j > j$. If the inequality is not strict (that is, $\sigma_j \geq j$), then j is called a *weak excedance*. The number of n-permutations with k excedances coincides with the number of n-permutations with k descents.

The number of n-permutations with exactly k left-to-right maxima is equal to the unsigned Stirling number of the first kind, $|s(n, k)|$ (see Chapter 4).

1.4.8. An *inversion* of a permutation σ is a pair (i, j) of positions where the entries of a permutation are in the opposite order: $i < j$ and $\sigma_i > \sigma_j$.

For example, the permutation $\sigma = 23154$ has three inversions: $(1, 3), (2, 3)$, and $(4, 5)$, for the pairs of entries $(2, 1)$, $(3, 1)$, and $(5, 4)$.

So, a descent is just an inversion at two adjacent positions.

Sometimes, an inversion is defined as the pair of values (σ_i, σ_j) whose order is reversed; this makes no difference to the number of inversions, and this pair (reversed) is also an inversion in the above sense for the inverse permutation σ^{-1}. The number of inversions is an important measure of the degree to which the entries of a permutation are out of order; it is the same for σ and for σ^{-1}.

To bring a permutation with k inversions into order (that is, transform it into the identity permutation) by successively applying (right-multiplication by) adjacent transpositions is always possible and requires a sequence of k such operations.

The number of permutations of n elements with k inversions is expressed by a *Mahonian number*; it is the coefficient $[x^k] \prod_{m=1}^{n} \sum_{i=0}^{m-1} x^i$ of x^k in the expansion of the product

$$\prod_{m=1}^{n} \sum_{i=0}^{m-1} x^i = 1(1+x)(1+x+x^2)\cdots(1+x+x^2+\cdots+x^{n-1}).$$

The triangle of Mahonian numbers starts with the elements $1, 1, 1, 1, 2, 2, 1, 1, 3, 5, 6, 5, 3, 1, 1, \ldots$ (sequence A008302 in the OEIS).

Alternating permutations

1.4.9. An *alternating permutation* (or *zigzag permutation*) of the set $\{1, 2, 3, \ldots, n\}$ is an arrangement of those numbers so that each entry is alternately greater or lesser than the previous entry.

Formally, a permutation $\sigma = \sigma_1 \cdots \sigma_n$ is said to be *alternating* if its entries alternately rise and descend. Thus, each entry other than the first and the last should be either larger or smaller than both of its neighbors. The *up-down* permutations are such permutations for which the conditions $\sigma_1 < \sigma_2 > \sigma_3 < \cdots$ hold, while the *down-up* permutations satisfy the conditions $\sigma_1 > \sigma_2 < \sigma_3 > \cdots$. Many authors use the term "alternating" to refer only to the *up-down* permutations.

For example, the five alternating (up-down) permutations of $\{1, 2, 3, 4\}$ are $1\,3\,2\,4$ (as $1 < 3 > 2 < 4$); $1\,4\,2\,3$ (as $1 < 4 > 2 < 3$);

2 3 1 4 (as 2 < 3 > 1 < 4); 2 4 1 3 (as 2 < 4 > 1 < 3); 3 4 1 2 (as 3 < 4 > 1 < 2). There is a simple one-to-one correspondence between the down-up and up-down permutations: replacing each entry σ_i with $n+1-\sigma_i$ reverses the relative order of the entries.

By convention, in any naming scheme, the unique permutations of length 0 (the permutation of the empty set) and 1 (the permutation consisting of a single entry 1) are taken to be alternating.

This type of permutation was first studied by Désiré André in the 19th century (see [MSY96] and [Andr79]).

The numbers A_n of *up-down* permutations on $\{1, 2, \ldots, n\}$ are known as *Euler zigzag numbers* (or *up-down numbers*). The first few values of A_n are $1, 1, 1, 2, 5, 16, 61, 272, 1385, 7936, 50521, \ldots$ (sequence A000111 in the OEIS).

The numbers Z_n count the permutations of $\{1, \ldots, n\}$, that are either up-down or down-up (or both, for $n < 2$). So, $Z_n = 2A_n$ for $n \geq 2$; the first few values of Z_n are 1, 1, 2, 4, 10, 32, 122, 544, 2770, 15872, 101042, ... (sequence A001250 in the OEIS).

Pattern-avoiding permutations

1.4.10. A *pattern* is a smaller permutation embedded in a larger permutation. More precisely, a permutation *sigma* of an n-set contains a permutation α of a k-set, $k \leq n$, as a pattern if we can find the k digits of σ that appear in the same relative order as the digits of α. Consider, for example, the permutation $\sigma = 18274653$. If, instead of looking at all eight digits, we focus on the digits 8, 4, and 6, we note that the largest of these three digits comes first, the smallest of these digits comes second, and the middle of these three digits comes last, just as in the permutation $\alpha = 312$. So, the digits 846 are called a 312-pattern inside σ. It is easy to see that the permutation σ also contains the patterns 123, 132, 231, and 321. On the other hand, it does not contain the pattern 213. In this case, we say that the permutation σ *avoids* the pattern 213.

Pattern-avoiding permutations have provided many counting problems. In particular, the number of permutations of an n-set that avoid the pattern 213 is equal to the nth Catalan number C_n.

Circular permutations

1.4.11. Permutations, when considered as arrangements, are sometimes referred to as linearly ordered arrangements. In these arrangements, there is a first element, a second element, and so on. If, however, the objects are arranged in a circular manner, this distinct ordering no longer exists, that is, there is no "first element" in the arrangement, and any element can be considered the start of the arrangement. The arrangements of objects in a circular manner are called *circular permutations*. These can be formally defined as equivalence classes of ordinary permutations of the objects for the equivalence relation generated by moving the final element of the linear arrangement to its front.

Two circular permutations are equivalent if one can be rotated into the other (that is, cycled without changing the relative positions of the elements). The four circular permutations 1324, 3241, 2413, and 4132 on four letters are considered to be the same.

The circular arrangements are to be read counterclockwise, so 1324 and 1423 are not equivalent since no rotation can bring one to the other.

The number of circular permutations of a set S with n elements is $(n-1)!$.

Enumeration of permutations

1.4.12. In the following table, we present the numbers of main types of permutations of a given n-set.

$|X|$	Any	Circular	k-cycles	Derangements	With k-cycles	With k-anscents	Alternating	123-avoid
0	1	1	1	1	1	1	1	1
1	1	1	0, 1	0	0, 1	1, 0	1	1
2	2	1	0, 2, 1	1	0, 1, 1	1, 1, 0	2	2
3	6	2	0, 3, 3, 2	2	0, 2, 3, 1	1, 4, 1, 0	4	5
4	24	6	0, 4, 6, 8, 6	9	0, 6, 11, 6, 1	1, 11, 11, 1, 0	10	14
⋮	⋮	⋮	⋮	⋮	⋮	⋮	⋮	⋮
n	$n!$	$(n-1)!$	$C_{n-1}^k(k-1)!$	$[\frac{n}{e}]$	$|s(n,k)|$	$E(n,k)$	Z_n	C_n
OEIS	A000142	A000142	A111492	A000166	A008275	A008292	A001250	A000108

Exercises

1. Find all n-permutations for $n = 1, 2, 3, 4$. Find all even n-permutations for $n = 1, 2, 3, 4$. Find all odd n-permutations for $n = 1, 2, 3, 4$.
2. Find the number of k-cycles among all n-permutations, $n = 1, 2, 3, 4$. Prove that the number of k-cycles among all n-permutations is $\binom{n}{k}(k-1)!$.
3. Find the numbers $E(n, k)$ for $n = 1, 2, 3, 4$ and $k = 1, 2, \ldots, n$.
4. Find the numbers $|s(n, k)|$ for $n = 1, 2, 3, 4$ and $k = 1, 2, \ldots, n$.
5. Find the numbers A_n for $n = 1, 2, 3, 4$.
6. For $n = 1, 2, 3, 4, 5, 6$, find the number $!n$ of *derangements*, i.e., the permutations of an n-set that have no fixed points. Prove that the number $!n$ is equal to the nearest integer to $\frac{n!}{e}$.

1.5. Recurrent Relations

Recurrent relations in combinatorics

1.5.1. To derive the solutions to many problems, we can use the method of moving from a given problem concerning n objects to a similar problem concerning a smaller number ($n-1$, $n-2$, etc.) of objects. Such a method is called the *recurrence method* (from Latin *recurrere*, meaning *return*), and we call *recurrent relations* such relationships that allow us to get information about n objects using information about $n-1$ ($n-2$, $n-3$, etc.) objects.

A *recurrent relation of order* k for a sequence $f_1, f_2, \ldots, f_n, \ldots$ is a relationship that allows us to express an element f_{n+k} of the sequence through k previous elements $f_n, f_{n+1}, \ldots, f_{n+k-1}$ of the sequence; a *linear recurrent relation of the order* k *with constant coefficients* is the relation

$$f_{n+k} = a_1 f_{n+k-1} + a_2 f_{n+k-2} + \cdots + a_k f_n, \quad a_1, \ldots, a_k \in \mathbb{R},$$

where $a_1, \ldots, a_k$ are arbitrary constants.

1.5.2. Many recurrent relations arise when we solve combinatorial problems. For example, from the theory of combinatorial configurations, it is known that $P_m = mP_{m-1}$, that is, the number of permutations of m elements can be calculated if we know the number of permutations of $m-1$ elements. Thus, knowing the initial condition $P_1 = 1$, one can find $P_2, P_3 \ldots, P_n, \ldots$ using the recurrence method.

Similarly, the formula $C_n^k = C_{n-1}^k + C_{n-1}^{k-1}$ gives a recurrent relation for the number C_n^k of combinations: the number C_n^k of combinations from n elements by k elements can be found by knowing the number of combinations from $n-1$ elements (by k elements and $k-1$ elements). Thus, using the initial conditions $C_0^0 = 1$, $C_n^0 = 1$, and $C_n^n = 1$, all values of C_n^k can be obtained through sequential calculations. Using this algorithm, we get a number triangle, called *Pascal's triangle*.

Recurrent relations and Fibonacci numbers

1.5.3. A widely known problem connected with recurrent relations is *the rabbit problem*, which was considered by Leonardo Pisano (circa 1170 – circa 1250), one of the most significant Western mathematicians of the Middle Ages. Pisano, an Italian merchant, authored the book, *Liber Abaci* in 1202. He later came to be known as Fibonacci. (See, for example, [DeMo10] and [Hogg69]).

The problem is as follows:

- *A newly born breeding pair of rabbits are put in a field; each breeding pair mates at the age of one month, and at the end of their second month they always produce another pair of rabbits; and rabbits never die, but continue breeding forever. How many pairs will be there in one year?*

Simple consideration gives us a recurrent relation:

$$u_n = u_{n-1} + u_{n-2}$$

with initial conditions as $u_1 = u_2 = 1$, which allows us to calculate sequentially all elements of the sequence $u_1, u_2, \ldots, u_n, \ldots$, which are called *Fibonacci numbers*.

It is easy to check that there exist infinitely many sequences satisfying the same recurrent relation $a_n = a_{n-1} + a_{n-2}$. For example, if $a_1 = 3, a_2 = 5$, then we obtain a sequence $3, 5, 8, 13, 21, 34, \ldots$, which is an subsequence of the Fibonacci sequence. If $a_1 = 1, a_2 = 3$, then we get the sequence of *Lucas numbers* $1, 3, 4, 7, 11, 18, 29, 47, 76, 123, \ldots$ (sequence A000032 in the OEIS), etc. But with the initial conditions $u_1 = 1, u_2 = 1$, the Fibonacci sequence $1, 1, 2, 3, 5, 8, 13, 21, 34, 55, \ldots$ (sequence A000045 in the OEIS), defined by the recurrent ration $u_n = u_{n-1} + u_{n-2}$, is unique.

Recurrent relations and closed formulas

1.5.4. If we have a recurrent relation of the kth order and the initial values $f_1, f_2, \ldots, f_k$ of a sequence, we can write out all the members of the considered sequence but cannot compute the next element without the previous ones. Sometimes, it is not convenient, and there is often a need for a formula that expresses the nth element f_n of the sequence only as a function of its index n.

A formula, expressing the nth element of a sequence $f_1, f_2, \ldots, f_n, \ldots$ as a function of its index n is called a *closed formula* for the elements of the sequence $f_1, f_2, \ldots, f_n, \ldots$. If the sequence $f_1, f_2, \ldots, f_n, \ldots$ is defined by a recurrent relation, then the closed formula for the elements of the sequence $f_1, f_2, \ldots, f_n, \ldots$ is also called a *solution* of this recurrent relation.

So, the formula $P_n = n!$, giving the sequence $1, 2, 6, 24, 120, 720, \ldots$ of factorials, is a solution of the recurrent relation $P_{n+1} = (n+1)P_n$; it is a closed formula for the number of permutations of n elements.

The formula $C_n^k = \frac{n!}{k!(n-k)!}$ is a closed formula for the number of combinations from n elements by k elements.

The formula $A_n^k = \frac{n!}{(n-k)!}$ is a closed formula for the number of partial permutations from n elements by k elements.

The well-known *Binet's formula* $u_n = \frac{\left(\frac{1+\sqrt{5}}{2}\right)^n - \left(\frac{1-\sqrt{5}}{2}\right)^n}{\sqrt{5}}$ is a closed formula for the nth Fibonacci number u_n.

Solution of recurrent relations

1.5.5. For solutions to recurrent relations, there are no general rules. However, there is a very large class of recurrent relations solved by a common method. These are the *linear recurrent relations with constant coefficients*: $f_{n+k} = a_1 f_{n+k-1} + a_2 f_{n+k-2} + \cdots + a_k f_n$. Their solution is based on a study of the solutions to the equation $x^k = a_1 x^{k-1} + a_2 x^{k-2} + \cdots + a_k$, called the *characteristic equation* of a given recurrent relation $f_{n+k} = a_1 f_{n+k-1} + a_2 f_{n+k-2} + \cdots + a_k f_n$.

For $k = 2$, the theorem concerning the solutions of the linear recurrent relation $f_{n+2} = a_1 f_{n+1} + a_2 f_n$ is as follows ([DeMo10]):

- If the characteristic equation $x^2 = a_1 x + a_2$ of a linear recurrent relation $f_{n+2} = a_1 f_{n+1} + a_2 f_n$ has two different roots r_1 and r_2, then the general solution of the recurrent relation $f_{n+2} = a_1 f_{n+1} + a_2 f_n$ has the form $f_n = C_1 r_1^{n-1} + C_2 r_2^{n-1}$, where C_1 and C_2 are arbitrary constants; if the equation $x^2 = a_1 x + a_2$ has two coinciding roots $r_1 = r_2 = r$, then the general solution of the recurrent relation $f_{n+2} = a_1 f_{n+1} + a_2 f_n$ has the form $f_n = (C_1 + C_2 n) r^{n-1}$, where C_1 and C_2 are arbitrary constants.

A solution of a liner recurrent relation $f_{n+k} = a_1 f_{n+k-1} + \cdots + a_k f_n$ of the kth order, $k > 2$, can be found similarly. If all k roots of the characteristic equation $x^k = a_1 x^{k-1} + \cdots + a_k$ of the considered recurrent relation are different, then the general solution has the form

$$f_n = C_1 r_1^{n-1} + C_2 r_2^{n-1} + \cdots + C_k r_k^{n-1}.$$

If we have s coinciding roots, $r_1 = r_2 = \cdots = r_s = r$, then to this value corresponds a contribution $r^{n-1}(C_1 + C_2 n + \cdots + C_s n^{s-1})$ to the general solution.

Consider, for example, a linear recurrent relation $f_{n+4} = 5 f_{n+3} - 6 f_{n+2} - 4 f_{n+1} + 8 f_n$ of the fourth order with constant coefficients. The characteristic equation of this relation has the form $x^4 - 5x^3 + 6x^2 + 4x - 8 = 0$. The roots of this equation are $r_1 = -1, r_2 = r_3 = r_4 = 2$. Therefore, the general solution of the considered recurrent

relation has the form
$$f_n = C_1(-1)^{n-1} + (C_2 + C_3 n + C_4 n^2) 2^{n-1}.$$

1.5.6. The recurrent relation for Fibonacci numbers has the form $f_n = f_{n-1} + f_{n-2}$. It is a linear recurrent relation of the second order with constant coefficients ($a_1 = a_2 = 1$). The characteristic equation of this relation has the form $x^2 - x - 1 = 0$. It has two distinct roots: $\alpha = \frac{1+\sqrt{5}}{2}$ and $\beta = \frac{1-\sqrt{5}}{2}$. Therefore, the general solution to this recurrent relation has the form $f_n = C_1(\frac{1+\sqrt{5}}{2})^{n-1} + C_2(\frac{1-\sqrt{5}}{2})^{n-1}$. Using the initial conditions $f_1 = f_2 = 1$, we can easily show that $C_1 = \frac{1+\sqrt{5}}{2\sqrt{5}}$ and $C_2 = 1 - C_1 = -\frac{1-\sqrt{5}}{2\sqrt{5}}$. Then, a solution, corresponding to the specified initial conditions, takes the form $f_n = \frac{1}{\sqrt{5}}(\frac{1+\sqrt{5}}{2})^n - \frac{1}{\sqrt{5}}(\frac{1-\sqrt{5}}{2})^n$.

So, we have proven *Binet's formula*:
$$u_n = \frac{\alpha^n - \beta^n}{\sqrt{5}}, \quad \text{where} \quad \alpha = \frac{1+\sqrt{5}}{2}, \quad \beta = \frac{1-\sqrt{5}}{2}.$$

Exercises

1. Find in the sequence $u_1, u_2, \ldots, u_{12}$ of the first 12 Fibonacci numbers: all prime numbers; all composite numbers; and all square free numbers. Find the value of $\frac{u_{n+1}}{u_n}, n \leq 11$.
2. Prove the following identities:

   (a) $u_1 + u_2 + \cdots + u_n = u_{n+2} - 1$;

   (b) $u_1 + u_3 + \cdots + u_{2n-1} = u_{2n}$;

   (c) $u_2 + u_4 + \cdots + u_{2n} = u_{2n+1} - 1$;

   (d) $u_1^2 + u_2^2 + \cdots + u_n^2 = u_n u_{n+1}$;

   (e) $u_{n+m} = u_{n-1} u_m + u_n u_{m+1}$;

   (f) $u_{2n} = u_{n+1}^2 + u_{n-1}^2$;

   (g) $u_{3n} = u_{n+1}^3 + u_n^3 - u_{n-1}^3$;

   (h) $u_n^2 = u_{n-1} u_{n+1} + (-1)^{n+1}$;

   (i) $u_a + u_{a+1} + \cdots + u_n = u_{n+2} - u_{a+1}$;

   (j) $u_{n+2} u_{n-1} = u_{n+1}^2 - u_n^2$.

3. Prove *Binet's formula*, $u_n = \frac{(\frac{1+\sqrt{5}}{2})^n - (\frac{1-\sqrt{5}}{2})^n}{\sqrt{5}}$, through induction by n.
4. Find a recurrent relation for the sequence 1, 3, 6, 10, 15, ... of *triangular numbers*. Find the closed formula for the sequence of triangular numbers.
5. Make sure that if k is fixed, the binomial coefficients C_n^k can be calculated using the recurrence formula $\binom{n}{k} = \frac{n}{n-k}\binom{n-1}{k}$ with initial value $\binom{k}{k} = 1$. Do the appropriate calculations for $k = 1, 2, 3, 4, 5, 6, 7, 8$.
6. Prove that, for a fixed n, in order to calculate the elements of the nth row of Pascal's triangle, we can use the "one-dimensional" recurrent relation $\binom{n}{k} = \frac{n+1-k}{k}\binom{n}{k-1}$ with the initial condition $\binom{n}{0} = 1$.

1.6. Generating Functions

Classical generating functions

1.6.1. The method of generating functions is closely connected with the recurrence method; it became widespread in combinatorics in the 19th century (see, for example, [FlSe09] and [Land02]).

A *generating function* of the sequence $a_0, a_1, a_2, \ldots, a_n, \ldots$ is the formal power series

$$a_0 + a_1 x + a_2 x^2 + \cdots + a_n x^n + \cdots.$$

If in some area this series converges to the function $f(x)$, then we call the function $f(x)$ the *generating function* of the sequence $a_0, a_1, \ldots, a_n, \ldots$.

We consider the following several important examples of generating functions:

- The function $f(x) = \frac{1}{1-x}$ is the generating function of the sequence $1, 1, \ldots, 1, 1, \ldots$, as, using the known formula of the sum of an infinite geometric progression, we can say that

$$\frac{1}{1-x} = 1 + 1 \cdot x + 1 \cdot x^2 + \cdots, \quad |x| < 1.$$

- Squaring both sides of the equality of the previous example, we get the equality

$$\frac{1}{(1-x)^2} = (1 + x + x^2 + \cdots)(1 + x + x^2 + \cdots)$$
$$= 1 + 2x + 3x^2 + \cdots, \quad |x| < 1;$$

thus, the function $f(x) = \frac{1}{(1-x)^2}$ is the generating function for the sequence $1, 2, 3, \ldots, n, \ldots$ of positive integers.

- Similarly, it holds that

$$\frac{1}{(1-x)^3} = (1 + 2x + 3x^2 + \cdots)(1 + x + x^2 + \cdots)$$
$$= 1 + 3x + 6x^2 + \cdots, \quad |x| < 1;$$

thus, the function $f(x) = \frac{1}{(1-x)^3}$ is the generating function for the sequence $1, 3, 6, \ldots, \frac{n(n+1)}{2}, \ldots$ of *triangular numbers*.

- The function $f(x) = (1+x)^n$ is the generating function for the sequence C_n^k, $k = 0, 1, \ldots, n$, of numbers of k-combinations of a given n-set for all possible k, as

$$(1+x)^n = C_n^0 + C_n^1 x + \cdots + C_n^n x^n$$

for all real numbers x. In other words, for a fixed integer $n \geq 0$, the function $(1+x)^n$ is the generating function for the sequence $\binom{n}{k}$, $k = 0, 1, 2, \ldots, n$, of *binomial coefficients*:

$$(1+x)^n = \binom{n}{0} + \binom{n}{1}x + \cdots + \binom{n}{n}x^n, \quad x \in \mathbb{R}.$$

- If we change in the previous example the number $n \in \mathbb{N}$ to the number $\alpha \in \mathbb{R}$ and the binomial coefficients $\binom{n}{k} = \frac{n(n-1)\cdots(n-k+1)}{k!}$ to the *generalized binomial coefficients* $\binom{\alpha}{k} = \frac{\alpha(\alpha-1)\cdots(\alpha-k+1)}{k!}$, we get the decomposition

$$(1+x)^\alpha = 1 + \alpha x + \frac{\alpha(\alpha-1)}{1 \cdot 2}x^2 + \frac{\alpha(\alpha-1)(\alpha-2)}{1 \cdot 2 \cdot 3}x^3 + \cdots$$
$$= \binom{\alpha}{0} + \binom{\alpha}{1}x + \cdots + \binom{\alpha}{n}x^n + \cdots, \quad |x| < 1.$$

Therefore, for a given $\alpha \in \mathbb{R}$, the function $f(x) = (1+x)^\alpha$ is the generating function for the infinite sequence $\binom{\alpha}{k} = \frac{\alpha(\alpha-1)\cdots(\alpha-k+1)}{k!}$, $k = 0, 1, 2, \ldots$, of generalized binomial coefficients.

1.6.2. In general, finding the generating function of a given sequence is a very difficult task. However, there is a class of sequences for which the generating function may be found using a common, relatively simple algorithm. These sequences are given by linear recurrent relations with constant coefficients.

The rule for finding the generating function in this case looks like this ([DeMo10]):

- The generating function of the sequence $c_0, c_1, \ldots, c_n, \ldots$, satisfying the linear recurrent relation

$$b_0 c_{m+k} + b_1 c_{m+k-1} + \cdots + b_m c_k = 0$$

of the mth order with constant coefficients $b_0, b_1, \ldots, b_m$ and the initial conditions $c_0 = C_0, \ldots, c_{m-1} = C_{m-1}$, where $C_0, \ldots, C_{m-1}$ are some constants, has the form $\frac{f(x)}{\varphi(x)}$, i.e., is a ratio of two polynomials $f(x)$ and $\varphi(x)$, where $\varphi(x) = b_0 + b_1 x + \cdots + b_m x^m$ and $f(x) = a_0 + a_1 x + \cdots + a_n x^n$, $n < m$; the coefficients $a_1, \ldots, a_n$ of $f(x)$ can be found from the following conditions:

$$\begin{cases} a_0 = & b_0 c_0, \\ a_1 = & b_0 c_1 + b_1 c_0, \\ \vdots \\ a_n = b_0 c_0 + b_1 c_{n-1} + \cdots + b_n c_0. \end{cases}$$

Moreover, it holds that

$$\frac{f(x)}{\varphi(x)} = c_0 + c_1 x + \cdots + c_n x^n + \cdots, \quad |x| < \min\{|x_1|, \ldots, |x_m|\},$$

where $x_1, \ldots, x_m$ are solutions to the equation $\varphi(x) = 0$.

For example, find the generating function of the Fibonacci sequence $1, 1, 3, 5, 8, \ldots$. Since the elements of this sequence satisfy the recurrent relation $c_{k+2} - c_{k+1} - c_k = 0$ with initial conditions

$c_0 = 1, c_1 = 1$, then, in our notation, $b_0 = 1, b_1 = -1, b_2 = 1$; therefore, it holds that $\varphi(x) = 1 - x - x^2$. Calculating $a_0 = b_0 \cdot c_0 = 1 \cdot 1 = 1$, $a_1 = b_0 \cdot c_1 + b_1 \cdot c_0 = 1 \cdot 1 - 1 \cdot 1 = 0$, we get that $f(x) \equiv 1$. So, the generating function of the sequence $c_0 = 1, c_1 = 1, c_2 = 2, \ldots, c_n, \ldots$ has the form $\frac{1}{1-x-x^2}$.

The roots of the equation $x^2 + x - 1 = 0$ are $x_1 = \frac{-1+\sqrt{5}}{2}, x_2 = \frac{-1-\sqrt{5}}{2}$. So, the decomposition $\frac{1}{1-x-x^2} = c_0 + c_1 x + \cdots + c_n x^n + \cdots$, or, which is the same, the decomposition $\frac{1}{1-x-x^2} = u_1 + u_2 x + \cdots + u_{n+1} x^n + \cdots$, holds for $|x| < \frac{\sqrt{5}-1}{2} = \frac{1}{\alpha}$, where $\alpha = \frac{1+\sqrt{5}}{2}$. Therefore, we get the following statement:

- *The generating function of the sequence of Fibonacci numbers is $\frac{1}{1-x-x^2}$, i.e., it holds that*

$$\frac{1}{1-x-x^2} = u_1 + u_2 x + u_3 x^2 + \cdots + u_{n+1} x^n + \cdots$$

$$= \frac{1}{1-x-x^2}, \quad |x| < \frac{1}{\alpha} = \frac{\sqrt{5}-1}{2} \approx 0,6.$$

Exponential generating functions

1.6.3. When we study sequences $\{a_n\}_n$, it is often more convenient to investigate (instead of the properties of the series $\sum_{n=0}^{\infty} a_n x^n$) a behavior of the series $\sum_{n=0}^{\infty} a_n \frac{x^n}{n!}$.

An *exponential generating function* of the sequence $a_0, a_1, a_2, \ldots, a_n, \ldots$ is a formal power series

$$a_0 + a_1 \frac{x}{1!} + a_2 \frac{x^2}{2!} + \cdots + a_n \frac{x^n}{n!} + \cdots.$$

If in some region this series converges to the function $g(x)$, then we consider the function $g(x)$ as the *exponential generating function* of the sequence $a_0, a_1, \ldots, a_n, \ldots$.

We consider some classical examples of exponential generating functions, as follows:

- The function e^x is the exponential generating function of the sequence $1, 1, \ldots, 1, 1, \ldots$; it follows from the above definition and the well-known decomposition

$$e^x = 1 + \frac{1}{n!}x + \frac{1}{2!}x^2 + \cdots + \frac{1}{n!}x^n + \cdots,$$

which is true for any $x \in \mathbb{R}$.

- Considering the decomposition $(1+x)^\alpha = 1 + \alpha x + \frac{\alpha(\alpha-1)}{1\cdot 2}x^2 + \frac{\alpha(\alpha-1)(\alpha-2)}{1\cdot 2\cdot 3}x^3 + \cdots$, occurring for $|x| < 1$, we can check that for any complex value of α, the function $(1+x)^\alpha$ is the exponential generating function for the sequence $1, \alpha, \alpha(\alpha-1), \ldots, \alpha(\alpha-1)\cdots(\alpha-n+1), \ldots$:

$$(1+x)^\alpha = 1 + \alpha\frac{x}{1!} + \alpha(\alpha-1)\frac{x^2}{2!} + \alpha(\alpha-1)(\alpha-2)\frac{x^3}{3!}$$

$$+ \cdots + \alpha(\alpha-1)(\alpha-2)\cdots(\alpha-n+1)\frac{x^n}{n!} + \cdots, \quad |x| < 1.$$

1.6.4. There are numerous such examples, which can be obtained by very simple observation: the notion of exponential generating function is closely related to the decomposition of functions into the *Taylor series* (in fact, into the *Maclaurin series*), which is an infinite sum of power functions.

A *Taylor series* (at a point a) for the real- or complex-valued function $f(x)$ that is infinitely differentiable in some neighborhood of the point a is a formal power series of the form

$$f(a) + \frac{f'(a)}{1!}(x-a) + \frac{f''(a)}{2!}(x-a)^2 + \cdots + \frac{f^{(n)}(a)}{n!}(x-a)^n + \cdots,$$

where $f^{(n)}(a)$ denotes the nth derivative of f, evaluated at the point a. (The derivative of order zero of f is defined to be f itself; $(x-a)^0$ and $0!$ are both defined to be 1.)

If $f(x)$ is equal to the sum of its Taylor series in some area, we write that

$$f(x) = f(a) + \frac{f'(a)}{1!}(x-a) + \frac{f''(a)}{2!}(x-a)^2$$

$$+ \frac{f'''(a)}{3!}(x-a)^3 + \cdots + \frac{f^{(n)}(a)}{n!}(x-a)^n + \cdots$$

for such values of the argument x.

For $a = 0$, we obtain a *Maclaurin series* (or *Taylor–Maclaurin series*):
$$f(0) + \frac{f'(0)}{1!}x + \frac{f''(0)}{2!}x^2 + \cdots + \frac{f^{(n)}(0)}{n!}x^n + \cdots.$$
Thus, the Maclaurin series for the function $f(x)$ is, by definition, an exponential generating function of the sequence $f(0), f'(0), f''(0), \ldots, f^{(n)}(0), \ldots$ of derivatives of the function $f(x)$ of the orders n, $n = 0, 1, 2, 3, \ldots$, evaluated at zero. If the Maclaurin series converges to the function $g(x)$ in some neighborhood of zero, then we write that
$$g(x) = f(0) + \frac{f'(0)}{1!}x + \frac{f''(0)}{2!}x^2 + \cdots + \frac{f^{(n)}(0)}{n!}x^n + \cdots$$
and state that the *exponential generating function* of the sequence $f(0), f'(0), f''(0), \ldots, f^{(n)}(0), \ldots$ is the function $g(x)$.

1.6.5. In our book, we repeatedly use the decompositions into the Maclaurin series of some classical functions. A list of the most famous decompositions of this kind is provided as follows.

Maclaurin series decompositions of classical functions

- The *exponential function* e^x (with base e) has the decomposition
$$e^x = 1 + \frac{x}{1!} + \frac{x^2}{2!} + \frac{x^3}{3!} + \cdots = \sum_{n=0}^{\infty} \frac{x^n}{n!}, \quad x \in \mathbb{C}.$$

- The *natural logarithm* (logarithm with base e) has the decompositions (called *Mercator series*, or *Newton–Mercator series*)
$$\log(1+x) = x - \frac{x^2}{2} + \frac{x^3}{3} - \cdots = \sum_{n=0}^{\infty} \frac{(-1)^n x^{n+1}}{(n+1)}$$
$$= \sum_{n=1}^{\infty} \frac{(-1)^{n-1} x^n}{n}, \quad -1 < x \leq 1;$$
$$\log(1-x) = -x - \frac{x^2}{2} - \frac{x^3}{3} - \cdots = -\sum_{n=0}^{\infty} \frac{x^{n+1}}{(n+1)}$$
$$= -\sum_{n=1}^{\infty} \frac{x^n}{n}, \quad -1 \leq x < 1.$$

- The *binomial series* is the power series $\left(\text{with } \binom{\alpha}{n} = \dfrac{\alpha(\alpha-1)\cdots(\alpha-n+1)}{n!}\right)$

$$(1+x)^\alpha = 1 + \sum_{n=1}^{\infty} \binom{\alpha}{n} x^n, \quad |x| < 1,\ \alpha \in \mathbb{C}.$$

- The *geometric series*, a special case of the binomial series with $\alpha = -1$, is

$$\frac{1}{1-x} = 1 + x + x^2 + x^3 + \cdots = \sum_{n=0}^{\infty} x^n, \quad |x| < 1.$$

- The *trigonometric function* $\sin x$ has the following Maclaurin series:

$$\sin x = x - \frac{x^3}{3!} + \frac{x^5}{5!} - \cdots = \sum_{n=0}^{\infty} \frac{(-1)^n}{(2n+1)!} x^{2n+1}, \quad x \in \mathbb{C}.$$

- The *trigonometric function* $\cos x$ has the following Maclaurin series:

$$\cos x = 1 - \frac{x^2}{2!} + \frac{x^4}{4!} - \cdots = \sum_{n=0}^{\infty} \frac{(-1)^n}{(2n)!} x^{2n}, \quad x \in \mathbb{C}.$$

1.6.6. In combinatorial problems, the question of the convergence of a formal series specifying a generating function of a given sequence is not discussed in detail. We write out the function to which this series converges and name it the generating function of the considered sequence based on the fact that the specified convergence is performed in some neighborhood of zero, but the exact properties of this neighborhood, as a rule, are not investigated. However, in several classic cases, when the information about the convergence of a particular series is well known, we indicate existing boundaries, referring to the corresponding sources.

Readers can themselves investigate the question of convergence of the decompositions presented in the book ([Ficht01]): a *convergence interval* of the power series $\sum_{n=0}^{\infty} a_n x^n$ is a set of values of x satisfying the condition $|x| < r$, and a *convergence radius* r in most cases can be calculated by the *Cauchy–Hadamar formula*:

$$\frac{1}{r} = \overline{\lim_{n \to \infty}} |a_n|^{1/n}.$$

Multidimensional generating functions

1.6.7. In some cases, even the exponential generating function cannot help in a considered situation. It happens when the sequence under investigation is "multidimensional," that is, when it contains not one but two or more parameters. In our arsenal, an example of such a sequence exists: it is the sequence of elements of Pascal's triangle C_n^k. We got out of the situation by describing by the function $(1+x)^n$ the behavior of elements $C_n^0, C_n^1, C_n^2, \ldots, C_n^n$ of a given row of the triangle (with the index n), that is, considering the finite one-dimensional subsequence of a two-dimensional object consisting of all possible numbers C_n^k, $n \geq 0, 0 \leq k \leq n$.

Of course, any "two-dimensional" set $c_{n,k}$ with non-negative integer indices n, k can be "expanded" into a line, but sometimes this is not convenient. Such sequences can be represented by the *two-dimensional generating function* $\sum_{n,k=0}^{\infty} c_{n,k} x^n y^k$. For example, for the elements of Pascal's triangle, such a function has the form $\sum_{n,k=0}^{\infty} C_n^k x^n y^k$. Simple reasoning allows us to obtain the following result:

$$\sum_{n,k=0}^{\infty} C_n^k x^k y^n = \sum_{n=0}^{\infty} \left(\sum_{k=0}^{\infty} C_n^k x^k \right) y^n = \sum_{n=0}^{\infty} \left(\sum_{k=0}^{n} C_n^k x^k \right) y^n$$

$$= \sum_{n=0}^{\infty} (1+x)^n y^n = \sum_{n=0}^{\infty} ((1+x)y)^n = \frac{1}{1-(1+x)y}.$$

By studying the resulting function $\frac{1}{1-(1+x)y}$, we can derive the properties of all binomial coefficients. In some situations, they turn out to be very useful.

Exercises

1. Let $f(x)$ and $g(x)$ be the generating functions of the sequences $\{a_n\}_n$ and $\{b_n\}_n$, respectively. Prove that $h(x)$ is the generating function of the sequence $\{c_n\}_n$ if:
   - $h(x) = \alpha f(x) + \beta g(x)$, $c_n = \alpha a_n + \beta b_n$;
   - $h(x) = f(x)g(x)$, $c_n = a_0 b_n + a_1 b_{n-1} + \cdots + a_n b_0$;

- $h(x) = tf(x)$, $c_0 = 0$, $c_n = a_{n-1}$, $n \geq 1$;
- $h(x) = f(\alpha x)$, $c_n = \alpha^n a_n$;
- $h(x) = tf'(x)$, $c_n = na_n$;
- $h(x) = \frac{f(x)}{1-x}$, $c_n = a_0 + a_1 + \cdots + a_n$.

2. Prove that the generating function of the sequence 1, 2, 3, ..., n, ... of positive integers has the form $\frac{1}{(1-x)^2}$, taking the derivative of the generating function $\frac{1}{1-x}$ of the sequence 1, 1, 1, ...

3. Find the generating function of the sequence $1^2, 2^2, 3^2, \ldots, n^2, \ldots$ of the squares of positive integers. Find the generating function of the sequence $1^3, 2^3, 3^3, \ldots, n^3, \ldots$ of the cubes of positive integers.

4. Prove that there are the following decompositions for the *square root function* and its inverse:

$$(1+x)^{\frac{1}{2}} = 1 + \tfrac{1}{2}x - \tfrac{1}{8}x^2 + \tfrac{1}{16}x^3 - \tfrac{5}{128}x^4 + \tfrac{7}{256}x^5 - \cdots$$

$$= \sum_{n=0}^{\infty} \frac{(-1)^{n-1}(2n)!}{4^n(n!)^2(2n-1)} x^n,$$

$$(1+x)^{-\frac{1}{2}} = 1 - \tfrac{1}{2}x + \tfrac{3}{8}x^2 - \tfrac{5}{16}x^3 + \tfrac{35}{128}x^4 - \tfrac{63}{256}x^5 + \cdots$$

$$= \sum_{n=0}^{\infty} \frac{(-1)^n(2n)!}{4^n(n!)^2} x^n.$$

5. Check that the generating function for the sequence $a_n = \alpha^n$, $n = 0, 1, 2, \ldots$, is $f(x) = (1 - \alpha x)^{-1}$.

1.7. Asymptotic Formulas

Infinite sums and generating functions

1.7.1. Generating functions can be used to solve some problems of finding infinite and finite sums.

For example, the generating function of the sequence of Fibonacci numbers has the form $\frac{1}{1-x-x^2}$, and the corresponding decomposition $\frac{1}{1-x-x^2} = u_1 + u_2 x + u_3 x^2 + \cdots$ holds for $|x| < \frac{1}{\alpha}$, $\alpha = \frac{1+\sqrt{5}}{2}$. Taking

$x = \frac{1}{2}$, we obtain on the right-hand side of the equality an infinite sum $u_1 + \frac{u_2}{2} + \frac{u_3}{4} + \cdots + \frac{u_n}{2^{n-1}} + \cdots$, and on the left-hand side, we have the value $\frac{1}{1-\frac{1}{2}-\frac{1}{4}} = \frac{4}{4-2-1} = 4$. Therefore, the following equality holds:

$$u_1 + \frac{u_2}{2} + \frac{u_3}{4} + \cdots + \frac{u_n}{2^{n-1}} + \cdots = \sum_{n=1}^{\infty} \frac{u_n}{2^{n-1}} = 4.$$

The generating function of the sequence $1, 1, \ldots, 1, \ldots$ has the form $\frac{1}{1-x}$, and the decomposition $\frac{1}{1-x} = 1 + x + x^2 + \cdots$ exists for $|x| < 1$. Taking $x = \frac{1}{2}$, we get the equality $\sum_{n=0}^{\infty} 2^{-n} = 2$. Taking $x = -\frac{1}{2}$, we get the equality $\sum_{n=0}^{\infty} (-2)^{-n} = \frac{2}{3}$.

The generating function of the finite sequence $C_n^0, C_n^1, \ldots, C_n^n$ has the form $(1+x)^n$. In this case, the equality $(1+x)^n = C_n^0 + C_n^1 x + \cdots + C_n^n x^n$ holds for all $x \in \mathbb{R}$. Taking $x = 1$, we get that $C_n^0 + C_n^1 + \cdots + C_n^n = 2^n$. Taking $x = -1$, we get that $C_n^0 - C_n^1 + \cdots + (-1)^n C_n^n = 0$. Taking $x = 2$, we get that $C_n^0 + 2C_n^1 + \cdots + 2^n C_n^n = 3^n$.

Big-O and little-o notations

1.7.2. In order to obtain from a given infinite sum the corresponding finite sum, we can usually use the so-called *asymptotic formulas*.

Let $f(x)$ and $g(x)$ be complex-valued functions of real argument such that $g(x) > 0$ for all large x, i.e., for all $x > a$, where a is a fixed positive real number. We say that $f(x) = O(g(x))$ if there exists a positive real number x_0 such that, for all $x \geq x_0$, the inequality $|f(x)| \leq C \cdot g(x)$ holds, where $C > 0$ is a positive constant.

Consider several simple examples:

- $\sin x = O(1)$ since for all $x \in \mathbb{R}$, it holds that $|\sin x| \leq 1$, i.e., an inequality $|\sin x| \leq Cg(x)$ holds for $C = 1$ and $g(x) \equiv 1$.
- $\frac{1}{x} = O(1)$ since for all $x \geq 1$, it holds that $|\frac{1}{x}| = \frac{1}{x} \leq 1$, i.e., the inequality $|\frac{1}{x}| \leq Cg(x)$ holds for $x_0 = 1, C = 1, g(x) \equiv 1$.
- $5x^2 - 2x + 1 = O(x^2)$ since for all $x \geq 1$, it holds that $|5x^2 - 2x + 1| \leq |5x^2| + |-2x| + 1 = 5|x^2| + 2|x| + 1 \leq 5|x^2| + 2|x^2| + |x^2| = 8|x^2| = 8x^2$, i.e., the inequality $|5x^2 - 2x + 1| \leq Cg(x)$ holds for $x_0 = 1, C = 8, g(x) = x^2$.

54   Catalan Numbers

In addition to the "O()"-symbol (*big-O notation*), the "o()"-symbol (*little-o notation*) is often used: we say that $f(x) = o(g(x))$ if $\lim_{x \to \infty} \frac{f(x)}{g(x)} = 0$.

In other words, writing $f(x) = O(g(x))$ means that at large x, the ratio $\frac{|f(x)|}{|g(x)|}$ is bounded from above, whereas the condition $f(x) = o(g(x))$ indicates that at large x, this ratio is small.

1.7.3. A classic example of using asymptotic formulas involves studying the behavior of partial sums $\sum_{n \leq x} \frac{1}{n}$ of the *harmonic series* $\sum_{n=1}^{\infty} \frac{1}{n}$.

- For the sum $\sum_{n \leq x} \frac{1}{n}$, the following asymptotic formula holds:

$$\sum_{n \leq x} \frac{1}{n} = \log x + O(1).$$

□ As $\frac{1}{n+1} \leq \int_{n}^{n+1} \frac{dt}{t} \leq \frac{1}{n}$, it holds that

$$\sum_{n=1}^{N} \frac{1}{n} - 1 + \frac{1}{N+1}$$

$$= \sum_{n=1}^{N} \frac{1}{n+1} \leq \sum_{n=1}^{N} \int_{n}^{n+1} \frac{dt}{t} = \log N + \int_{N}^{N+1} \frac{dt}{t} \leq \sum_{n=1}^{N} \frac{1}{n}.$$

As $\frac{1}{N+1} \leq \int_{N}^{N+1} \frac{dt}{t} \leq \frac{1}{N}$, we get that

$$0 < \frac{1}{N+1} \leq \int_{N}^{N+1} \frac{dt}{t} \leq \sum_{n=1}^{N} \frac{1}{n} - \log N \leq \frac{1}{N} + 1 - \frac{1}{N+1} \leq 2.$$

So, it holds that

$$\sum_{n=1}^{N} \frac{1}{n} - \log N = O(1). \qquad \square$$

1.7.4. Using more serious reasoning, the previous theorem can be clarified ([DeMo10]):

- For the sum $\sum_{n \leq n} \frac{1}{n}$, there exists an asymptotic formula

$$\sum_{n \leq x} \frac{1}{n} = \log x + \gamma + O(x^{-1}),$$

where $\gamma = 0.5772215\ldots$ is the Euler–Mascheroni constant: $\gamma = \lim_{n \to \infty} (\sum_{k=1}^{n} \frac{1}{k} - \log n)$..

Asymptotic equivalence

1.7.5. Our results allow us to state that, for large x, the sum $\sum_{n \leq x} \frac{1}{n}$ behaves itself as $\log x$. In fact, since for $x \to \infty$ the value $O(1)$ is bounded by a positive constant, the ratio $\frac{O(1)}{\log x}$ goes to zero, and we have that

$$\lim_{x \to \infty} \frac{\sum_{n \leq x} \frac{1}{n}}{\log x} = \lim_{x \to \infty} \frac{\log x + O(1)}{\log x} = \lim_{x \to \infty} \left(1 + O\left(\frac{1}{\log x}\right)\right) = 1.$$

For an indication of the asymptotic coincidence of two functions, one uses a special symbol.

Let $f(x)$ and $g(x)$ be complex-valued functions of real argument. If $\lim_{x \to \infty} \frac{f(x)}{g(x)} = 1$, we say that the functions $f(x)$ and $g(x)$ are asymptotically equal and write $f(x) \sim g(x)$.

So, we can write that

$$\sum_{n \leq x} \frac{1}{n} \sim \log x.$$

Consider several examples:

- $5x^2 - 2x + 1 \sim 5x^2$, as $\lim_{x \to \infty} \frac{5x^2 - 2x + 1}{5x^2} = \lim_{x \to \infty} (1 - \frac{2x-1}{5x^2}) = 1$.
- $C_n^2 \sim \frac{n^2}{2}$, as $\lim_{n \to \infty} \frac{C_n^2}{n^2/2} = \lim_{n \to \infty} \frac{n(n-1)/2}{n^2/2} = \lim_{n \to \infty} \frac{n^2-n}{n^2} = \lim_{n \to \infty} (1 - \frac{1}{n}) = 1$.
- $\sin x \sim x$ for $x \to 0$, as $\lim_{x \to 0} \frac{\sin x}{x} = 1$.

Stirling's formula

1.7.6. For estimations and asymptotic formulas for the majority of combinatorial numbers, in particular, binomial coefficients, we should use *Stirling's formula* (or *Stirling's approximation*, *Moivre formula*), which is a formula for the approximative calculation of factorials (see, for example, [GKP94], [Gelf59], and [Kara83]).

It is a good approximation that leads to accurate results even for small values of n. It is named after J. Stirling, though a related but less precise result was first stated by A. de Moivre.

1.7.7. The weak version of the estimation of $n!$ is as follows:

- For all $n \geq 2$, it holds that

$$\log n! = n \log n + O(n).$$

☐ Let us show that

$$n \log n - n < \log n! < n \log n.$$

In fact, the inequality $\log n! < n \log n$ is a consequence of the trivial inequality $n! < n^n$.

On the other hand, a Riemann sum approximation for $\int_1^n \log x\, dx$ using the right endpoints is $\log 2 + \cdots + \log n = \log n!$, which overestimates, so

$$\log n! > \int_1^n \log x\, dx = n \log n - n + 1.$$

So, we get the following inequalities:

$$n \log n - n + 1 < \log n! < n \log n, \quad -n + 1 < \log n! - n \log n < 0,$$
$$|\log n! - n \log n| < n - 1 < n.$$

Therefore, we obtain that $\log n! - n \log n = O(n)$ and $\log n! = n \log n + O(n)$. ☐

Dividing by $n \log n$ both sides of the last equality, we obtain that $\frac{\log n!}{n \log n} = 1 + O(\frac{1}{\log n})$. So, it is proven that

$$\log n! \sim n \log n.$$

1.7.8. The simplest variant of Stirling's formula has the form
$$\log n! = n \log n - n + O(\log n).$$

□ As before, a Riemann sum approximation for $\int_1^n \log x\, dx$ using the right endpoints has the form $\log 2 + \cdots + \log n = \log n!$, and we get
$$\log n! > \int_1^n \log x\, dx = n \log n - n + 1.$$

On the other hand, a Riemann sum approximation for $\int_1^n \log x\, dx$ using left endpoints is $\log 1 + \cdots + \log(n-1) = \log n! - \log n$, and we get
$$\log n! - \log n < \int_1^n \log x\, dx = n \log n - n + 1,$$

or, equivalently,
$$\log n! < n \log n - n + \log n + 1.$$

So, we get the inequalities
$$n \log n - n + 1 < \log n! < n \log n - n + \log n + 1,$$
$$1 < \log n! - n \log n - n < \log n + 1,$$
$$|\log n! - n \log n - n| < \log n + 1 \leq 2 \log n.$$

Therefore, we obtain that $\log n! - n \log n + n = O(\log n)$ and $\log n! = n \log n - n + O(\log n)$. □

In other words, we have proven that
$$n! = e^{n \log n - n + O(\log n)}.$$

As $e^{n \log n - n} = n^n \cdot e^{-n} = (\frac{n}{e})^n$, we get the equality
$$n! = \left(\frac{n}{e}\right)^n \cdot O(n).$$

1.7.9. The error term $O(\log n)$ in the above asymptotic formula for $\log n!$ can be expressed more precisely as $\frac{1}{2}\log(2\pi n) + O(\frac{1}{n})$, giving

the asymptotic formula
$$\log n! = n \log n - n + \tfrac{1}{2} \log(2\pi n) + O(\tfrac{1}{n}).$$
So, a more exact approximation of $n!$ is
$$\lim_{n \to \infty} \frac{n!}{\sqrt{2\pi n} \left(\frac{n}{e}\right)^n} = 1,$$
which is equivalent to the classical variant of *Stirling's formula*:
$$n! \sim \sqrt{2\pi n} \left(\frac{n}{e}\right)^n.$$
Note that a formula for approximation of $n!$ was first discovered by A. de Moivre in the form
$$n! \sim [\text{constant}] \cdot n^{n+\frac{1}{2}} e^{-n}.$$
Stirling's contribution consisted of showing that the constant is precisely $\sqrt{2\pi}$.

1.7.10. Stirling's formula is the first approximation of the decomposition of the factorial into the Stirling series (see sequences A001163 and A001164 in the OEIS):
$$n! = \sqrt{2\pi n} \left(\frac{n}{e}\right)^n$$
$$\cdot \left(1 + \frac{1}{12n} + \frac{1}{288n^2} - \frac{139}{51840n^3} - \frac{571}{2488320n^4} + O\left(n^{-5}\right)\right).$$

However, the following version of the bounds for $n!$ holds for all $n \geq 1$, rather than only asymptotically (H. Robbins, 1955; [Robb55]):
$$\sqrt{2\pi n} \left(\frac{n}{e}\right)^n e^{\frac{1}{12n+1}} < n! < \sqrt{2\pi n} \left(\frac{n}{e}\right)^n e^{\frac{1}{12n}}.$$

Exercises

1. Find the following sums:
- $u_1 + \frac{u_2}{3} + \frac{u_3}{9} + \cdots + \frac{u_n}{3^{n-1}} + \cdots;$
- $\frac{u_1}{5} + \frac{u_2}{25} + \frac{u_3}{125} + \cdots + \frac{u_n}{5^n} + \cdots;$
- $5 + \frac{5}{4} + \frac{5}{16} + \cdots + \frac{5}{4^n} + \cdots;$
- $\frac{3}{8} + \frac{9}{16} + \frac{27}{32} + \cdots + \frac{3^n}{2^{n+2}} + \cdots.$

Give other examples of convergent infinite sums; find their values.
2. Prove that $\cos 2x = O(1)$; $\frac{1}{1+x} = O(x^{-1})$; $5x^7 - 98x_1^4 = O(x^7)$; $(n+1)^3 = O(n^3)$; $(n+1)^3 = o(n^4)$. Give other examples that use the $O()$-symbol.
3. Prove the following properties of the $O()$-symbol:
   (a) $C \cdot O(f(x)) = O(f(x))$;
   (b) $O(-f(x)) = O(f(x))$;
   (c) $O(f(x)) + O(f(x)) = O(f(x))$;
   (d) $O(f(x)) \cdot O(g(x)) = O(f(x)g(x))$;
   (e) $O(O(f(x))) = O(f(x))$;
   (f) $O(f(x)) + o(f(x)) = O(f(x))$.
4. Prove that $e^x = 1 + x + \frac{x^2}{2} + O(x^3)$.
5. Prove that $\sum_{n \leq x} \frac{1}{n} = \log x + \gamma + O(\frac{1}{x})$, where $\gamma = 0,5772215\ldots$ is the Euler–Mascheroni constant.
6. Prove that for $x \to 0$, the relation $\cos x \sim 1$ holds.
7. Check that the relation $f(x) \sim g(x)$ means that $f(x) = g(x) + o(g(x))$.

References

[AbSt72], [Abra74], [Aign82], [Ande03], [BaCo87], [Bern99], [Berg71], [BeSl95], [BiBa70], [Bond93], [Bron01], [Buch09], [Comt74], [CoRo96], [Dave99], [Dede63], [DeKo13], [Deza18], [DeMo10], [Dick05], [Erus00], [Eule48], [Ficht01], [FlSe09], [Gelf98], [Gelf59], [GGL95Û], [GuJa90], [Hagg04], [Hall70], [Hara03], [GaSa92], [HaWr79], [Hogg69], [Hons91], [IrRo90], [Jablo01], [Kara83], [KoKo77], [KuAD88], [Land94], [Land02], [Lege79], [LiWi92], [Lips88], [Mark0], [Mazu10], [Mazu10], [Ore48], [Pasc54], [Plou92], [Rior79], [Rior80], [RoTe09], [Robb55], [Rybn82], [Smit84], [Stra16], [Stru87], [Stan97], [Uspe76], [Vile14], [VVV06], [Voro61], [VoKu84], [Went99].

Chapter 2
Pascal's Triangle

2.1. Construction of Pascal's Triangle

A history of the question

2.1.1. *Pascal's triangle* is a triangular array arising in many problems of probability theory, combinatorics, and algebra.

Known to much of the Western world as Pascal's triangle, it is named after the French mathematician Blaise Pascal (1623–1662), although other mathematicians studied it centuries before him in Persia, India, China, Germany, and Italy.

The triangle is constructed by the following law: the top row (zeroth row) of the triangle contains only one unity. The next (first) row contains two unities. In the second row, the rightmost and leftmost elements are again equal to unities, and the internal element is obtained as the sum of the two elements of the first row, located above it, that is, it is equal to $1 + 1 = 2$. In general, the rightmost and leftmost elements of any row of Pascal's triangle are equal to 1, and any internal element, if it exists, is obtained as the sum of the elements located directly above it in the previous row.

Through sequential calculations, we get the following picture:

$$\begin{array}{ccccccccc} & & & & 1 & & & & \\ & & & 1 & & 1 & & & \\ & & 1 & & 2 & & 1 & & \\ & 1 & & 3 & & 3 & & 1 & \\ 1 & & 4 & & 6 & & 4 & & 1 \\ \vdots & \vdots & \vdots & \vdots & \vdots & \vdots & \vdots & \vdots & \vdots \end{array}$$

It is natural to call such a representation of the triangle the *isosceles Pascal's triangle*.

The Persian mathematician Al-Karaji (953–1029) wrote a now-lost book which contained the first description of Pascal's triangle. It was later reproduced by Omar Khayyam (1048–1131), another Persian mathematician; thus, the triangle is also referred to as *Khayyam's triangle* in Iran. Several theorems related to the triangle were known, including the binomial theorem. O. Khayyam used a method of finding the nth roots based on the binomial expansion and, therefore, the binomial coefficients.

Pascal's triangle was known in China during the early 11th century as a result of the work of a Chinese mathematician, Jia Xian (1010–1070). During the 13th century, Yang Hui (1238–1298) presented the triangle, and hence it is still known as *Yang Hui's triangle* in China.

2.1.2. In Europe, Pascal's triangle appeared for the first time in the *Arithmetic* of Jordanus de Nemore in the 13th century.

Petrus Apianus (1495–1552) published the full triangle on the frontispiece of his book on business calculations in 1527.

Michael Stifel (1487–1567) published a portion of the triangle (from the second to the middle column in each row) in 1544, describing it as a table of figurate numbers.

In Italy, Pascal's triangle is referred to as *Tartaglia's triangle*, named after the Italian algebraist Niccolò Fontana Tartaglia (1500–1577), who published six rows of the triangle in 1556.

Gerolamo Cardano (1501–1576) also published the triangle as well as the additive and multiplicative rules for its construction in 1570.

However, the triangle was named after Blaise Pascal (1623–1662), a French mathematician, physicist, inventor, and philosopher. B. Pascal's *Treatise on Arithmetical Triangle* (*Traité du triangle arithmétique*) was published in 1665 (or, according to other sources, in 1663). In this work, B. Pascal collected several results then known about the triangle and employed them to solve problems in probability theory.

In 1708, the triangle was named after B. Pascal by the French mathematician Pierre Raymond de Montmort (1678–1719), who called it *Pascal's table for combinations* (*Table de M. Pascal pour les combinaisons*). In 1730, the triangle was again named after B. Pascal by the French mathematician Abraham de Moivre (1667–1754), who called it *Pascal's arithmetic triangle* (*Triangulum Arithmeticum Pascalianum*), which became the basis of the modern Western name.

The triangular form of the considered two-dimensional number sequence, represented in the *Treatise on the Number Triangle* by B. Pascal, differs from the above shape of an isosceles triangle. In fact, in this work, a triangular table was published, in which the zeroth row and zeroth column consist of ones, and each internal element is equal to the sum of the previous number in the same row and the previous number in the same column:

```
1  1  1   1  1  1
1  2  3   4  5
1  3  6  10
1  4  10
1  5
1
```

Thus, our isosceles triangle differs from the triangle considered by B. Pascal himself; it is turned by 45 degrees. Let us call B. Pascal's form of representation the *base Pascal's triangle*.

In the same form, only not in the form of a triangle but in the form of a rectangular table, the described numbers appeared a hundred years earlier in the work of N. Tartaglia, *General Treatise on the Number and Measure* (1556–1560):

```
1  1   1   1    1    1
1  2   3   4    5    6
1  3   6  10   15   21
1  4  10  20   35   56
1  5  15  35   70  126
1  6  21  56  126  252
```

It is natural to call this table *Tartaglia's rectangle*.

Recurrent construction of Pascal's triangle

2.1.3. Consider the isosceles Pascal's triangle.

$$\begin{array}{ccccccc}
 & & & 1 & & & \\
 & & 1 & & 1 & & \\
 & 1 & & 2 & & 1 & \\
 1 & & 3 & & 3 & & 1 \\
1 & 4 & & 6 & & 4 & 1 \\
\vdots & \vdots & \vdots & \vdots & \vdots & \vdots & \vdots
\end{array}$$

Denoting the kth element of the nth row of the isosceles Pascal's triangle as T_n^k, $n = 0, 1, 2, \ldots$, $k = 0, 1, 2, \ldots, n$, we can write down the above definition in algebraic form:

$$T_n^k = T_{n-1}^k + T_{n-1}^{k-1}, \text{ and } T_0^0 = T_n^0 = T_n^n = 1.$$

This recurrent relation is called *Pascal's rule* (or *Pascal's formula*).

Placing the initial elements of each row of the triangle in the same column, we obtain the following rectangular shape:

$$\begin{array}{ccccc}
1 & & & & \\
1 & 1 & & & \\
1 & 2 & 1 & & \\
1 & 3 & 3 & 1 & \\
1 & 4 & 6 & 4 & 1 \\
\vdots & \vdots & \vdots & \vdots & \vdots
\end{array}$$

It is natural to call such a representation the *rectangular Pascal's triangle*.

The algebraic notation for the rule of construction of this rectangular triangle is not different from that of the previous one: if the kth element of the nth row of the rectangular Pascal's triangle is indicated as T_n^k, then $T_n^k = T_{n-1}^k + T_{n-1}^{k-1}$, and $T_0^0 = T_n^0 = T_n^n = 1$.

2.1.4. In each of these triangles, the *horizontals* are clearly visible: they correspond to its rows and consist of the elements

$$T_n^0, T_n^1, T_n^2, \ldots, T_n^{n-2}, T_n^{n-1}, T_n^n, \quad n = 0, 1, 2, \ldots.$$

From Pascal's rule, it immediately follows that:

- the sum of the elements of each row of Pascal's triangle, starting with the first row, is twice the sum of the elements of the previous row.

☐ In fact, it holds that

$$T_{n+1}^0 + T_{n+1}^1 + T_{n+1}^2 + \cdots + T_{n+1}^n + T_{n+1}^{n+1}$$
$$= T_n^0 + (T_n^0 + T_n^1) + (T_n^1 + T_n^2) + \cdots + (T_n^{n-1} + T_n^n) + T_n^n$$
$$= 2(T_n^0 + T_n^1 + T_n^2 + \cdots + T_n^{n-1} + T_n^n). \qquad \square$$

Therefore, we can prove that:

- *the sum of the elements of the nth row of Pascal's triangle is equal to* 2^n:

$$T_n^0 + T_n^1 + T_n^2 + \cdots + T_n^n = 2^n.$$

☐ This is not difficult to prove by induction, using the previous property and the fact that the sum of the elements of the zeroth row is 1. ☐

Looking closely at the rows of Pascal's triangle, we can note that they are symmetrical:

- *The elements of a given row, equidistant from the ends of the row, are equal to each other:*

$$T_n^k = T_n^{n-k}.$$

☐ The proof can be obtained by induction. In fact, the first few rows of the triangle are symmetrical. Suppose now that the nth row of the triangle is symmetrical; we prove the symmetry of the $(n+1)$th row. It follows from the construction that the statement is true for $k = 0$:

$$T_{n+1}^0 = T_{n+1}^{n+1} = 1.$$

For a positive integer k, we use the induction's assumption: if the nth row of Pascal's triangle is symmetrical, then

$$T_{n+1}^k = T_n^{k-1} + T_n^k = T_n^{n-(k-1)} + T_n^{n-k} = T_n^{((n+1)-k)-1} + T_n^{(n+1)-k}$$
$$= T_{n+1}^{(n+1)-k}. \qquad \square$$

2.1.5. In the rectangular Pascal's triangle, the *verticals* are also clearly visible. The corresponding lines in the isosceles Pascal's triangle are called *right-descending diagonals*. They consist of the elements

$$T_n^n, T_{n+1}^n, T_{n+2}^n, T_{n+3}^n, \ldots, \quad n = 0, 1, 2, \ldots.$$

Considering the verticals of the rectangular Pascal's triangle, it is not difficult to note that:

- *the zero-vertical* $1, 1, 1, 1, 1, \ldots$ *consists of unities.*

☐ $T_n^0 = 1$ by the construction of the triangle. ☐

Moreover, it holds that:

- *the first vertical of the rectangular Pascal's triangle as well as the first right-descending diagonal of the isosceles Pascal's triangle consist of consecutive positive integers*:

$$T_n^1 = n.$$

☐ This fact can be proven by induction. Indeed, $T_1^1 = 1$ by definition, and if $T_n^1 = n$, then $T_{n+1}^1 = T_n^1 + T_n^0 = n + 1$. ☐

A pattern of construction of the second vertical $1, 3, 6, 10, 15, \ldots$ is more complicated, but readers familiar with number theory will easily guess that it consists of *triangular numbers*.

Recall that the *triangular numbers* $1, 3, 6, 10, 15, 21, 28, 36, 45, 55, \ldots$ (sequence A000217 in the OEIS) correspond to the numbers of balls, with which we can lay out a triangle; the nth triangular number $S_3(n)$ is calculated using the formula $\frac{n(n+1)}{2}$ (see [DeDe12]). It turns out that:

- *the second vertical* $1, 3, 6, 15, 21, \ldots$ *of the rectangular Pascal's triangle as well as the second right-descending diagonal of the isosceles Pascal's triangle consist of triangular numbers*:

$$T_n^2 = S_3(n-1) = \frac{n(n-1)}{2}, \quad n \geq 2.$$

☐ This can be proved by induction too, using the previous result. Indeed, $T_2^2 = 1 = \frac{2 \cdot (2-1)}{2} = S_3(1)$, and if $T_n^2 = \frac{n(n-1)}{2}$, then

$$T_{n+1}^2 = T_n^2 + T_n^1 = \frac{n(n-1)}{2} + n = \frac{n(n+1)}{2} = S_3(n). \qquad \square$$

The *tetrahedral numbers* $1, 4, 10, 20, 35, 56, 84, 120, 165, 220, \ldots$ (sequence A000292 in the OEIS) are the three-dimensional analog of two-dimensional triangular numbers. They correspond to the minimum number of balls with which we can construct regular triangular pyramids; the nth tetrahedral number $S_3^3(n)$ can be calculated using the formula $\frac{n(n+1)(n+2)}{6}$ (see [DeDe12]). It is easy to see that:

- *the third vertical of the rectangular Pascal's triangle as well as the third right=descending diagonal of the isosceles Pascal's triangle consist of tetrahedral numbers:*

$$T_n^3 = S_3^3(n-2) = \frac{n(n-1)(n-2)}{6}, \quad n \geq 3.$$

□ This result can be checked according to the same scheme. For $n = 3$, one has $T_3^3 = 1 = \frac{1 \cdot 2 \cdot 3}{6} = S_3^3(1)$. Going from n to $n+1$, one obtains

$$T_{n+1}^3 = T_n^3 + T_n^2 = S_3^3(n-2) + S_3(n-1)$$
$$= \frac{(n-2)(n-1)n}{6} + \frac{(n-1)n}{2} =$$
$$= \frac{(n-1)n}{6} \cdot (n+1) = \frac{(n-1)n(n+1)}{6} = S_3^3(n-1). \quad \blacksquare$$

The fourth vertical of the rectangular Pascal's triangle as well as the fourth right-descending diagonal of the isosceles Pascal's triangle consist of numbers corresponding to a four-dimensional simplex; they are the four-dimensional analog of the two-dimensional triangular and three-dimensional tetrahedral numbers.

The next verticals (right-descending diagonals) contain sequences of the less-known multidimensional analog of the two-dimensional triangular and three-dimensional tetrahedral numbers.

The *left-descending diagonals* for both the isosceles and rectangular Pascal's triangles are defined similarly. They consist of the elements

$$T_n^0, T_{n+1}^1, T_{n+2}^2, T_{n+3}^3, \ldots, \quad n = 0, 1, 2, \ldots.$$

Both types of descending diagonals are lines parallel to the sides of the triangle, which, in turn, can be seen as a special case of descending diagonals. Since the rows of Pascal's triangle are symmetrical, in the isosceles Pascal's triangle, the nth left-descending diagonal consists of the same elements as the nth right-descending diagonal.

2.1.6. In the rectangular Pascal's triangle, we can easily see the lines which can be called *ascending diagonals*. They consist of the elements

$$T_n^0, T_{n-1}^1, T_{n-2}^2, T_{n-3}^3, \ldots, \; n = 0, 1, 2, \ldots,$$

which are now easy to find and in the isosceles Pascal's triangle.

It turns out that they are associated with the sequence $1, 1, 2, 3, 5, \ldots$ of *Fibonacci numbers* (see Chapter 1). Thus:

- *the sum of the elements of the nth ascending diagonal of Pascal's triangle is the $(n+1)$th Fibonacci number:*

$$T_n^0 + T_{n-1}^1 + T_{n-2}^2 + T_{n-3}^3 + \cdots = u_{n+1}.$$

☐ It is easy to prove, using Pascal's rule and the recurrent relation, $u_{n+1} = u_n + u_{n-1}$ for Fibonacci numbers. Indeed, $T_0^0 = 1 = u_1$, $T_1^0 = 1 = u_2$, $T_2^0 + T_1^1 = 1 + 1 = 2 = u_3$. If $T_{n-1}^0 + T_{n-2}^1 + T_{n-3}^2 + T_{n-4}^3 + \cdots = u_n$ and $T_n^0 + T_{n-1}^1 + T_{n-2}^2 + T_{n-3}^3 + \cdots = u_{n+1}$, then

$$T_{n+1}^0 + T_n^1 + T_{n-1}^2 + T_{n-2}^3 + \cdots = T_n^0 + (T_{n-1}^0 + T_{n-1}^1)$$
$$+ (T_{n-2}^1 + T_{n-2}^2) + (T_{n-3}^2 + T_{n-3}^3) + \cdots$$
$$= (T_n^0 + T_{n-1}^1 + T_{n-2}^2 + T_{n-3}^3 + \cdots)$$
$$+ (T_{n-1}^0 + T_{n-2}^1 + T_{n-3}^2 + T_{n-4}^3 + \cdots)$$
$$= u_n + u_{n-1} = u_{n+1}. \quad \square$$

2.1.7. Finally, in the isosceles Pascal's triangle, it is natural to consider the *principal diagonal*, consisting of the elements

$$T_0^0, T_2^1, T_4^2, \ldots, T_{2n}^n, \ldots .$$

This imaginary line is also associated with special numbers. In fact, by dividing the element T_{2n}^n by $n+1$, we get the nth *Catalan number* C_{n+1} (see Chapter 3).

Reformulating the already proven property $T_n^k = T_n^{n-k}$, we can state that:

- Pascal's triangle is symmetrical with respect to the principal diagonal.

2.1.8. Consider now Pascal's representation of the constructed two-dimensional number sequence:

```
1 1 1  1  1 1
1 2 3  4  5
1 3 6  10
1 4 10
1 5
1
```

In this triangular table, called the *base Pascal's triangle*, the zeroth row and the zeroth column consist of unities, and each internal element is equal to the sum of the previous number in the same row and the previous number in the same column.

Tartaglia's rectangle is constructed by the same law but in the form of a rectangular table:

```
1 1  1  1   1   1
1 2  3  4   5   6
1 3  6  10  15  21
1 4  10 20  35  56
1 5  15 35  70  126
1 6  21 56  126 252
```

Algebraically, the rule for the construction of the base Pascal's triangle (or Tarataglia's rectangle) has the following form: if the kth element of the nth row of the base Pascal's triangle (Tarataglia's rectangle) is indicated as A_n^k, then $A_0^0 = A_n^0 = A_0^n = 1$, and for positive integers n and k, it holds that

$$A_n^k = A_n^{k-1} + A_{n-1}^k.$$

These representations allow us to emphasize the above-mentioned features of Pascal's triangle.

Thus, in these constructions, the nth column coincides with the nth row (we invite the reader to prove this by induction using the formal definition). In other words, the ascending diagonals, as well as the entire table, are symmetrical with respect to the *principal descending diagonal* $A_0^0, A_1^1, \ldots, A_n^n, \ldots$, the bisector of the right angle. The sequences listed above form the columns (and, hence, the rows) of the table: the first column is a sequence of unities $1, 1, 1, 1, 1, \ldots$; the second column presents a set of natural numbers $1, 2, 3, 4, 5, \ldots$; the third column contains the sequence $1, 3, 6, 10, 15, \ldots$ of triangular numbers; in the fourth column, we can find the sequence $1, 4, 10, 20, 35 \ldots$, of tetrahedral numbers, etc.

So, at the moment, we built using a simple recurrent relation, containing only the operation of addition, a number triangle which has several obvious properties. But what does the built object have to do with combinatorics? We show it in the following section.

Pascal's triangle and combinations

2.1.9. As was defined in Chapter 1, the *k-combinations* of a given n-set are unordered sets, consisting of k different elements selected from n different elements of a given n-set. The number of k-combinations of an n-set is indicated by the symbol C_n^k. It turns out that:

- the number C_n^k of k-combinations of an n-set satisfies the recurrent relation

$$C_n^k = C_{n-1}^k + C_{n-1}^{k-1}.$$

□ In order to prove this identity, it is enough to note that when fixing one element, for example a_1, from a given set of n different elements $a_1, a_2, \ldots, a_n$, any k-combination of the considered n-set either does or does not contain the fixed element a_1. The number of k-combinations of the first type is equal to C_{n-1}^k: select all k elements from the set $a_2, \ldots, a_n$. The number of k-combinations of the second type is equal to C_{n-1}^{k-1}: having the element a_1, we choose the remaining $k-1$ elements from the set $a_2, \ldots, a_n$. By the combinatorial rule of sum, we obtain that the total number C_n^k of k-combinations of

n elements is equal to the sum of C_{n-1}^k and C_{n-1}^{k-1}, which completes the proof. □

In addition, it is obvious that from the n elements, we can choose zero elements in only one way, that is, $C_n^0 = 1$. Similarly, from n elements, we can choose n elements in only one way, that is, $C_n^n = 1$. It is reasonable to assume that from zero objects, we can select zero objects in only one way, that is, $C_0^0 = 1$.

The obtained recurrent relation coincides with the recurrent relation $T_n^k = T_{n-1}^k + T_{n-1}^{k-1}$ for the elements T_n^k of Pascal's triangle. Moreover, the same are the initial conditions: $T_0^0 = T_n^0 = T_n^n = 1$. So, it can be stated that, for all possible values of n and k, the numbers C_n^k and T_n^k are equal: $T_n^k = C_n^k$.

- Pascal's triangle consists of the numbers C_n^k of k-combinations of an n-set, where $n = 0, 1, 2, \ldots, k = 0, 1, 2, \ldots, n$:

$$C_0^0 = 1$$
$$C_1^0 = 1 \qquad C_1^1 = 1$$
$$C_2^0 = 1 \qquad C_2^1 = C_1^0 + C_1^1 = 1 + 1 = 2 \qquad C_2^2 = 1$$
$$C_3^0 = 1 \qquad C_3^1 = 1 + 2 = 3 \qquad C_3^2 = 1 + 2 = 3 \qquad C_3^3 = 1$$
$$\vdots \qquad \vdots \qquad \vdots \qquad \vdots$$

Exercises

1. Construct the first 20 rows of Pascal's triangle.
2. Construct the first 20 rows (each of length 20) of Tartaglia's rectangle.
3. Prove that the second element of each row of Pascal's triangle corresponds to its index and the third element of each row is equal to the sum of the indices of all the previous rows.
4. Consider the following combinatorial situation. At point A, a group of 2^n people start. Half of them go southwest, while the other half go southeast. Reaching the first intersection, each group splits again: half go southeast, and half go southwest. The same separation occurs at each intersection. How many people will come to each intersection of the nth row?

2.2. Pascal's Triangle and Binomial Coefficients

2.2.1. *Binomial theorem* gives a formula for the decomposition of a positive integer power of a binomial (i.e., a sum of two monomials, for example, $a + b$), having the form

$$(a+b)^n = \sum_{k=0}^{n} \binom{n}{k} a^{n-k} b^k = \binom{n}{0} a^n + \binom{n}{1} a^{n-1} b + \cdots$$
$$+ \binom{n}{k} a^{n-k} b^k + \cdots + \binom{n}{n} b^n, \ n = 0, 1, 2, \ldots.$$

The numbers $\binom{n}{k}$ involved in this decomposition are called the *binomial coefficients*.

For a long time, it was believed that for natural values of the exponent n, this formula was invented by B. Pascal, who described it in the 17th century. However, the formula was known much earlier, in particular, it was known to Indian mathematicians of the 10th–11th centuries, the Chinese mathematician Yang Hui, who lived in the 13th century, as well as Islamic mathematicians al-Tusi (13th century) and al-Kashi (15th century).

Isaac Newton (1643–1727), around 1676, generalized this formula for the case where the exponent is an arbitrary real (or even complex) number. In this case, we have an infinite series:

$$(1+x)^\alpha = \sum_{n=0}^{\infty} \binom{\alpha}{n} x^n, \ |x| \leq 1.$$

The coefficients of this decomposition are given by the formula

$$\binom{\alpha}{n} = \frac{1}{n!} \prod_{k=0}^{n-1} (\alpha - k) = \frac{\alpha(\alpha-1)(\alpha-2) \cdots (\alpha - (n-1))}{n!},$$

where α, as was stated above, can be not only a non-negative integer but also a negative, an arbitrary real, or even a complex number. On the basis of this binomial decomposition by I. Newton, the theory of infinite series was built.

It turns out that, for non-negative integers n, the binomial coefficients $\binom{n}{k}$, $k = 0, 1, 2, \ldots, n$, form the corresponding row of Pascal's triangle. In other words,

- the elements of Pascal's triangle coincide with the binomial coefficients, i.e., for any non-negative integer n, the numbers T_n^k, $k = 0, 1, 2, \ldots, n$, are equal to the coefficients $\binom{n}{k}$, $k = 0, 1, 2, \ldots, n$, of the decomposition

$$(1+x)^n = \binom{n}{0} + \binom{n}{1}x + \cdots + \binom{n}{k}x^k + \cdots + \binom{n}{n}x^n.$$

□ The statement is valid for $n = 0$: we have the equality $(1+x)^0 = 1$, that is, $\binom{0}{0} = T_0^0 = 1$. It is also true for $n = 1$: we have that $(1+x)^1 = 1 + 1 \cdot x$, i.e., $\binom{1}{0} = T_1^0 = 1$, and $\binom{1}{1} = T_1^1 = 1$. In addition,

$$(1+x)^n = (1+x)^{n-1}(1+x)$$

$$= \left(\binom{n-1}{0} + \binom{n-1}{1}x + \cdots + \binom{n-1}{n-1}x^{n-1}\right)(1+x)$$

$$= \binom{n-1}{0} + \left(\binom{n-1}{1} + \binom{n-1}{0}\right)x + \left(\binom{n-1}{2} + \binom{n-1}{1}\right)$$

$$\times x^2 + \cdots + \left(\binom{n-1}{n-1} + \binom{n-1}{n-1}\right)x^{n-1} + \binom{n-1}{n-1}x^n.$$

So, $\binom{n}{0} = \binom{n-1}{0}$, $\binom{n}{n} = \binom{n-1}{n-1}$, and $\binom{n}{k} = \binom{n-1}{k} + \binom{n-1}{k-1}$, $1 \leq k \leq n-1$. Therefore, the binomial coefficients $\binom{n}{k}$ satisfy the same recurrent relation $\binom{n}{k} = \binom{n-1}{k} + \binom{n-1}{k-1}$, $1 \leq k \leq n-1$, as the elements of Pascal's triangle: $T_n^k = T_{n-1}^k + T_{n-1}^{k-1}$, $1 \leq k \leq n-1$. The initial conditions $\binom{n}{n} = \binom{n-1}{n-1} = \cdots = \binom{1}{1} = \binom{0}{0} = 1$, $\binom{n}{0} = \binom{n-1}{0} = \cdots = \binom{1}{0} = \binom{0}{0} = 1$ also coincide with the initial conditions for the elements of the Pascal's triangle: $T_n^n = T_{n-1}^{n-1} = \cdots = T_1^1 = T_0^0 = 1$, $T_n^0 = T_{n-1}^0 = \cdots = T_1^0 = T_0^0 = 1$. It holds that $\binom{n}{k} = T_n^k$, $0 \leq k \leq n$. □

So, we have proven that

- *Pascal's triangle consists of the binomial coefficients:*

$$\binom{0}{0} = 1$$
$$\binom{1}{0} = 1 \qquad \binom{1}{1} = 1$$
$$\binom{2}{0} = 1 \qquad \binom{2}{1} = 2 \qquad \binom{2}{2} = 1$$
$$\binom{3}{0} = 1 \qquad \binom{3}{1} = 3 \qquad \binom{3}{2} = 3 \qquad \binom{3}{3} = 1$$
$$\vdots \qquad \vdots \qquad \vdots \qquad \vdots$$

As was proven above, any element of Pascal's triangle is a number of combination. So, we obtain that

- *the binomial coefficients coincide with the numbers of combinations:*

$$\binom{n}{k} = C_n^k, 0 \le k \le n.$$

In other words, we have proven the following statement:

$$(1+x)^n = C_0^n + C_1^n x + \cdots + C_n^n x^n.$$

2.2.2. As the nth row of Pascal's triangle gives the binomial decomposition's coefficients for $(1+x)^n$, then, in order to obtain this decomposition, it is enough to build the corresponding number of rows of the triangle, or simply use already available data. Actually, this is the main aspect of the practical purpose of Pascal's triangle: it is simply a table of binomial coefficients, which is convenient to use in computational practice and, if necessary, expand to the desired size using an elementary additive algorithm that requires for its implementation only the operation of addition.

Al-Tusi (13th century) had a table of binomial coefficients up to $n = 12$ and gave the general rule for obtaining them, which, in modern language, coincides with Pascal's rule. Binomial coefficient tables were also known to others, including Chinese, Arab, and European mathematicians (such as P. Appian, 1527; M. Stiefel, 1544; and N. Tartaglia, 1556). Pascal himself (and many of his predecessors), as previously described, depicted these tables in a different shape, with "elevated" columns and a right angle at the top. In such a table, the numbers, corresponding to the binomial decomposition of $(1+x)^n$, are placed not along the same row but along the same ascending diagonal.

Exercises

1. Find the coefficients of the decomposition of the binomial $(1+x)^n$, $n = 5, 6, \ldots, 15$, using Pascal's triangle.
2. In a medieval mathematical text that was in circulation in Western Europe in the 15th century, binomial coefficients are calculated very clearly by exponentiating the number 10001 and are given by the following table:

$$100090036008401260126008400360009 0001$$
$$10008002800560070005600280008 0001$$
$$100070021003500350021000700 01$$
$$1000600150020001500060001$$
$$10005001000100005 0001$$
$$100040006000 40001$$
$$1000300030001$$
$$100020001$$
$$10001$$

Use this table to obtain the binomial coefficients for $n = 2, 3, 4, 5, 6, 7, 8, 9$. Continue with the construction of the table. Formulate this method of calculation of the binomial coefficients.

3. Check that the numbers $1, 11, 121, 1331, 14641$, derived from the nth row of Pascal's triangle, $n = 0, 1, 2, 3, 4$, by "gluing" them into the usual decimal number, are the nth powers of number 11. Is this statement true for $n = 5, 6, 7, 8$?

2.3. Closed Formula for Elements of Pascal's Triangle

2.3.1. So, we know how to compute the elements of Pascal's triangle using the simple additive recurrent relation. As it turned out, the numbers T_n^k coincide with the numbers of k-combinations C_n^k, or, which is the same but significantly more important for practical applications, with the coefficients $\binom{n}{k}$ of the binomial decomposition for $(1+x)^n$.

However, in some cases, it can be convenient and even necessary to have the ability to find an element C_n^k of Pascal's triangle without preliminary counting of all previous elements, but only focusing on

76   Catalan Numbers

the parameters n and k of C_n^k. A formula that allows us to find the value of C_n^k as a function $f(n, k)$ of its parameters n and k is called a *closed formula*.

For the numbers C_n^k, this formula is well known.

- It holds that

$$C_k^n = \binom{n}{k} = \frac{n!}{k!(n-k)!},$$

where $n!$ is the factorial of n: $n! = 1 \cdot 2 \cdot 3 \cdot \ldots \cdot n$ for a positive integer n and $0! = 1$.

□ In Chapter 1, we proved this formula through combinatorial reasoning. Let us try to derive a proof on the basis of using the recurrent relations.

Let $F_n^k = \frac{n!}{k!(n-k)!}$. Make sure that the numbers F_n^k satisfy the same recurrent relation with the same initial conditions as the numbers of k-combinations of an n-set. In fact,

$$F_{n-1}^k + F_{n-1}^{k-1} = \frac{(n-1)!}{k!((n-1)-k)!} + \frac{(n-1)!}{(k-1)!((n-1)-(k-1))!}$$

$$= \frac{(n-1)!}{(k-1)!(n-k-1)!}\left(\frac{1}{k} + \frac{1}{n-k}\right)$$

$$= \frac{(n-1)!n}{(k-1)!k(n-k-1)!(n-k)} = \frac{n!}{k!(n-k)!} = F_n^k.$$

Moreover, it holds that $F_0^0 = \frac{0!}{0!(0-0)!} = 1$, $F_n^0 = \frac{n!}{0!(n-0)!} = 1$, and $F_n^n = \frac{n!}{n!(n-n)!} = 1$. So, the value F_n^k satisfies the same recurrent relation $F_n^k = F_{n-1}^k + F_{n-1}^{k-1}$ with the same initial conditions $F_0^0 = F_n^0 = F_n^n = 1$ as the numbers C_n^k; therefore, F_n^k coincides with C_n^k for all possible values of the parameters n, k. □

2.3.2. Of course, we can calculate all the binomial coefficients for any n by direct multiplication of the n factors $1+x$, opening the brackets, and combining similar terms. However, the multiplicative methods of calculations, in most cases, are significantly more time-consuming

Pascal's Triangle   77

than the construction of Pascal's triangle additive methods. Moreover, mathematicians of antiquity and the Middle Ages, some of whom knew the closed formula for the calculation of binomial coefficients, obtained in this subsection, could not use it in practice because of a lack of convenient algebraic symbolism.

Exercises

1. Find the coefficients of the decomposition of $(1+x)^n$, for $n = 3, 4, 5, 6, 7, 8$, by constructing the corresponding row of Pascal's triangle and by using the closed formula $\binom{k}{n} = \frac{n!}{k!(n-k)!}$. Which method is easier? Why?

2. Find the value C_n^3, $n = 6, 7, 8$, and specify the maximal power of two that divides the resulting number. Do the calculations by building the appropriate row of Pascal's triangle and by using the closed formula $\binom{k}{n} = \frac{n!}{k!(n-k)!}$. Which method is easier? Why?

3. Check that if k is fixed, the binomial coefficients $\binom{n}{k}$ can be calculated using a recurrent formula $\binom{n}{k} = \frac{n}{n-k}\binom{n-1}{k}$ with the initial value $\binom{k}{k} = 1$. Perform the appropriate calculations for $k = 1, 2, 3, 4, 5, 6, 7, 8$.

4. Find the first few elements of the sequence $s_n = \prod_{k=0}^{n}\binom{n}{k} = \prod_{k=0}^{n}\frac{n!}{k!(n-k)!}$, $n \geq 0$, of products of elements of the nth row of Pascal's triangle. Find the first few members of the sequence $ss_n = \frac{s_{n+1}}{s_n}$ of ratios of the consecutive terms of the sequence s_n, $n \geq 0$. Prove the formula $ss_n = \frac{(n+1)^n}{n!}$. Find the first few members of the sequence $sss_n = \frac{ss_{n+1}}{ss_n}$ of ratios of the consecutive terms of the sequence ss_n, $n \geq 0$. Make sure that $sss_n = \left(\frac{n+1}{n}\right)^n$, $n \geq 1$. Using the second wonderful limit $e = \lim_{n \to \infty}\left(1 + \frac{1}{n}\right)^n$, formulate the relationship rule between the sequence of products of the elements of the nth row of Pascal's triangle and the number e.

5. Prove the following rule: $\binom{n}{k} = \frac{n}{k}\binom{n-1}{k-1}$.

6. Prove that $\binom{n-1}{k} - \binom{n-1}{k-1} = \frac{n-2k}{n}\binom{n}{k}$.

7. Prove the following rule: $\binom{n}{m}\binom{m}{k} = \binom{n}{k}\binom{n-k}{m-k}$. Make sure that this rule can be written as $\binom{n}{h}\binom{n-h}{k} = \binom{n}{k}\binom{n-k}{h}$.

78 Catalan Numbers

8. Prove that, for a fixed n, in order to calculate the elements of the nth row of Pascal's triangle, we can use a "one-dimensional" recurrent relation $\binom{n}{k} = \frac{n+1-k}{k}\binom{n}{k-1}$ with the initial condition $\binom{n}{0} = 1$.

9. Prove that the maximal power α, such that p^α, $p \in P$, divides C_n^k, has the form

$$\alpha = \left(\left\lfloor\frac{n}{p}\right\rfloor + \left\lfloor\frac{n}{p^2}\right\rfloor + \cdots + \left\lfloor\frac{n}{p^t}\right\rfloor + \cdots\right)$$
$$- \left(\left\lfloor\frac{k}{p}\right\rfloor + \left\lfloor\frac{k}{p^2}\right\rfloor + \cdots + \left\lfloor\frac{k}{p^t}\right\rfloor + \cdots\right)$$
$$- \left(\left\lfloor\frac{n-k}{p}\right\rfloor + \left\lfloor\frac{n-k}{p^2}\right\rfloor + \cdots + \left\lfloor\frac{n-k}{p^t}\right\rfloor + \cdots\right).$$

10. Using the closed formula for T_n^k, prove that the second diagonal of Pascal's triangle is formed by triangular numbers and the third diagonal of Pascal's triangle is formed by tetrahedral numbers. Find the closed formula for the m-dimensional analog of two-dimensional triangular numbers.

2.4. Generating Functions of the Sequences of Elements of Pascal's Triangle

2.4.1. Now, we try to find the generating function of the sequence of elements of Pascal's triangle. It turns out that this problem is extremely simple. Since, as it was previously proven,

$$(1+x)^n = \binom{n}{0} + \binom{n}{1}x + \binom{n}{2}x^2 + \cdots + \binom{n}{n}x^n,$$

it can be stated that:

- for a fixed n, the generating function $f(x)$ of the sequence $\binom{n}{0}, \binom{n}{1}, \binom{n}{2}, \ldots, \binom{n}{n}, 0, 0, \ldots$ of binomial coefficients has the form $f(x) = (1+x)^n$:

$$(1+x)^n = \binom{n}{0}x^0 + \binom{n}{1}x^1 + \binom{n}{2}x^2 + \binom{n}{n}x^n + 0 \cdot x^{n+1} + \cdots, \quad x \in \mathbb{R}.$$

2.4.2. As the values $\binom{n}{k}$ have two parameters n and k, the question of finding the generating function can be considered in a different way. Namely:

- for a fixed k, the generating function $g(x)$ of the sequence $\binom{0}{k}, \binom{1}{k}, \binom{2}{k}, \ldots, \binom{k}{k}, \binom{k+1}{k}, \ldots$ of binomial coefficients has the form $g(x) = \frac{y^k}{(1-y)^{k+1}}$:

$$\frac{y^k}{(1-y)^{k+1}} = \binom{0}{k} + \binom{1}{k}y + \binom{2}{k}y^2 + \cdots + \binom{k}{k}y^k + \binom{k+1}{k}y^{k+1} + \cdots$$

$$= \binom{k}{k}y^k + \binom{k+1}{k}y^{k+1} + \cdots + \binom{k+n}{k}y^{k+n} + \cdots, \quad |y| < 1.$$

□ This result follows from the decomposition $(1+z)^\alpha$ (see Chapter 1) with $z = -y$, and $\alpha = -(k+1)$. ■

2.4.3. In Chapter 1, we proved that:

- the two-dimensional generating function of the sequence of binomial coefficients $\binom{n}{k}$ for all n, k is the function $\frac{1}{1-y-xy}$:

$$\frac{1}{1-y-xy} = \sum_{n,k} \binom{n}{k} x^k y^n.$$

Other approaches to defining generating functions for certain sequences of elements of Pascal's triangle are presented in the following exercises.

Exercises

1. Using the generating function, find the sum of elements of the nth row of Pascal's triangle.
2. Using the generating function, find the alternating sum of elements of the nth row of Pascal's triangle.
3. Prove the following identities:
   (a) $C_n^0 + 2C_n^1 + 4C_n^2 + 8C_n^3 + \cdots + 2^n C_n^n = 3^n$;
   (b) $C_n^0 + 3C_n^1 + 9C_n^2 + 27C_n^3 + \cdots + 3^n C_n^n = 4^n$;
   (c) $C_n^0 + kC_n^1 + k^2 C_n^2 + k^3 C_n^3 + \cdots + k^n C_n^n = (k+1)^n$.

80 Catalan Numbers

4. Prove the following equalities:
   (a) $\frac{C_3^3}{8} + \frac{C_4^3}{16} + \cdots + \frac{C_n^3}{2^n} + \cdots = 2$;
   (b) $\frac{C_4^4}{16} + \frac{C_5^4}{32} + \cdots + \frac{C_n^4}{2^n} + \cdots = 2$;
   (c) $\frac{C_k^k}{2^k} + \frac{C_{k+1}^k}{2^{k+1}} + \cdots + \frac{C_n^k}{2^n} + \cdots = 2$;
   (d) $\frac{C_3^3}{27} + \frac{C_4^3}{81} + \cdots + \frac{C_n^3}{3^n} + \cdots = \frac{3}{16}$;
   (e) $\frac{C_4^4}{81} + \frac{C_5^4}{243} + \cdots + \frac{C_n^4}{3^n} + \cdots = \frac{3}{16}$;
   (f) $\frac{C_k^k}{3^k} + \frac{C_{k+1}^k}{3^{k+1}} + \cdots + \frac{C_n^k}{3^n} + \cdots = \frac{3}{16}$.

5. Check that the generating function of the sequence of binomial coefficients $\binom{n+k}{k}$ for integers n, k is a symmetric function $\frac{1}{1-x-y}$, i.e., it holds that $\sum_{n,k} \binom{n+k}{k} x^k y^n = \frac{1}{1-x-y}$.

6. Check that the exponential generating function of the sequence of binomial coefficients $\binom{n+k}{k}$ for all n, k has the form e^{x+y}, i.e., $e^{x+y} = \sum_{n,k} \frac{1}{(n+k)!} \binom{n+k}{k} x^k y^n$.

2.5. Properties of Pascal's Triangle

Simplest properties of Pascal's triangle

2.5.1. When we discussed the recurrent scheme of building Pascal's triangle, a number of its obvious properties were already described. In this section, we give, together with those already known, a large list of other properties of the triangle. In most cases, for already known facts, other proofs are given.

All properties are formulated for the isosceles Pascal's triangle in terms of k-combinations.

- *Pascal's triangle is symmetrical with respect to its principal diagonal:*

$$C_n^k = C_n^{n-k}.$$

□ In order to prove this property, it is enough to note that

$$C_n^k = \frac{n!}{k!(n-k)!} = \frac{n!}{(n-k)!(n-(n-k))!} = C_n^{n-k}. \qquad \blacksquare$$

- The sum of the elements of the nth row of Pascal's triangle is equal to 2^n:
$$C_n^0 + C_n^1 + C_n^2 + \cdots + C_n^n = 2^n.$$

□ Using the binomial decomposition of $(1+x)^n$ for $x = 1$, we get
$$2^n = (1+1)^n = C_n^0 + C_n^1 + C_n^2 + \cdots + C_n^n. \qquad \blacksquare$$

- The alternating sum of the elements of the nth row of Pascal's triangle, $n \geq 1$, is equal to zero:
$$C_n^0 - C_n^1 + C_n^2 - \cdots + (-1)^n C_n^n = 0.$$

□ To prove this fact, we use the binomial decomposition of $(1+x)^n$, taking $x = -1$:
$$0 = (1-1)^n = C_n^0 - C_n^1 + C_n^2 - \cdots + (-1)^n C_n^n. \qquad \blacksquare$$

- The sum of elements of the nth row of Pascal's triangle, $n \geq 1$, is twice the sum of the elements of the previous row of the triangle; the alternating sum of the elements of the nth row of Pascal's triangle, $n \geq 2$, is twice the alternating sum of the elements of the previous row of the triangle.

□ The first statement follows from the fact that the sum of the elements of the $(n-1)$th row is equal to 2^{n-1}, and the sum of the elements of the nth row is 2^n; the second statement is trivial. $\qquad \blacksquare$

- For each row of Pascal's triangle, starting with the first row, the sum of the elements located at even positions is equal to the sum of the elements located at odd positions:
$$C_n^1 + C_n^3 + C_n^5 + \cdots = C_n^0 + C_n^2 + C_n^4 + \cdots.$$

□ As we have proven that
$$C_n^0 - C_n^1 + C_n^2 - \cdots + (-1)^n C_n^n = 0,$$
it holds that
$$C_n^1 + C_n^3 + C_n^5 + \cdots = C_n^0 + C_n^2 + C_n^4 + \cdots. \qquad \blacksquare$$

Since the sum of all the elements of a given row is equal to the power of two, we get another property:

82   Catalan Numbers

- The sum of the elements of the nth row of Pascal's triangle, $n \geq 1$, located at even positions, is equal to 2^{n-1}; the sum of the elements of the nth row of Pascal's triangle, $n \geq 1$, located at even positions, is also 2^{n-1}.

We can also consider more complex sums, including those of the elements of a given row of Pascal's triangle:

- The sum of the elements of the nth row of Pascal's triangle with weights $0, 1, 2, 3, \ldots, n$ is equal to $n2^{n-1}$:

$$C_n^1 + 2C_n^2 + 3C_n^3 + \cdots + nC_n^n = n2^{n-1}, \quad n \geq 1.$$

□ We easily get this formula by differentiating the expansion of $(1+x)^n$ and then taking the value $x = 1$:

$$((1+x)^n)' = \left(C_n^0 + C_n^1 x + C_n^2 x^2 + \cdots + C_n^n x^n\right)';$$
$$n(1+x)^{n-1} = C_n^1 + 2C_n^2 x + \cdots + nC_n^n x^{n-1};$$
$$n2^{n-1} = n(1+1)^{n-1} = C_n^1 + 2C_n^2 + \cdots + nC_n^n.$$

However, the same result can be obtained if we note that

$$kC_n^k = \frac{n!}{k!(n-k)!} = n\frac{(n-1)!}{(k-1)!((n-1)-(k-1))!} = nC_{n-1}^{k-1}.$$

Then, using the relationship $C_{n-1}^0 + C_{n-1}^1 + \cdots + C_{n-1}^{n-1} = 2^{n-1}$, we obtain that

$$C_n^1 + 2C_n^2 + 3C_n^3 + \cdots + nC_n^n$$
$$= nC_{n-1}^0 + nC_{n-1}^1 + nC_{n-1}^2 + \cdots + nC_{n-1}^{n-1}$$
$$= n(C_{n-1}^0 + C_{n-1}^1 + \cdots + C_{n-1}^{n-1}) = n2^{n-1}. \quad \blacksquare$$

- The sum of the elements of the nth row of Pascal's triangle with weights $0, 1^2, 2^2, 3^2, \ldots, n^2$ is equal to $(n + n^2)2^{n-2}$:

$$C_n^1 + 4C_n^2 + 9C_n^3 + \cdots + n^2 C_n^n = (n + n^2)2^{n-2}, \quad n \geq 1.$$

□ To prove this property, let us once again differentiate the binomial decomposition of $(1+x)^n$ and, after carrying out simple transformations, take $x = 1$:

$$nx(1+x)^{n-1} = C_n^1 x + 2C_n^2 x^2 + \cdots + nC_n^n x^n;$$

$$(nx(1+x)^{n-1})' = (C_n^1 x + 2C_n^2 x^2 + \cdots + nC_n^n x^n)';$$

$$n(n-1)x(1+x)^{n-2} + n(1+x)^{n-1} = C_n^1 + 4C_n^2 x + \cdots + n^2 C_n^n x^{n-1};$$

$$(n+n^2)2^{n-2} = n(n-1)(1+1)^{n-2} + n(1+1)^{n-1} = C_n^1 + 4C_n^2 + \cdots + n^2 C_n^n.$$

□

- *The sum of the squares of elements of the nth row of Pascal's triangle is equal to the central element of the (2n)th row of the triangle:*

$$(C_n^0)^2 + (C_n^1)^2 + \cdots + (C_n^n)^2 = C_{2n}^n.$$

□ Using the symmetry of Pascal's triangle, this property can be rewritten as

$$C_n^0 C_n^n + C_n^1 C_n^{n-1} + \cdots + C_n^n C_n^0 = C_{2n}^n;$$

it is a special case of a more general combinatorial identity,

$$C_n^0 C_m^k + C_n^1 C_m^{k-1} + \cdots + C_n^k C_m^0 = C_{m+n}^k, \quad n \geq m \geq k,$$

which is called the *Vandermond convolution*. We can prove this identity using the method of constructing the corresponding combinatorial situation. In particular, we make a team of k persons consisting of n men and m women. Since there are exactly $n+m$ people, we can choose from them k members of the team in C_{n+m}^k ways. On the other hand, if the team should have exactly i men and $i-k$ women, $i = 0, 1, 2, \ldots, k$, then the number of ways of selecting such a team is $C_n^i C_m^{k-i}$. Summing up all the possibilities, we get the result. □

2.5.2. Consider now the diagonals of Pascal's triangle.

- *The sum of the elements of the $(n+1)$th ascending diagonal of Pascals' triangle, $n \geq 1$, is equals to the sum of the elements of two previous ascending diagonals of the triangle.*

☐ Indeed, any number on the $(n + 1)$th diagonal, except the first and last unities, is the sum of the two numbers on two previous diagonals, nth and $(n - 1)$th. Moreover, different elements of the $(n + 1)$th diagonal forming pairs of elements of the two previous diagonals have no common terms, and all such pairs are involved in the formation of the elements of the $(n + 1)$th diagonal. Therefore, the sum of elements of the $(n + 1)$th diagonal is the sum of the elements of the nth diagonal added to the sum of the elements of the $(n - 1)$th diagonal. Algebraically, the sum X_{n+1} of the elements of the $(n + 1)$th ascending diagonal satisfies the relation

$$\begin{aligned} X_{n+1} &= C_{n+1}^0 + C_n^1 + C_{n-1}^2 + C_{n-2}^3 + \cdots \\ &= C_n^0 + (C_{n-1}^0 + C_{n-1}^1) + (C_{n-2}^1 + C_{n-2}^2) + (C_{n-3}^2 + C_{n-3}^3) + \cdots \\ &= (C_n^0 + C_{n-1}^1 + C_{n-2}^2 + C_{n-3}^3 + \cdots) \\ &\quad + (C_{n-1}^0 + C_{n-2}^1 + C_{n-3}^2 + C_{n-4}^3 + \cdots) = X_n + X_{n-1}. \quad \square \end{aligned}$$

- *The sum of the elements of the nth ascending diagonal of Pascal's triangle, $n \geq 2$, is the sum of all the elements of Pascal's triangle located above the $(n - 1)$th ascending diagonal increased by one.*

☐ This property is easy to derive from the previous one by noting that the sum X_n of the elements of the nth ascending diagonal satisfies the relation

$$\begin{aligned} X_n &= X_{n-2} + X_{n-1} = X_{n-2} + (X_{n-3} + X_{n-2}) \\ &= X_{n-2} + X_{n-3} + (X_{n-4} + X_{n-3}) = \cdots \\ &= X_{n-2} + X_{n-3} + X_{n-4} + \cdots + X_1 + (X_0 + X_1) \\ &= X_{n-2} + X_{n-3} + X_{n-4} + \cdots + X_1 + X_0 + 1. \quad \square \end{aligned}$$

As was previously observed, the diagonals of Pascal's triangle are related to various special numbers. So, the last two properties can be rewritten in terms of *Fibonacci numbers*.

- *The sum of the elements of the $(n - 1)$th ascending diagonal of Pascal's triangle, $n \geq 1$, is the nth Fibonacci number:*

$$C_{n-1}^0 + C_{n-2}^1 + C_{n-3}^2 + \cdots = u_n.$$

□ This property immediately follows from the previous one if we note that the sum of the elements of the zeroth diagonal is $u_1 = 1$, and the sum of the elements of the first ascending diagonal is equal to $u_2 = 1$. □

- *The sum of all the elements of Pascal's triangle located above the nth ascending diagonal, $n \geq 1$, increased by unity is equal to the $(n+2)$th Fibonacci number:*

$$C_0^0 + C_1^0 + (C_2^0 + C_1^1) + \cdots + (C_{n-1}^0 + C_{n-2}^1 + C_{n-3}^2 + \cdots) + 1 = u_{n+2}.$$

□ This property follows from the previous property and from the well-known identity

$$u_1 + u_2 + \cdots + u_n = u_{n+2} - 1.$$ □

- *Along the descending diagonals of Pascal's triangle, triangular numbers are arranged, as well as their generalizations to the case of spaces of all dimensions.*

□ We discussed this question in detail earlier. The second left- (right-) descending diagonal consists of the numbers $C_2^2, C_3^2, C_4^2, \ldots, C_n^2, \ldots$, that is, from the triangular numbers $1, 3, 6, 10, 15, \ldots$. The third left- (right-) descending diagonal consists of the numbers $C_3^3, C_4^3, C_5^3, \ldots, C_n^3, \ldots$, that is, from the tetrahedral numbers $1, 4, 10, 20, 35, \ldots$. The numbers $C_4^4, C_5^4, C_6^4, \ldots, C_n^4, \ldots$ of the fourth diagonal form the sequence $1, 5, 15, 35, 70, \ldots$ of four-dimensional simplicial numbers; they are the four-dimensional analogs of two-dimensional triangular and three-dimensional tetrahedral numbers. In general, the numbers $C_k^k, C_{k+1}^k, C_{k+2}^k, \ldots, C_n^k, \ldots$, forming the kth diagonal, are the k-dimensional analog of triangular and tetrahedral numbers. The sequence $1, 2, 3, 4, 5, \ldots$ of the elements $C_1^1, C_2^1, C_3^1, \ldots$ of the first diagonal can be considered a one-dimensional analog of triangular and tetrahedral numbers, and the unities, forming the sides of the triangle, as their zero-dimensional analog. From this point of view, we can state that *Pascal's triangle is formed only from the figurate numbers*. □

By definition, the nth triangular number is equal to the sum of the first n natural numbers, the nth tetrahedral number is the sum of the first n triangular numbers, and, in general, the nth k-dimensional simplicial number $S_3^k(n)$ is the sum $S_3^{k-1}(1)+S_3^{k-1}(2)+\cdots+S_3^{k-1}(n)$ of the first n $(k-1)$-dimensional simplicial numbers. Using this approach, we can formulate another property of the diagonals of Pascal's triangle, which is often called the *club rule*.

- *The sum of the first n elements of the kth left- (right-) descending diagonal of Pascal's triangle is equal to the elements of the triangle located immediately below and to the right (to the left) of the last item to be summed:*

$$C_k^k + C_{k+1}^k + C_{k+2}^k + \cdots + C_n^k = C_{n+1}^{k+1}, \quad n \geq k.$$

□ We have already proved this property using multidimensional simplicial numbers. However, it can be checked directly:

$$\begin{aligned}
C_{n+1}^{k+1} &= C_n^k + C_n^{k+1} = C_n^k + (C_{n-1}^k + C_{n-1}^{k+1}) \\
&= C_n^k + C_{n-1}^k + (C_{n-2}^k + C_{n-2}^{k+1}) \\
&= C_n^k + C_{n-1}^k + C_{n-2}^k + \cdots (C_{k+1}^k + C_{k+1}^{k+1}) \\
&= C_n^k + C_{n-1}^k + C_{n-2}^k + \cdots + C_{k+1}^k + C_k^k.
\end{aligned}$$ □

After consideration of two or three examples, it will become clear to the reader why we use in this case the word "club."

Divisibility of rows of Pascal's triangle

2.5.3. In this section, we consider the properties of Pascal's triangle, related to the divisibility of binomial coefficients. It's easy to see that all the numbers of the 5th row of the triangle, except the rightmost and leftmost unities, are divided by 5; all the numbers of the 7th row, except the rightmost and leftmost unities, are divided by 7. Which rows have the same property? Obviously, they are the 2nd and 3rd rows. All other rows, up to and including the 10th row, have no such property. Which condition combines the numbers 2, 3, 5, and 7 and distinguishes them from other numbers up to 10? Yes, they are prime numbers.

For the convenience of further presentation, we say that *the nth row of Pascal's triangle is divisible by n* if all the internal elements of this row are divisible by n. We can prove that:

- *the nth row of Pascal's triangle is divisible by n if and only if n is a prime number.*

In fact, there is a more general result:

- *The nth row of Pascal's triangle is divisible by m if and only if m is a prime and n is the natural power of that prime.*

In other words, the main result of this section can be formulated as follows:

- *The greatest common divisor of all internal elements of the nth row of Pascal's triangle is equal to a prime number p if $n = p^k$, $p \in P, k \in \mathbb{N}$, and is equal to unity in all other cases.*

We prove this statement in several steps. First of all, note that:

- *if the nth row of Pascal's triangle is divisible by m, then n is divisible by m.*

□ This fact immediately follows from the fact that in the nth row, there is an element equal to n: $C_n^1 = n$. □

Further considerations use the following fundamental fact:

- *the binomial coefficient C_p^k, $1 \le k \le p-1$, is divisible by the prime number p.*

□ We use the fundamental property of divisibility by a prime: if a and b are positive integers and the product ab is divisible by a prime p, then at least one of the numbers a and b is divisible by p.

The equality $C_p^k = \frac{p!}{k!(p-k)!}$ can be rewritten for $0 < k < p$ as the equality

$$1 \cdot 2 \cdots k \cdot 1 \cdot 2 \cdot (p-k) C_p^k = p!.$$

Since the right-hand side, $p!$, is divisible by p, the left-hand side must be divisible by p. However, for $0 < k < p$, neither $k!$ nor $(k-p)!$

contain the number p; therefore, none of the left-hand side factors except C_p^k can be divisible by p. Since p is a prime, C_p^k must be divisible by p. The statement is proven. □

So, we have proven the first basic statement:

- *if p is a prime number, then the pth row of Pascal's triangle is divisible by p.*

For further consideration, we should use *Lucas' theorem*, well known as the number theory statement about the remainder of the division of the binomial coefficient C_n^m by the prime p, obtained by the French mathematician François Édouard Anatole Lucas (1842–1891) in 1878:

- *For a prime number p and positive integers n and m, there exists the following congruence:*

$$C_n^m \equiv \prod_{i=0}^{k-1} C_{n_i}^{m_i} \pmod{p},$$

where $n = \overline{(n_{k-1} \cdots n_0)}_p = n_{k-1}p^{k-1} + n_{k-2}p^{k-2} + \cdots + n_1 p + n_0$, *and* $m = \overline{(m_{k-1} \cdots m_0)}_p = m_{k-1}p^{k-1} + m_{k-2}p^{k-2} + \cdots + m_1 p + m_0$, *are the base-p expansions of the numbers n and m.*

□ For a proof, consider the coefficient at x^m in the polynomial $(x+1)^n$. On the one hand, it is equal to C_n^m. On the other hand, writing $n = n_{k-1}p^{k-1} + n_{k-2}p^{k-2} + \cdots + n_1 p + n_0$ in the base p and using the fact that C_p^k is divisible by p, $1 \le k \le p-1$, we get that

$$(x+1)^n = (x+1)^{n_{k-1}p^{k-1} + n_{k-2}p^{k-2} + \cdots + n_1 p + n_0}$$

$$= \prod_{i=0}^{k-1}(x+1)^{n_i p^i} = (x+1)^{n_0} \prod_{i=1}^{k-1}((x+1)^p)^{n_i p^{i-1}}$$

$$= (x+1)^{n_0} \prod_{i=1}^{k-1}(x^p + C_p^1 x^{p-1} + \cdots + C_p^{p-1} x + 1)^{n_i p^{i-1}}$$

$$\equiv (x+1)^{n_0} \prod_{i=1}^{k-1}(x^p + 1)^{n_i p^{i-1}}$$

$$\equiv (x+1)^{n_0}(x^p+1)^{n_1}\prod_{i=2}^{k-1}((x^p+1)^p)^{n_ip^{i-2}}$$

$$= (x+1)^{n_0}(x^p+1)^{n_1}\prod_{i=2}^{k-1}((x^p)^p+C_p^1(x^p)^{p-1}$$

$$+\cdots+C_p^{p-1}(x^p)^1+1)^{n_ip^{i-2}}$$

$$\equiv (x+1)^{n_0}(x^p+1)^{n_1}\prod_{i=2}^{k-1}(x^{p^2}+1)^{n_ip^{i-2}}$$

$$\equiv (x+1)^{n_0}(x^p+1)^{n_1}(x^{p^2}+1)^{n_2}\prod_{i=3}^{k-1}(x^{p^2}+1)^{n_ip^{i-3}}$$

$$\equiv \cdots \equiv \prod_{i=0}^{k-1}(x^{p^i}+1)^{n_i} \pmod{p}.$$

In order to obtain the coefficient of x^m from the last product, we should take from the zeroth factor the coefficient of x^{m_0}, we should take from the first factor the coefficient of x^{m_1p}, and, in general, we should take from the ith factor the coefficient of $x^{m_ip^i}$. So, it holds that

$$\prod_{i=0}^{k-1}(x^{p^i}+1)^{n_i} = \prod_{i=0}^{k-1}(C_{n_i}^0(x^{p^i})^0+C_{n_i}^1(x^{p^i})^1+C_{n_i}^2(x^{p^i})^2+\cdots+C_{n_i}^{n_i}(x^{p^i})^{n_i})$$

$$= \prod_{i=0}^{k-1}(C_{n_i}^0 x^{0\cdot p^i}+C_{n_i}^1 x^{1\cdot p^i}+C_{n_i}^2 x^{2\cdot p^i}+\cdots+C_{n_i}^{n_i} x^{n_i\cdot p^i})$$

$$= \prod_{i=0}^{k-1}\sum_{t_i=0}^{n_i} C_{n_i}^{t_i} x^{t_i\cdot p^i} = \sum_{\substack{t_0,\ldots,t_{k-1}, \\ 0\le t_i\le n_i}}\prod_{i=0}^{k-1} C_{n_i}^{t_i} x^{t_i\cdot p^i}$$

$$= \sum_{\substack{t_0,\ldots,t_{k-1}, \\ 0\le t_i\le n_i}} C_{n_0}^{t_0} C_{n_1}^{t_1}\cdots C_{n_{k-1}}^{t_{k-1}} x^{t_0+t_1p+\cdots+t_{k-1}p^{k-1}}.$$

Comparing the coefficients of x^m, we get the congruence

$$C_n^m \equiv \prod_{i=0}^{k-1} C_{n_i}^{m_i} \pmod{p}.$$ □

Now, it is easy to prove that:

- the nth row of Pascal's triangle is divisible by a prime number p if and only if its number n is a power of p, i.e., $n = p^\alpha$, $\alpha \in \mathbb{N}$.

□ In fact, if the number n of a row is p^α, $p \in P$, $\alpha \in \mathbb{N}$, i.e., if it holds that $n = 0 + 0 \cdot p + 0 \cdot p^2 + \cdots + 1 \cdot p^\alpha$, then, for $0 < m < n$, we have $m = m_0 + m_1 p + \cdots + m_{\alpha-1} p^{\alpha-1} + 0 \cdot p^\alpha$, and

$$C_n^m \equiv C_0^{m_0} \cdot C_0^{m_1} \cdot C_0^{m_2} \cdots \cdot C_1^0 \pmod{p}.$$

As at least one of the values $m_0, m_1, \ldots, m_{\alpha-1}$ is not zero, the right-hand side of the congruence gives zero: $m_i \neq 0 \Rightarrow C_0^{m_i} = 0$; hence, the number C_n^m, $0 < n < m$, is divisible by p.

In order to better understand our reasons, we check that, in this situation, the rightmost and leftmost elements of the row are not divisible by p. Therefore, these coefficients correspond to the values C_n^m with $m = 0$ and $m = p^\alpha$. In the first case, we get $m = 0 + 0 \cdot p + \cdots + 0 \cdot p^{\alpha-1} + 0 \cdot p^\alpha$ and $C_n^m \equiv C_0^0 \cdot C_0^0 \cdot C_0^0 \cdots \cdot C_1^0 \equiv 1 \pmod{p}$. In the second case, it holds that $m = 0 + 0 \cdot p + \cdots + 0 \cdot p^{\alpha-1} + 1 \cdot p^\alpha$ and $C_n^m \equiv C_0^0 \cdot C_0^0 \cdot C_0^0 \cdots \cdot C_1^1 \equiv 1 \pmod{p}$. □

Now, we prove the opposite statement:

- If n is not equal to a positive integer power p^α, $\alpha \in \mathbb{N}$, of a prime number p, then the nth row of Pascal's triangle is not divisible by p.

□ Indeed, in this case, the representation of n in base p has the form $n = n_0 + n_1 p + n_2 p^2 + \cdots + n_k p^k$, where $n_k \neq 0$.

Let us discard the trivial case of $n = n_0$. In this case, $n < p$; therefore, already, the first internal element n of the nth row is not divisible by p: $C_n^1 = n$, p does not divide n.

If $n \geq p$ and n is not a power of p, then $k \geq 1$, and there exists at least one coefficient n_i in the decomposition $n = n_0 + n_1 p + n_2 p^2 + \cdots + n_k p^k$ other than n_k and not equal to zero. In this case, we can always find a binomial coefficient C_n^m, $0 < m < n$, which is not congruent to zero modulo p. Now, it is enough to take $m = n_0 + n_1 p + \cdots + n_{i-1} p^{i-1} + 0 \cdot p^i + n_{i+1} p^{i+1} + \cdots + n_k p^k$. In this case, $C_n^m \equiv C_{n_0}^{n_0} \cdot C_{n_1}^{n_1} \cdot C_{n_i}^{0} \cdots C_{n_k}^{n_k} \equiv 1 (mod\ p)$. The statement is proven. □

It is possible to prove a more general statement. It turns out that the rows of Pascal's triangle can be divisible only by prime numbers. So, no row of the triangle is divisible by p^2 as well as by the product of two primes pq.

First, we prove that:

- *if m has at least two different prime divisors, then no row of Pascal's triangle is divisible by m.*

□ Indeed, if some row is divisible by m and m has two prime divisors p and q, then this row is divisible by the primes p and q; therefore, according to the previous statement, its number is both the power of p and the power of q, which is not possible. □

Now, let us show that:

- *no row of Pascal's triangle is divisible by p^2, where p is a prime.*

□ In fact, if some row of Pascal's triangle is divisible by p^2, then it is divisible by p, and its number n has the form p^k, $k \geq 2$. Then, the element $C_{p^k}^{p^{k-1}}$ of this row with the number p^{k-1} must be divisible by p^2.

However, it is not difficult to prove that $C_{p^k}^{p^{k-1}}$ is not divisible by p^2. To do this, just recall that the power of p in the integer factorization of $n!$ has the form $\lfloor \frac{n}{p} \rfloor + \lfloor \frac{n}{p^2} \rfloor + \cdots + \lfloor \frac{n}{p^t} \rfloor + \cdots$; therefore, the power α of p in the integer factorization of the number $C_{p^k}^{p^{k-1}} = \frac{(p^k)!}{(p^{k-1})!(p^k - p^{k-1})!}$ is

$$\alpha = \left(\left\lfloor \frac{p^k}{p} \right\rfloor + \left\lfloor \frac{p^k}{p^2} \right\rfloor + \cdots + \left\lfloor \frac{p^k}{p^k} \right\rfloor \right)$$

$$- \left(\left\lfloor \frac{p^{k-1}}{p} \right\rfloor + \left\lfloor \frac{p^{k-1}}{p^2} \right\rfloor + \cdots + \left\lfloor \frac{p^{k-1}}{p^{k-1}} \right\rfloor \right)$$

$$- \left(\left\lfloor \frac{p^k - p^{k-1}}{p} \right\rfloor + \left\lfloor \frac{p^k - p^{k-1}}{p^2} \right\rfloor + \cdots + \left\lfloor \frac{p^k - p^{k-1}}{p^{k-1}} \right\rfloor \right)$$

$$= \left(p^{k-1} + p^{k-2} + \cdots + p + 1 \right) - \left(p^{k-2} + p^{k-3} + \cdots + p + 1 \right)$$

$$- \left((p^{k-1} - p^{k-2}) + (p^{k-2} - p^{k-3}) + \cdots + (p-1) \right)$$

$$= p^{k-1} - (p^{k-1} - 1) = 1.$$

So, the number $C_{p^k}^{p^{k-1}}$, $k \geq 2$, is not divisible by p^2 – a contradiction. □

Thus, we have proven the main result of this section:

- *The nth row of Pascal's triangle is divisible by m if and only if m is a prime number and n is a positive integer power of this prime.*

In other words:

- *the largest common divisor of all the internal elements of the nth row of Pascal's triangle is equal to the prime number p if $n = p^k$, $k \in \mathbb{N}$, and is equal to unity in all other cases.*

Pascal's triangle and fractals

2.5.4. Using the divisibility properties of the rows of Pascals' triangle and Lukas's theorem, we can obtain many beautiful structures, connected to the triangle and directly related to fractals.

In particular, on the basis of proven properties, it is possible to paint Pascal's triangle in different ways. If we mark by one color the elements of the triangle divisible by some positive integer m and mark by the other color the elements of the triangle not divisible by m, then unexpected patterns can be obtained.

Let us start with the prime number 2. Because the numbers divisible by 2 are called *even* and the numbers not divisible by 2 are called *odd*, we study the parity of the elements of Pascal's triangle. It turns out that:

- the number of odd numbers in the nth row of Pascal's triangle is 2^k, where k is the number of unities in the binary representation of the number n.

☐ This follows from Lukas's theorem. Indeed, if we want to get an odd number C_n^t, then, for a given binary representation $\overline{(n_{k-1}n_{k-2}\cdots n_1 n_0)_2}$ of n, in order to construct a binary representation $\overline{(t_{k-1}t_{k-2}\cdots t_1 t_0)_2}$ of the number t, we have no choice for the indices i corresponding to zero n_i: in these cases, it is necessary to take $t_i = 0$; for indices j, corresponding to the values $n_j = 1$, we can take t_j equal to both, 0 or 1. Hence, by the combinatorial rule of sum, we get exactly $2 \cdot \cdots \cdot 2 = 2^k$ possibilities to "build" the corresponding binary decomposition $\overline{(t_{k-1}t_{k-2}\cdots t_1 t_0)_2}$ of the number t and, therefore, for a selection of an odd number C_n^t. ☐

Note that we have also proved another important property:

- The binomial coefficient $\binom{n}{k}$ is odd if and only if in the binary representation of the number k, the unities are not placed in those positions where, in the binary representation of the number n, zeros are placed.

2.5.5. Now, we cast Pascal's triangle as follows: instead of the value of the number, in its position, a black circle for odd values should be drawn and a white circle for even values should be drawn.

We know that all the internal elements of a row of Pascal's triangle are even if and only if the number of this row is a power of two, i.e., it has the form 2^n. Then, using the recurrent construction of the triangle, we make sure that in the next row, there are two odd leftmost elements and two odd rightmost elements (as the sum of the odd unity and its even neighbor), and all the other internal elements are even. In the next row, we go back to even numbers as the two neighbors of the extreme unities, while the length of the central even part will decrease by another two unities. Continuing our consideration, we get an inverted central triangle with a base containing $2^n - 1$

numbers and a height $2^n - 1$ consisting of only even numbers, while on the sides there will be additional smaller "even" triangles.

Based on the facts we already have, we can make sure that, in such a coloring, we get a set of white triangles that form a fractal self-similar structure. If the number n of the considered rows is large, then the resulting structure is close to the well-known *Sierpiński triangle*, and for $n \to \infty$, this structure coincides with it.

The *Sierpiński triangle* (or *Sierpiński gasket*, *Sierpiński sieve*) is a plane fractal with the overall shape of an equilateral triangle, subdivided recursively into smaller equilateral triangles. Originally constructed as a curve, this is one of the basic examples of self-similar sets. It was named after the Polish mathematician Wacław Sierpiński (1882–1969) but appeared as a decorative pattern many centuries before the work of Sierpiński.

The Sierpiński triangle may be constructed from an equilateral triangle by repeated removal of triangular subsets: starting with an equilateral triangle, subdivide it into four smaller congruent equilateral triangles, and remove the central triangle; repeat such subdivision with each of the remaining smaller triangles infinitely.

2.5.6. Using similar algorithms, we can build many different types of coloring of Pascal's triangle. The triangle that is constructed "with respect to" the number 3 — the elements that do not divisible by 3 are depicted in black and those divisible by 3 in white — is similar to the already constructed triangle "with respect to" the number 2: the sequence of central inverted white triangles is accompanied by a set of smaller triangles, appearing symmetrically on both sides of the central structure. The same outlines have triangles constructed with respect to the number 5, with respect to the number 7, and, in general, with respect to a prime number p.

For more detailed calculations, it is convenient to use the following simple consequences of Lukas's theorem:

- *The binomial coefficient $\binom{n}{k}$ is divisible by a prime number p if and only if at least one digit in the base p representation of the number k exceeds the corresponding digit in the base-p representation of the number n.*

In other words, we can state that:

- *the binomial coefficient $\binom{n}{k}$ is not divisible by a prime p if and only if, in the base-p representation of the number k, all digits do not exceed the corresponding digits in the base-p representation of the number n.*

At the same time, it is easy to show that:

- *the number of binomial coefficients $\binom{n}{0}$, $\binom{n}{1}$, ..., $\binom{n}{n}$, not divisible by a prime number p is $(n_1+1)\cdots(n_k+1)$, where the numbers n_1, ..., n_n are digits of the base-p representation of the number n and $k = \lfloor \log_p n \rfloor + 1$ is its length.*

The situation becomes more complicated if we try to build a triangle with respect to the number 4. In this case, the central triangles will be replaced by smaller structures as none of the rows of Pascal's triangle are divisible by 4. We can see a similar situation for a triangle constructed with respect to any other composite number. At the same time, the symmetry of the obtained structure as well a its fractality will remain.

Estimations and asymptotic formulas for binomial coefficients

2.5.7. In order to obtain classical estimates and asymptotic formulas for the binomial coefficients, we use the *Stirling formula* for factorials (see Chapter 1):

$$n! \sim \sqrt{2\pi n}\left(\frac{n}{e}\right)^n.$$

However, we begin our presentation with a proof of rougher but much simpler factorial estimates:

- *There exist the inequalities*

$$\left(\frac{n}{e}\right)^n \leq n! \leq 2\cdot\left(\frac{n}{2}\right)^n.$$

□ To prove this fact, let us consider the sequence $\left\{\left(1+\frac{1}{n}\right)^n\right\}_n, n \in \mathbb{N}$, and use the following well-known result: $\lim_{n\to\infty}(1+\frac{1}{n})^n = e$.

In fact, using for $(1+\frac{1}{n})^n$ the binomial theorem, we get that

$$\left(1+\frac{1}{n}\right)^n = 1 + \frac{n}{1}\cdot\frac{1}{n} + \frac{n(n-1)}{1\cdot 2}\cdot\frac{1}{n^2} + \frac{n(n-1)(n-2)}{1\cdot 2\cdot 3}\cdot\frac{1}{n^3}$$
$$+\cdots+\frac{n(n-1)(n-2)\cdots(n-(n-1))}{1\cdot 2\cdot 3\cdots\cdots n}\cdot\frac{1}{n^n}$$
$$= 1 + 1 + \frac{1}{1\cdot 2}\cdot\left(1-\frac{1}{n}\right) + \frac{1}{1\cdot 2\cdot 3}\cdot\left(1-\frac{1}{n}\right)\cdot\left(1-\frac{2}{n}\right)$$
$$+\cdots+\frac{1}{1\cdot 2\cdot 3\cdots\cdots n}\cdot\left(1-\frac{1}{n}\right)\cdot\left(1-\frac{2}{n}\right)\cdots\cdots\left(1-\frac{n-1}{n}\right).$$

From this equality, it follows that, with an increase in n, the number of positive terms on the right-hand side increases. In addition, with an increase in n, the number of $\frac{1}{n}$ decreases, so the values $\left(1-\frac{1}{n}\right), \left(1-\frac{2}{n}\right),\ldots$ increase. Therefore, the sequence $\left\{\left(1+\frac{1}{n}\right)^n\right\}_n, n \in \mathbb{N}$, is an increasing sequence; therefore, it holds that $2 \leq \left(1+\frac{1}{n}\right)^n \leq e$.

Now, we can prove the above double inequality for the factorial. Let us do it by induction. For $n = 1$, everything is correct:

$$\left(\frac{1}{e}\right)^1 \leq 1! \leq 2\cdot\left(\frac{1}{2}\right)^1.$$

Going from n to $n+1$, we are using, on the one hand, the inequality $e \geq \left(1+\frac{1}{n}\right)^n$:

$$(n+1)! = n!\cdot(n+1) \geq \left(\frac{n}{e}\right)^n\cdot(n+1) \geq \left(\frac{n+1}{e}\right)^{n+1}\cdot e\cdot\left(\frac{n+1}{n}\right)^{-n}$$
$$\geq \left(\frac{n+1}{e}\right)^{n+1}\cdot e\cdot\left(\left(1+\frac{1}{n}\right)^n\right)^{-1} \geq \left(\frac{n+1}{e}\right)^{n+1}.$$

On the other hand, we are using the inequality $\left(1+\frac{1}{n}\right)^n \geq 2$:

$$(n+1)! = n!\cdot(n+1) \leq 2\cdot\left(\frac{n}{2}\right)^n\cdot(n+1) \leq \left(\frac{n+1}{2}\right)^{n+1}\cdot 2\cdot\left(\frac{n+1}{n}\right)^{-n}$$
$$\leq \left(\frac{n+1}{2}\right)^{n+1}\cdot 2\cdot\left(\left(1+\frac{1}{n}\right)^n\right)^{-1} \leq \left(\frac{n+1}{2}\right)^{n+1}.$$

□

2.5.8. Now, we have all the information to get the following simplest estimations of the binomial coefficient $\binom{n}{k}$.

- *There exist the inequalities*

$$\left(\frac{n}{k}\right)^k \leq \binom{n}{k} \leq \left(\frac{n \cdot e}{k}\right)^k, \quad 1 \leq k \leq n.$$

□ Indeed, a direct verification shows that, for a given k, for all $i < k$, the following inequality holds: $\frac{n-i}{k-i} \geq \frac{n}{k}$. So, we get

$$\binom{n}{k} = \frac{n!}{k!(n-k)!} = \frac{n(n-1)\cdots(n-k+1)}{1 \cdot 2 \cdots k}$$

$$= \frac{n}{k} \cdot \frac{n-1}{k-1} \cdot \frac{n-2}{k-2} \cdots \frac{n-k+1}{1} \geq \left(\frac{n}{k}\right)^k.$$

On the other hand, as $k! \geq \left(\frac{k}{e}\right)^k$,

$$\binom{n}{k} = \frac{n!}{k!(n-k)!} = \frac{n(n-1)\cdots(n-k+1)}{k!} \leq \frac{n^k}{k!} \leq \left(\frac{n \cdot e}{k}\right)^k. \quad □$$

2.5.9. When n is large and k is sufficiently less than n, we can prove that

$$\binom{n}{k} = \frac{n(n-1)\cdots(n-k+1)}{k!} \sim \frac{(n-\frac{k}{2})^k}{k^k e^{-k}\sqrt{2\pi k}} = \frac{(\frac{n}{k}-\frac{1}{2})^k e^k}{\sqrt{2\pi k}};$$

therefore,

$$\log \binom{n}{k} \sim k \log \left(\frac{n}{k} - \frac{1}{2}\right) + k - \frac{1}{2} \log(2\pi k).$$

If we need more accuracy, we can approximate $\log(n(n-1)\cdots(n-k+1))$ by the integral, having obtained an estimation

$$\log \binom{n}{k} \sim (n+0.5) \log \frac{n+0.5}{n-k+0.5} + k \log \frac{n-k+0.5}{k} - 0.5 \log(2\pi k).$$

So, for example, for $n = 20$ and $k = 10$, it holds that $\log \binom{n}{k} \approx 12{,}127$, while our approximations give the values $12{,}312$ and $12{,}133$, respectively.

2.5.10. For large values of n, we can get an approximation of the binomial coefficient of the form

$$\binom{n}{k} \sim \frac{2^n}{\sqrt{0.5n\pi}} e^{-\frac{(k-0.5n)^2}{0.5n}},$$

related to the *normal distribution* ([Went99]).

If both parameters n and k are significantly larger than unity, the Stirling formula gives the following approximation of the binomial coefficient $\binom{n}{k}$, exponential at n:

$$\log\binom{n}{k} \sim nH\left(\frac{k}{n}\right),$$

where $H(x) = -x\log(x) - (1-x)\log(1-x)$ is the *binary entropy*.

2.5.11. From Stirling's formula, it is again easy to obtain a well-known approximation of the *central binomial coefficient* $\binom{2n}{n}$:

$$\binom{2n}{n} \sim \frac{2^{2n}}{\sqrt{\pi n}}.$$

In fact, as $n! \sim \sqrt{2\pi n}\left(\frac{n}{e}\right)^n$,

$$\binom{2n}{n} = \frac{(2n)!}{(n!)^2} \sim \frac{\sqrt{2\pi(2n)}\left(\frac{2n}{e}\right)^{2n}}{(\sqrt{2\pi n}\left(\frac{n}{e}\right)^n)^2} \sim \frac{2^{2n}}{\sqrt{\pi n}}.$$

The lower estimations of the binomial coefficients, based on Stirling's formula, are also well known. So, for the central binomial coefficient, there is the inequality

$$\sqrt{n}\binom{2n}{n} \geq 2^{2n-1}.$$

In general, it holds that

$$\sqrt{n}\binom{mn}{n} \geq \frac{m^{m(n-1)+1}}{(m-1)^{(m-1)(n-1)}}, \quad m \geq 2, n \geq 1.$$

2.5.12. Now, we consider some estimations of the sums of binomial coefficients:

- One of the simplest estimations of the sum $\sum_{m=0}^{k}\binom{n}{m}$ of binomial coefficients has the form

$$\sum_{m=0}^{k}\binom{n}{m} \leq 1 + n^k.$$

□ Let us prove this result by induction. For $k = 0$ and $k = 1$, the statement is true: $\binom{n}{0} \leq 1$ and $\binom{n}{0} + \binom{n}{1} \leq 1 + n$. Going from k to $k+1$, we get:

$$\sum_{m=0}^{k+1}\binom{n}{m} = \sum_{m=0}^{k}\binom{n}{m} + \binom{n}{k+1} \leq 1 + n^k + \binom{n}{k+1}$$

$$= 1 + n^k + \frac{n(n-1)\cdots(n-k)}{(k+1)!} \leq 1 + n^k + n(n-1)\cdots$$

$$(n-k) \leq 1 + n^k + n^k(n-1) = 1 + n^{k+1}. \qquad \square$$

Chebyshev's inequality is well known:

$$\sum_{k=0}^{m}\binom{n}{k} \leq \frac{n}{(\frac{n}{2}-m)^2} 2^{n-3}, \quad m < \frac{n}{2}.$$

The *entropy estimation*

$$\sum_{k=0}^{m}\binom{n}{k} \leq t 2^{nH(\frac{m}{n})}, \quad m \leq \frac{n}{2},$$

where $H(x) = -x\log_2 x - (1-x)\log_2(1-x)$ is the *binary entropy* of x, already appeared in a slightly different form above.

The *Chernov inequality* has the form

$$\sum_{k=0}^{n/2-\lambda}\binom{n}{k} \leq 2^n e^{-2\lambda^2/n}.$$

Exercises

1. Check that the *polarity property* of the even rows of Pascal's triangle holds for small values of n: we get zero if we take the central element of the nth row of Pascal's triangle with an even

number n, subtract from it the two elements closest to the center, then add the two next closest to the center elements, etc. For example, for the fourth row $1, 4, 6, 4, 1$, we get the formula $6 - (4+4) + (1+1) = 0$, and for the sixth row $1, 6, 15, 20, 15, 6, 1$, we get the formula $20 - (15+15) + (6+6) - (1+1) = 0$. Prove this property.

2. For small values of n, check the following statement: *if in the nth row of Pascal's triangle, n is odd, we add all the numbers with indices congruent to 0 modulo 3, 1 modulo 3, and 2 modulo 3, i.e., with indices of the from $3n, 3n+1, 3n+2$, then the first two sums will be equal, and the third sum is the previous one minus 1.* Prove this statement.

3. Check that the *Vandermond convolution* $C_{n+p}^k = \sum_{m=0}^{M} C_n^m C_p^{k-m}$, where $M = \min\{k, n-1\}$, can be proved using the relationships

$$C_{n+p}^k = C_{n+p-1}^k + C_{n+p-1}^{k-1}$$
$$= C_{n+p-2}^k + C_{n+p-2}^{k-1} + C_{n+p-2}^{k-1} + C_{n+p-2}^{k-3}$$
$$= C_{n+p-2}^k + 2C_{n+p-2}^{k-1} + C_{n-2}^{k-3}$$
$$= C_{n-3}^k + 3C_{n-3}^{k-1} + 3C_{n-3}^{k-2} + C_{n-3}^{k-3} = \cdots.$$

4. Prove the *Shu–Vandermond identity* $\sum_{j=0}^{k} \binom{m}{j}\binom{n-m}{k-j} = \binom{n}{k}$ using the following two decompositions: $(1+x)^m (1+x)^{n-m}$ and $(1+x)^n$. Make sure that it works for any complex m and n and for any integer non-negative k.

5. Using the relation $\sum_{j=0}^{m} \binom{m}{j}^2 = \binom{2m}{m}$, prove the formulas:

  (a) $\sum_{i=0}^{n} i \binom{n}{i}^2 = \frac{n}{2}\binom{2n}{n}$;   (b) $\sum_{i=0}^{n} i^2 \binom{n}{i}^2 = n^2 \binom{2n-2}{n-1}$.

6. Prove that $\binom{n}{0}\binom{a}{a} - \binom{n}{1}\binom{a+1}{a} + \cdots + (-1)^n \binom{n}{n}\binom{a+n}{a} = (-1)^n \binom{a}{n}$.

7. Check that $\sum_{k=1}^{m} \frac{(-1)^{k-1}}{k}\binom{m}{k} = H_m$, where $H_m = \sum_{k=1}^{m} \frac{1}{k}$ is the mth harmonic number.

8. Prove that

$$\sum_{i+j=m} \binom{n}{j}\binom{n}{i}(-1)^j = \begin{cases} (-1)^{m/2}\binom{n}{m/2}, & \text{if } m \equiv 0 \pmod{2}, \\ 0, & \text{if } m \equiv 1 \pmod{2}. \end{cases}$$

2.6. Polynomials Related to Pascal's Triangle

2.6.1. The simplest polynomial, connected with the elements of Pascal's triangle is, obviously, the polynomial $(1+x)^n$ in variable x of degree n; its coefficients are defined by the nth row of Pascal's triangle:

$$(1+x)^n = \sum_{k=0}^{n} \binom{n}{k} x^k = \binom{n}{0} x^0$$
$$+ \binom{n}{1} x^1 + \binom{n}{2} x^2 + \cdots + \binom{n}{n-1} x^{n-1} + \binom{n}{n} x^n.$$

In general, when a binomial, $x+y$, is raised to a positive integer power, n, the expression expands into

$$(x+y)^n = \sum_{k=0}^{n} \binom{n}{k} x^{n-k} y^k = \binom{n}{0} x^0 y^n + \binom{n}{1} x^1 y^{n-1} + \binom{n}{2} x^2 y^{n-2}$$
$$+ \cdots + \binom{n}{n-1} x^{n-1} y^1 + \binom{n}{n} x^n y^0$$
$$= \binom{n}{0} x^n y^0 + \binom{n}{1} x^{n-1} y^1 + \binom{n}{2} x^{n-2} y^2$$
$$+ \cdots + \binom{n}{n-1} x^1 y^{n-1} + \binom{n}{n} x^0 y^n.$$

The last equality can be considered a special property of the polynomial sequence $f_0(x) = x^0, f_1(x) = x^1, \ldots, f_n(x) = x^n, \ldots$. In fact, it holds that

$$f_n(x+y) = \sum_{k=0}^{n} \binom{n}{k} f_k(x) f_{n-k}(y).$$

In mathematics, a *polynomial sequence*, i.e., a sequence f_n, $n = 0, 1, 2, \ldots$, of polynomials indexed by non-negative integers in which the index of each polynomial equals its degree, is said to be of *binomial type* if it satisfies the sequence of identities

$$p_n(x+y) = \sum_{k=0}^{n} \binom{n}{k} p_k(x) p_{n-k}(y).$$

Many such sequences exist. We list several examples as follows:

- The sequence $\{x^n, n = 0, 1, 2, \ldots\}$ is of binomial type.
- The sequence $\{x^{\underline{n}} = x(x-1)(x-2)\cdots(x-n+1), n = 0, 1, 2, \ldots\}$ of *falling factorials* is of binomial type.
- The sequence $\{x^{\overline{n}} = x(x+1)(x+2)\cdots(x+n-1), n = 0, 1, 2, \ldots\}$ of *rising factorials* is of binomial type.
- The sequence $\{p_n(x) = x(x-an)^{n-1}, n = 0, 1, 2, \ldots\}$ of *Abel polynomials* forms a polynomial sequence of binomial type.
- The sequence $\{T_n(x) = \sum_{k=0}^{n} S(n,k) x^k, n = 0, 1, 2, \ldots\}$ of *Touchard polynomials*, where $S(n,k)$ are *Stirling numbers of the second kind*, forms a polynomial sequence of binomial type.

Every sequence of binomial type may be expressed in terms of the *Bell polynomials* ([Deza24]).

2.6.2. There is a natural relationship between the binomial coefficients and one of the classes of polynomials with rational coefficients. Thus, for any non-negative integer k, the expression $\binom{t}{k}$ can be considered a polynomial $t(t-1)(t-2)\cdots(t-k+1)$ in variable t of degree k with integer coefficients divided by $k!$:

$$\binom{t}{k} = \frac{t(t-1)(t-2)\cdots(t-k+1)}{k!} = \frac{t(t-1)(t-2)\cdots(t-k+1)}{k(k-1)(k-2)\cdots 2 \cdot 1}.$$

So, we get the polynomial

$$p_k(t) = \frac{t(t-1)(t-2)\cdots(t-k+1)}{k(k-1)(k-2)\cdots 2 \cdot 1}$$

in one variable t of degree k with rational coefficients. This polynomial can be defined for any real or even complex values of t.

If n is a non-negative integer, then we get the binomial coefficient $\binom{n}{k}$ as a value of the polynomial $p_k(t)$ at the point $t = n$:

$$p_k(n) = \binom{n}{k}.$$

For negative integer, real, or complex values of the variable t, we get in this way the so-called *generalized binomial coefficients*. For a given $\alpha \in \mathbb{C}$, the *generalized binomial coefficient* $\binom{\alpha}{k}$ is defined as $p_k(\alpha)$:

$$\binom{\alpha}{k} = p_k(\alpha) = \frac{\alpha(\alpha-1)(\alpha-2)\cdots(\alpha-k+1)}{k!}$$

$$= \frac{\alpha(\alpha-1)(\alpha-2)\cdots(\alpha-k+1)}{k(k-1)(k-2)\cdots 2 \cdot 1}.$$

In this case, we obtain the following infinite series:

$$(1+x)^\alpha = 1 + p_1(\alpha)x + p_2(\alpha)x^2 + \cdots + p_n(\alpha)x^n + \cdots$$

$$= 1 + \binom{\alpha}{1}x + \binom{\alpha}{2}x^2 + \cdots + \binom{\alpha}{n}x^n + \cdots, \quad |x| \leq 1.$$

The polynomial $p_k(t)$ is naturally related to the *falling factorial* $t^{\underline{k}} = t(t-1)\cdots(t-k+1)$.

- We have the relation

$$p_k(t) = \frac{t^{\underline{k}}}{k^{\underline{k}}}.$$

☐ In fact,

$$p_k(t) = \frac{t(t-1)(t-2)\cdots(t-k+1)}{k!}$$

$$= \frac{t(t-1)(t-2)\cdots(t-k+1)}{k(k-1)(k-2)\cdots 2 \cdot 1} = \frac{t^{\underline{k}}}{k^{\underline{k}}}. \quad \square$$

- The sequence $p_0(t), p_1(t), \ldots, p_k(t), \ldots$ satisfies the following recurrent relation:

$$p_{k+1}(t) = p_k(t)\frac{n+k}{k+1}, \quad p_0(t) \equiv 1.$$

□ In fact, we have $p_0(t) = \binom{t}{0} \equiv 1$, and

$$p_{k+1}(t) = \frac{t(t-1)(t-2)\cdots(t-k+1)(t-k)}{(k+1)!}$$
$$= \frac{t(t-1)(t-2)\cdots(t-k+1)}{k!} \cdot \frac{t-k}{k+1}$$
$$= p_k(t) \cdot \frac{t-k}{k+1}. \qquad \square$$

For a given non-negative integer k, we have the equalities $p_k(0) = p_k(1) = \cdots = p_k(k-1) = 0$, $p_k(k) = 1$. This set of values allows us to formulate the following property:

- *For each non-negative integer k, the polynomial $p_k(t)$ is characterized as the only polynomial of degree k satisfying the conditions*

$$p_k(0) = p_k(1) = \cdots = p_k(k-1) = 0, \quad p_k(k) = 1.$$

□ This statement follows from the fact that any polynomial of degree k is uniquely determined by setting $k+1$ of its values: recall that only one straight line passes through two points, only one parabola passes through three points on a plane, etc. □

The coefficients of the polynomial $p_k(t)$ can be expressed in terms of the *Stirling numbers of the first kind* $s(n,k)$.

- *There exists a decomposition*

$$p_k(t) = \frac{s(k,0)}{k!} + \frac{s(k,1)}{k!}t + \frac{s(k,2)}{k!}t^2 + \cdots + \frac{s(k,k)}{k!}t^k.$$

□ As it holds that (see Chapter 4)

$$t^{\underline{k}} = s(k,0) + s(k,1)t + s(k,2)t^2 + \cdots + s(k,k)t^k,$$

$$\frac{t^{\underline{k}}}{k!} = p_k(t) = \frac{s(k,0)}{k!} + \frac{s(k,1)}{k!}t + \frac{s(k,2)}{k!}t^2 + \cdots + \frac{s(k,k)}{k!}t^k.$$

However, this fact can be checked directly by induction: since $s(0,0) = 1$, $p_0(t) \equiv 1 = \frac{s(0,0)}{0!}t^0$. As $s(1,0) = 1$ and $s(1,1) = 1$,

$p_1(t) = t = \frac{s(1,0)}{0!}t^0 + \frac{s(1,1)}{1!}t^1$. As $s(k,0) = 0$, $s(k,k) = 1$ for $k \geq 1$, and $s(k+1,i) = s(k,i-1) - ks(k,i)$, then using the formula $p_{k+1}(t) = p_k(t)\frac{t-k}{k+1}$, we obtain

$$p_{k+1}(t) = p_k(t)\frac{t-k}{k+1} = \left(\sum_{i=0}^{k} \frac{s(k,i)}{k!}t^i\right)\frac{t-k}{k+1}$$

$$= \sum_{i=0}^{k} \frac{s(k,i)}{(k+1)!}t^{i+1} - \sum_{i=0}^{k} \frac{ks(k-1,i)}{(k+1)!}t^i$$

$$= \frac{ks(k,0)}{(k+1)!}t^0 + \sum_{i=0}^{k-1} \frac{s(k,i-1) - ks(k,i)}{(k+1)!}t^i + \frac{s(k,k)}{(k+1)!}t^{k+1}$$

$$= \frac{s(k+1,0)}{(k+1)!}t^0 + \sum_{i=1}^{k-1} \frac{s(k+1,i)}{(k+1)!}t^i + \frac{s(k+1,k+1)}{(k+1)!}t^{k+1}$$

$$= \sum_{i=0}^{k+1} \frac{s(k+1,i)}{(k+1)!}t^i.$$

□

The derivative of the polynomial $p_k(t)$ can be obtained using logarithmic differentiation.

- *The derivative of the polynomial $p_k(t)$ has the following form:*

$$p'_k(t) = p_k(t)\left(\frac{1}{t} + \frac{1}{t-1} + \frac{1}{t-2} + \cdots + \frac{1}{t-k+1}\right).$$

□ In fact, it holds that

$$\log p_k(t) = \log \frac{t(t-1)(t-2)\cdots(t-k+1)}{k!}$$

$$= \log t + \log(t-1) + \log(t-2) + \cdots + \log(t-k+1) - \log k!,$$

$$\log' p_k(t) = \left(\log \frac{t(t-1)(t-2)\cdots(t-k+1)}{k!}\right)'$$

$$= (\log t + \log(t-1) + \log(t-2) + \cdots + \log(t-k+1) - \log k!)'.$$

As $\log' p_k(t) = \frac{p_k'(t)}{p_k(t)}$ and $\log'(t-i) = \frac{1}{t-i}$, $i = 0, 1, 2, \ldots, k-1$, we get that

$$\frac{p_k'(t)}{p_k(t)} = \frac{1}{t} + \frac{1}{t-1} + \frac{1}{t-2} + \cdots + \frac{1}{t-k+1}.$$
□

- Each polynomial $p_k(t)$ is an integer polynomial, that is, it takes integer values for all integer values of the argument.

□ For the non-negative integers t, this result follows from the fact that $\binom{t}{k}$ is a number of combinations and, therefore, is a positive integer. For negative integer values of t, the situation is similar except for the sign. □

2.6.3. Over any field of characteristic zero (that is, any field that contains the rational numbers), each polynomial $p(t)$ of degree at most d is uniquely expressible as a linear combination $\sum_{k=0}^{d} a_k \binom{t}{k}$ of binomial polynomials $p_k(t) = \binom{t}{k}$. The coefficient a_k is the kth difference of the sequence $p(0), p(1), \ldots, p(k)$. Explicitly,

$$a_k = \sum_{i=0}^{k} (-1)^{k-i} \binom{k}{i} p(i).$$

Thus, we can state that

- the polynomials $p_k(t)$, $k = 0, 1, 2, \ldots$, form a basis for the space of polynomials.

In particular, for any integer polynomial $p(t)$, the coefficients a_k will be integers, so the polynomials $p_k(t)$, $k = 0, 1, 2, \ldots$, form a basis for the space of integer polynomials, in which any integer polynomial has a uniquely linear representation, $\sum_{k=0}^{d} a_k \binom{t}{k}$, with integer coefficients. For example, the integer polynomial $\frac{3t(3t+1)}{2}$ has the decomposition $9p_2(t) + 6p_1(t)$.

At the same time, the standard basis $1, x, x^2, \ldots$ cannot express all integer polynomials if only integer coefficients are used since $\binom{x}{2} = \frac{x^2}{2} - \frac{x}{2}$ already has fractional coefficients in the powers of x.

This result can be generalized to polynomials of many variables. Namely, if the polynomial $R(x_1, \ldots, x_m)$ of degree k has real coefficients and takes integer values with integer values of variables, then

$$R(x_1, \ldots, x_m) = P\left(\binom{x_1}{1}, \ldots, \binom{x_1}{k}, \ldots, \binom{x_m}{1}, \ldots, \binom{x_m}{k}\right),$$

where P is an integer polynomial.

Exercises

1. Expand by powers t the polynomial $p_k(t)$ for $k = 0, 1, 2, 3, 4, 5, 6, 7, 8, 9$.
2. Write the Taylor series decompositions for $(1+x)^\alpha$, where $\alpha = -1, -2, -3, -4, -5$.
3. Write the Taylor series decompositions for $(1+x)^t$, where $t = 1/2, -1/2$.
4. Check that the polynomial $p_k(t)$ takes integer values at integers t, $t \in \{0, \pm 1, \pm 2, \ldots, \pm 9\}$, if $k = 0, 1, 2, 3, 4, 5, 6, 7, 8, 9$.
5. Present as a linear combination of binomial coefficients the polynomials $1 + t^3$, $1 + t^4$, $t + t^3$, and $t + t^3 + t^4$.
6. Prove that for any polynomial $P(x)$ of degree less than n, it holds that $\sum_{j=0}^{n}(-1)^j \binom{n}{j} P(j) = 0$.
7. Check that, for $k = 0, \ldots, n-1$, the identity $\sum_{j=0}^{k}(-1)^j \binom{n}{j} = (-1)^k \binom{n-1}{k}$ is a special case of the previous problem.
8. Prove that for a polynomial $P(x)$ of degree $k \leq n$, it holds that $\sum_{j=0}^{n}(-1)^j \binom{n}{j} P(n-j) = n! a_n$, where a_n is the coefficient at x^n in $P(x)$.

2.7. Pascal's Triangle in the Family of Special Numbers

2.7.1. In the previous sections, we have reviewed in detail the occurrence of various special numbers in Pascal's triangle. The numbers mentioned include figurative numbers, Fibonacci numbers, Catalan numbers, and perfect numbers. In this section, we consider the possibilities of generalizing Pascal's triangle and getting in this way other interesting special combinatorial numbers.

Pascal's triangle and figurate numbers

2.7.2. A *figurate number* is a number that can be represented by a regular geometrical arrangement of equally spaced points. If the arrangement forms a regular polygon, the number is called a *polygonal number*. So, there exist triangular numbers, square numbers, pentagonal numbers, hexagonal numbers, etc. Figurate numbers can also form other shapes, such as centered polygons and three-dimensional solids.

In particular, by adding to a point two, three, four, etc., points, then organizing the points in the form of an equilateral triangle, and counting the number of points in each such triangle, we can obtain the sequence $1, 3, 6, 10, 15, 21, 28, 36, 45, 55, \ldots$ of *triangular numbers* (sequence A000217 in the OEIS).

The above construction implies that $S_3(n) = 1 + 2 + \cdots + n$, and we obtain the following recurrent formula for triangular numbers:

$$S_3(n+1) = S_m(n) + (n+1), \quad S_3(1) = 1.$$

Since the sum $1 + 2 + \cdots + n$ of the first n positive integers is equal to $\frac{n(n+1)}{2}$, we obtain the following closed formula for the nth triangular number:

$$S_3(n) = \frac{n(n+1)}{2}.$$

Placing points in some special order in the space, instead of on the plane, we obtain *space figurate numbers*. The most commonly known such numbers are *pyramidal numbers*, corresponding to triangular, square, pentagonal, hexagonal, heptagonal, and, in general, m-gonal pyramids. They can be represented as sums of the corresponding polygonal numbers.

In particular, *tetrahedral numbers* (or *triangular pyramidal numbers*) are space figurate numbers which correspond to placing dots in the configuration of a tetrahedron. The nth tetrahedral number

$S_3^3(n)$ is the sum of the first n consecutive triangular numbers; it has the form
$$S_3^3(n) = \frac{n(n+1)(n+2)}{6}.$$

So, the sequence of tetrahedral numbers is obtained by consecutive summation of the sequence $1, 3, 6, 10, 15, 21, 28, 36, 45, 55, \ldots$ (sequence A000217 in the OEIS) of triangular numbers and begins with elements $1, 4, 10, 20, 35, 56, 84, 120, 165, 220, \ldots$ (sequence A000292 in the OEIS).

Formally, figurate numbers can be constructed in any dimension k, but, for $k \geq 4$, such construction loses its usual physical sense.

Pentatope numbers (or *hypertetrahedral numbers, triangulo-triangular numbers*) are figurate number which represent a four-dimensional hypertetrahedron. They are the four-dimensional analog of three-dimensional tetrahedral numbers and two-dimensional triangular numbers.

The nth pentatope number $S_3^4(n)$ is given as the sum of the first n tetrahedral numbers:
$$S_3^4(n) = S_3^3(1) + S_3^3(1) + \cdots + S_3^3(n).$$

Since $S_3^3(n) = \frac{n(n+1)(n+2)}{6}$, we get the following recurrent formula for the sequence of pentatope numbers:
$$S_3^4(n+1) = S_3^4(n) + \frac{(n+1)(n+2)(n+3)}{6}, \quad S_3^4(1) = 1.$$

It yields the closed formula for the nth pentatope number:
$$S_3^4(n) = \frac{n(n+1)(n+2)(n+3)}{24}.$$

So, the first few pentatope numbers are $1, 5, 15, 35, 70, 126, 210, 330, 495, 715, \ldots$ (sequence A000332 in the OEIS).

Five-dimensional hypertetrahedron numbers start with the elements $1, 6, 21, 56, 126, 252, 462, 792, 1287, 2002, \ldots$ (sequence A000389 in the OEIS). They are obtained as the partial sums of the sequence $1, 5, 15, 35, 70, 126, 210, 330, 495, 715, \ldots$ (sequence A000332 in the OEIS) of pentatore numbers.

Six-dimensional hypertetrahedron numbers start with the elements 1, 7, 28, 84, 210, 462, 924, 1716, 3003, 5005, ... (sequence A000579). They are obtained as the partial sums of the sequence 1, 7, 28, 84, 210, 462, 924, 1716, 3003, 5005, ... (sequence A000579 in the OEIS) of five-dimensional hypertetrahedron numbers.

By definition, we get the following recurrent formula for the sequence of k-dimensional hypertetrahedron numbers:

$$S_3^k(n+1) = S_3^k(n) + S_3^{(k-1)}(n+1), \quad S_3^k(1) = 1.$$

Now, it is easy to obtain the closed formula for the nth k-dimensional hypertetrahedron number, which has the form

$$S_3^k(n) = \frac{n(n+1)(n+2)\cdots(n+(k-1))}{k!}, \quad \text{or} \quad S_3^k(n) = \frac{n^{\overline{k}}}{k!},$$

where $n^{\overline{k}} = n(n+1)(n+2)\cdots(n+(k-1))$ is a *rising factorial*.

2.7.3. One can easily find the family of figurate numbers in Pascal's triangle. In fact, any triangular number can be written in the form

$$S_3(n) = \binom{n+1}{2},$$

i.e., it is a *binomial coefficient*. Therefore:

- triangular numbers form the third diagonal 1, 3, 6, 10, 15, ... of Pascal's triangle.

In other words, triangular numbers are found in the third position either from the left to right or from the right to left in Pascal's triangle.

Similarly,

$$S_3^3(n) = \binom{n+2}{3},$$

i.e., any tetrahedral number is a binomial coefficient. The tetrahedral numbers form the fourth diagonal 1, 4, 10, 20, 35, ... of Pascal's triangle.

Pentatope numbers also coincide with binomial coefficients:
$$S_3^4(n) = \binom{n+3}{4}.$$
They form the fifth diagonal $1, 5, 15, 35, 70, \ldots$ of Pascal's triangle.

In general, for a given dimension k, $k \geq 2$, the nth k-dimensional hypertetrahedron number $S_3^k(n)$, representing a k-dimensional simplex, has the form
$$S_3^k(n) = \frac{n^{\overline{k}}}{k!} = \binom{n+k-1}{k}.$$
Hence, the k-dimensional hypertetrahedron numbers form the $(k+1)$th diagonal of Pascal's triangle, i.e., they can be found at the $(k+1)$th position either from the left to right or from the right to left in Pascal's triangle.

Moreover, the *linear numbers* $1, 2, 3, 4, 5, \ldots$ can be considered the one-dimensional analog of the two-dimensional triangular and three-dimensional tetrahedral numbers. The nth linear number $S_3^1(n) = n$ is equal to the binomial coefficient $\binom{n}{1}$, and such numbers form the second diagonal $1, 2, 3, 4, 5, \ldots$ of Pascal's triangle.

At last, the sequence $1, 1, 1, 1, 1, \ldots$ can be seen as the zero-dimensional analog of linear, triangular, and tetrahedral numbers. Thus, the nth element $S_3^0(n) = 1$ of this sequence coincides with the binomial coefficient $\binom{n-1}{0}$. These zero-dimensional figurate numbers form the first diagonal $1, 1, 1, 1, 1, \ldots$ of Pascal's triangle.

So, viewing from the opposite perspective, any element $\binom{n}{k}$, $n \geq 0$, $0 \leq k \leq n$, of Pascal's triangle is a figurate number, representing a k-dimensional analog of triangular numbers. In fact,
$$\binom{n}{k} = S_3^k(n - k + 1),$$
and we can rewrite Pascal's triangle in the form

$$
\begin{array}{ccccccccc}
 & & & & S_3^0(1) & & & & \\
 & & & S_3^0(2) & & S_3^1(1) & & & \\
 & & S_3^0(3) & & S_3^1(2) & & S_3^2(1) & & \\
 & S_3^0(4) & & S_3^1(3) & & S_3^2(2) & & S_3^3(1) & \\
S_3^0(5) & & S_3^1(4) & & S_3^2(3) & & S_3^3(2) & & S_3^4(1) \\
\vdots & \vdots & \vdots & \vdots & \vdots & \vdots & \vdots & \vdots & \vdots
\end{array}
$$

In other words, one can rewrite the binomial theorem in the form

$$(1+x)^n = S_3^0(n+1) + S_3^1(n)x + S_3^2(n-1)x^2 + \cdots + S_3^{n-1}(2)x^{n-2} + S_3^n(1)x^n.$$

2.7.4. The above consideration shows that Pascal's triangle consists only of figurate numbers. Moreover, the triangle, written as a square grid and padded with zeros, as was considered by Jakob Bernoulli (1654–1705) [Smit84], is called the *figurate number triangle*:

$$\begin{matrix} 1 & 0 & 0 & 0 & 0 & \cdots \\ 1 & 1 & 0 & 0 & 0 & \cdots \\ 1 & 2 & 1 & 0 & 0 & \cdots \\ 1 & 3 & 3 & 1 & 0 & \cdots \\ 1 & 4 & 6 & 4 & 1 & \cdots \\ \vdots & \vdots & \vdots & \vdots & \vdots & \ddots \end{matrix}$$

This triangle has entries $a_{ij} = \binom{i}{j}$, where $i = 0, 1, 2, 3, \ldots$ is the row number and $j = 0, 1, 2, 3, \ldots$ is the column number. In this construction, starting with the zeroth column, consisting of 1, we obtain the linear numbers in the first column, the triangular numbers in the second column, the tetrahedral numbers in the third column, etc. In general, the k-dimensional hypertetrahedron numbers form the kth column of the figurate number triangle.

Pascal's triangle and Mersenne numbers

2.7.5. A *Mersenne number* is a positive integer of the form

$$\mathcal{M}_n = 2^n - 1, \ n \in \mathbb{N}.$$

The first few Mersenne numbers are $1, 3, 7, 15, 31, 63, 127, 255, 511, 1023, 2047, \ldots$ (sequence A000225 in OEIS).

It is obvious that:

- the sum of all the entries of the first n rows of Pascal's triangle is equal to the nth Mersenne number $\mathcal{M}_n$.

☐ In fact, it is well known that the sum of all the elements of the ith row of Pascal's triangle is 2^i, $i = 0, 1, 2, 3, \ldots$. Then, the sum of all the elements of the first n rows of Pascal's triangle is equal to $2^0 + 2^1 + \cdots + 2^{n-1} = 2^n - 1$. That concludes the proof. □

2.7.6. It was proven in Section 2.5 that, for a given prime p, the *pth row of Pascal's triangle is divisible by p*. It means that in the pth row of Pascal's triangle, if p is a prime number, all the terms except unities are multiples of p.

So, for any Mersenne prime $\mathcal{M}_p$, we get the following property:

- *For a given Mersenne prime $\mathcal{M}_p$, the $\mathcal{M}_p$th row of Pascal's triangle is divisible by $\mathcal{M}_p$.*

In fact, the third row $1, 3, 3, 1$ is divisible by $\mathcal{M}_2 = 2^2 - 1 = 3$, the seventh row $1, 7, 21, 35, 35, 21, 7, 1$ is divisible by $\mathcal{M}_3 = 2^3 - 1 = 7$, etc.

Pascal's triangle and perfect numbers

2.7.7. A *perfect number* is a positive integer that is equal to the sum of its proper divisors, i.e., positive integer divisors, excluding the number itself.

For instance, the number 6 has three proper divisors, 1, 2 and 3, and $1 + 2 + 3 = 6$, so 6 is a perfect number. The number 28 has five proper divisors, 1, 2, 4, 7, and 14, and $1 + 2 + 4 + 7 + 14 = 28$, so 28 is a perfect number. Similarly, the number 496 has nine proper divisors, 1, 2, 4, 8, 16, 31, 62, 124, and 238; it is perfect since $1 + 2 + 4 + 8 + 16 + 31 + 62 + 124 + 248 = 496$. The number 8128 has 13 proper divisors, 1, 2, 4, 8, 16, 32, 64, 127, 254, 508, 1016, 2032, and 4064; it is perfect since $1 + 2 + 4 + 8 + 16 + 32 + 64 + 127 + 254 + 508 + 1016 + 2032 + 4064 = 8128$.

It is not known whether there are any odd perfect numbers or whether infinitely many perfect numbers exist. The first few perfect numbers are $6, 28, 496, 8128, 33550336, 8589869056, 137438691328,$ $2305843008139952128, 2658455991569831744654692615953842176, \ldots$ (sequence A000396 in the OEIS).

The set of even perfect numbers is completely characterized by the *Euclid–Euler theorem*, partially proved by Euclid and completed by L. Euler.

- An even natural number is perfect if and only if $n = 2^{k-1}(2^k - 1)$, where $2^k - 1 \in P$.

In terms of even perfect numbers, we can say now that:

- the even part of any even perfect number is the sum 2^{p-1} of all the elements of the $(p-1)$th row of Pascal's triangle, while the odd part of this even perfect number is the sum of all the elements of the first p (up to the $(p-1)$th) rows of Pascal's triangle, where p is a prime number, giving the Mersenne prime $\mathcal{M}_p = 2^p - 1$.

In fact, for the first even perfect number 6, its even part, 2, is the sum of the elements $1, 1$ of the first row of Pascal's triangle, while its odd part, 3, is the sum of the elements $1; 1, 1$ of the first two rows. For the second even perfect number 28, its even part, 4, is the sum of the elements $1, 2, 1$ of the second row of Pascal's triangle, while its odd part, 7, is the sum of the elements $1; 1, 1; 1, 2, 1$ of the first three rows. For the third even perfect number 496, its even part, 16, is the sum of the elements $1, 4, 6, 4, 1$ of the fourth row of Pascal's triangle, while its odd part, 31, is the sum of the elements $1; 1, 1; 1, 2, 1; 1, 3, 3, 1; 1, 4, 6, 4, 1$ of the first five rows, etc.

Considering Pascal's triangle modulo 2, we can find in this construction all even perfect numbers.

In fact, the nth row of Pascal's triangle is divisible by 2 if and only if $n = 2^k$ (see Section 2.5). In this case, all $2^k - 1$ internal entries of the row are even. So, exactly $2^k - 2$ central entries of the next $(2^k + 1)$th row will be even; exactly $2^k - 3$ central entries of the $(2^k + 2)$th row will be even, ..., exactly 1 central element of the $(2^{k+1} - 2)$th row will be even. In this way, we obtain in Pascal's triangle modulo 2 a black subtriangle which contains

$$1 + 2 + \cdots + (2^k - 1) = \frac{(2^k - 1)2^k}{2} = 2^{k-1}(2^k - 1)$$

elements. So, for a Mersenne prime $\mathcal{M}_p$, the number of elements in the corresponding black "even subtriangle" will be a perfect number.

For some additional information see, for example, [Bond93], [Buch09], [DeKo13], [Deza17], [Deza18], [Uspe76], and [Wiki24].

Pascal's triangle and Fermat numbers

2.7.8. A *Fermat number* F_n is a positive integer of the form

$$F_n = 2^{2^n} + 1, \ n \in \mathbb{Z}, \ n \geq 0.$$

The first few Fermat numbers are $3, 5, 17, 257, \ 65537, 4294967297,$ $18446744073709551617, \ldots$ (sequence A000215 in the OEIS).

The *Gauss–Wantzel theorem* states that *an n-sided regular polygon can be constructed with a compass and a straightedge if and only if n is the product of a power of 2 and distinct Fermat primes.*

Since there are exactly five known Fermat primes, we know exactly 31 numbers that are the products of distinct Fermat primes and hence 31 constructible odd-sided regular polygons.

They are 3, 5, 15, 17, 51, 85, 255 , 257, 771, 1285, 3855, 4369, 13107, 21845, 65535, 65537, 196611, 327685, 983055, 1114129, 3342387, 5570645, 16711935, 16843009, 50529027, 84215045, 252645135, 286331153, 858993459, 1431655765, and 4294967295 (sequence A045544 in the OEIS).

These numbers, when written in binary form, are equal to the first 32 rows of the modulo 2 Pascal's triangle, minus the top row, which corresponds to a monogon. Because of this, the unities in such a list form an approximation to the Sierpiński triangle.

This pattern breaks down after this, as the next Fermat number is composite. In the following, one can see the algorithm of this construction:

```
                              1
                          1       1
                       1     2       1
                    1     3     3       1
                 1     4     6     4       1
                 ⋮     ⋮     ⋮     ⋮       ⋮
```

modulo 2 →

```
                              1
                          1       1
                       1     0       1
                    1     1     1       1
                 1     0     0     0       1
                 ⋮     ⋮     ⋮     ⋮       ⋮
```

base 2 →

$1_2 = 1$
$11_2 = 1 \cdot 2 + 1 = 3$
$101_2 = 1 \cdot 2^2 + 0 \cdot 2 + 1 = 5$
$1111_2 = 1 \cdot 2^3 + 1 \cdot 2^2 + 1 \cdot 2 + 1 = 15$
$10001_2 = 1 \cdot 2^4 + 0 \cdot 2^3 + 0 \cdot 2^2 + 0 \cdot 2 + 1 = 17$
$\ldots$

Several additional rows of Pascal's triangle modulo 2 are presented as follows:

$$
\begin{array}{ccccccccc}
 & & & & 1 & & & & \\
 & & & 1 & & 1 & & & \\
 & & 1 & & 0 & & 1 & & \\
 & 1 & & 1 & & 1 & & 1 & \\
1 & & 0 & & 0 & & 0 & & 1 \\
\end{array}
\quad
\begin{array}{l}
= 1 \\
= 3 \\
= 5 \\
= 15 \\
= 17 \\
= 51 \\
= 85 \\
= 255
\end{array}
$$

For some additional information, see, for example, [Bond93], [Deza17], [Deza18], [DeMo10], [KLS01], [Tsan10], [Sier64], and [Uspe76].

Elements of Pascal's triangle with negative indices

2.7.9. We built Pascal's triangle by looking at its elements T_n^k for all non-negative integers n and all non-negative integers k, $k \leq n$.

The first possible expansion of the set of indices is completely natural. In particular, let us assume that, for any fixed non-negative integer n, the values T_n^k are defined for all integer values of k: they are equal to zero for all negative integers k and for all positive integers k, $k > n$:

$$T_n^k = 0, \ n = 0, 1, 2, 3, \ldots, k = \ldots, -3, -2, -1, n+1, n+2, n+3, \ldots \ .$$

This is fully consistent with the combinatorial interpretation of the elements of Pascal's triangle and with the recurrent scheme of the construction of the triangle. Let us imagine a digital plane filled with zeros. We place the unity in the position T_0^0. Then, using the recurrent relation $T_n^k = T_{n-1}^k + T_{n-1}^{k-1}$, we build the elements $T_1^0 = T_0^0 + T_0^{-1} = 1 + 0 = 1$ and $T_1^1 = T_0^1 + T_0^0 = 0 + 1 = 1$, the elements $T_2^0 = T_1^0 + T_1^{-1} = 1 + 0 = 1$, $T_2^1 = T_1^1 + T_1^0 = 1 + 1 = 2$, and $T_2^2 = T_1^2 + T_1^1 = 0 + 1 = 1$, etc. In fact, for this construction, it is sufficient to add to the recurrent relation $T_n = T_{n-1}^k + T_{n-1}^{k-1}$ the initial conditions $T_0^0 = 1$ and $T_0^k = 0$ for any non-zero integer k.

In this case, $T_1^k = T_0^k + T_0^{k-1}$, that is, $T_1^0 = T_1^1 = 1$, and all other $T_1^k = 0$. Similarly, $T_2^0 = 1, T_2^1 = 2, T_2^2 = 1$, and all other $T_2^k = 0$, etc.

Let us see what happens with negative values of n, if we assume that, for negative n and negative k, it holds that $T_n^k = 0$. Formally, $T_0^0 = T_{-1}^0 + T_{-1}^{-1}$, and hence, $T_{-1}^0 = T_0^0 - T_{-1}^{-1} = 1 - 0 = 1$. Furthermore, $T_0^1 = T_{-1}^1 + T_{-1}^0$, and hence, $T_{-1}^1 = T_0^1 - T_{-1}^0 = 0 - 1 = -1$. Similarly, $T_0^2 = T_{-1}^2 + T_{-1}^1$, and $T_{-1}^2 = T_0^2 - T_{-1}^1 = 0 + 1 = 1$. In general, $T_0^k = T_{-1}^k + T_{-1}^{k-1}$, and if $T_{-1}^{k-1} = (-1)^{k-1}$, then $T_{-1}^k = T_0^2 - T_{-1}^1 = 0 - (-1)^{k-1} = (-1)^k$.

Consider the next row. As $T_{-1}^k = T_{-2}^k + T_{-2}^{k-1}$, $T_{-2}^k = T_{-1}^k - T_{-2}^{k-1}$, and we have that $T_{-2}^0 = T_{-1}^0 - T_{-2}^{-1} = 1 - 0 = 1$, $T_{-2}^1 = T_{-1}^1 - T_{-2}^0 = -1 - 1 = -2$, $T_{-2}^2 = T_{-1}^2 - T_{-2}^1 = 1 + 2 = 3$, and $T_{-2}^k = (-1)^k (k+1)$.

Continue the consideration. As $T_{-2}^k = T_{-3}^k + T_{-3}^{k-1}$, $T_{-3}^k = T_{-2}^k - T_{-3}^{k-1}$, and we obtain that $T_{-3}^0 = T_{-2}^0 - T_{-3}^{-1} = 1 - 0 = 1$, $T_{-3}^1 = T_{-2}^1 - T_{-3}^0 = -2 - 1 = -3$, $T_{-3}^2 = T_{-2}^2 - T_{-3}^1 = 3 + 3 = 6$, and $T_{-2}^k = (-1)^k \frac{(k+1)(k+2)}{2}$.

In general, any T_{-n}^k with negative index $-n$ and with non-negative index k can be obtained using the recurrent relation $T_{-n}^k = T_{-n+1}^k - T_{-n}^{k-1}$, with initial conditions $T_{-n}^{-1} = 0$.

So, we obtain the following table:

n	T_n^{-1}	T_n^0	T_n^1	T_n^2	T_n^3	T_n^4	T_n^5	T_n^6	T_n^7	T_n^8	T_n^9	T_n^{10}
-4	0	1	-4	10	-20	35	-56	84	-120	165	-220	286
-3	0	1	-3	6	-10	15	-21	28	-36	45	-55	66
-2	0	1	-2	3	-4	5	-6	7	-8	9	-10	11
-1	0	1	-1	1	-1	1	-1	1	-1	1	-1	1
0	0	1	0	0	0	0	0	0	0	0	0	0
1	0	1	1	0	0	0	0	0	0	0	0	0
2	0	1	2	1	0	0	0	0	0	0	0	0
3	0	1	3	3	1	0	0	0	0	0	0	0
4	0	1	4	6	4	1	0	0	0	0	0	0

If we study the resulting table, then we see that, excluding the sign, above the usual Pascal's triangle, there is Tartaglia's rectangle. It is obtained for negative values of the index n.

It is easy to trace the following pattern: for negative integer values of the index $-n$ and for non-negative integer values of the index k,

there is a relation
$$T^k_{-n} = (-1)^k T^k_{n+k-1}, \ k \geq 0, \ n \geq 1.$$

On the one hand, this relation can be obtained by induction, using the recurrent scheme, analyzed above. We invite the reader to do this work themselves.

On the other hand, the question can be solved in a much easier way using the already discussed possibility of assigning to the parameter n in $\binom{n}{k}$ any non-negative integer, real, or even complex values. For a positive integer n, we get that

$$\binom{-n}{k} = \frac{(-n) \cdot (-n-1) \cdot (-n-2) \cdots (-n-k+2) \cdot (-n-k+1)}{k!}$$

$$= (-1)^k \frac{n \cdot (n+1) \cdot (n+2) \cdots (n+k-2) \cdot (n+k-1)}{k!}$$

$$= (-1)^k \binom{n+k-1}{k}.$$

For the special case of $n = -1$, we get that

$$\binom{-1}{k} = (-1)^k \binom{k}{k} = (-1)^k.$$

For example, if $n = -4$ and $k = 7$, then

$$\binom{-4}{7} = \frac{(-10) \cdot (-9) \cdot (-8) \cdot (-7) \cdot (-6) \cdot (-5) \cdot (-4)}{1 \cdot 2 \cdot 3 \cdot 4 \cdot 5 \cdot 6 \cdot 7}$$

$$= (-1)^7 \frac{4 \cdot 5 \cdot 6 \cdot 7 \cdot 8 \cdot 9 \cdot 10}{1 \cdot 2 \cdot 3 \cdot 4 \cdot 5 \cdot 6 \cdot 7} = (-1)^7 \binom{10}{7} = \binom{-1}{7}\binom{10}{7}.$$

Thus, for generalized binomial coefficients, there is the identity

$$\binom{-n}{k} = \binom{-1}{k}\binom{n+k-1}{k} = (-1)^k \binom{n+k-1}{k},$$

which is called the *upper index inverse*.

This operation is directly related to another combinatorial structure. We already noted that the binomial coefficient $\binom{n}{k}$ is closely related to the falling factorial $n^{\underline{k}} = n(n-1)(n-2)\cdots(n-k+1)$:

$$\binom{n}{k} = \frac{n^{\underline{k}}}{k!}.$$

And what happens if we replace the falling factorial by the rising factorial? Let us build the value $\left(\binom{n}{k}\right)$, replacing in the formula $\binom{n}{k}$ the falling factorial by the rising factorial:

$$\left(\binom{n}{k}\right) = \frac{n^{\overline{k}}}{k!} = \frac{n(n+1)(n+2)\cdots(n+k-1)}{k!}.$$

The constructed numbers (with non-negative integer indices n and non-negative indices k, $k \leq n$) have a natural combinatorial interpretation. They are equal to the number of k-combinations with repetitions of a given n-element set:

$$\left(\binom{n}{k}\right) = \overline{C}_n^k.$$

Thus, it can be argued that for a positive integer n and for a non-negative integer k, we have the relationship

$$\binom{-n}{k} = (-1)^k \left(\binom{n}{k}\right),$$

or, which is the same, the relationship

$$C_{-n}^k = (-1)^k \overline{C}_n^k.$$

Generalized Pascal's triangles

2.7.10. Let us move to the consideration of certain number triangles related to Pascal's triangle but built using a more complicated recurrent scheme: the so-called *generalized Pascal's triangles*.

Consider the generalized Pascal's triangle of the third order. In particular, the *generalized Pascal's triangle of the third order* is a number triangle in which at the vertex and on the sides are unities, and each internal element is the sum of the three numbers placed above it in the previous row. As a rule, the generalized Pascal's triangle of the third order is written as an isosceles triangle:

$$
\begin{array}{ccccccccc}
 & & & & 1 & & & & \\
 & & & 1 & 1 & 1 & & & \\
 & & 1 & 2 & 3 & 2 & 1 & & \\
 & 1 & 3 & 6 & 7 & 6 & 3 & 1 & \\
1 & 4 & 10 & 16 & 19 & 16 & 10 & 4 & 1 \\
\vdots & \vdots & \vdots & \vdots & \vdots & \vdots & \vdots & \vdots & \vdots
\end{array}
$$

With this construction, it can be assumed that before the start of the triangle's construction, all positions of the digital plane are filled with zeros. Then, it is enough to start with unity, located at the vertex of the triangle: each element of any next row, including the leftmost and rightmost elements, is obtained as the sum of the three numbers located directly above it. Of course, we can also construct the right-generalized Pascal's triangle, as follows:

$$
\begin{array}{ccccccccc}
1 & & & & & & & & \\
1 & 1 & 1 & & & & & & \\
1 & 2 & 3 & 2 & 1 & & & & \\
1 & 3 & 6 & 7 & 6 & 3 & 1 & & \\
1 & 4 & 10 & 16 & 19 & 16 & 10 & 4 & 1 \\
\vdots & \vdots & \vdots & \vdots & \vdots & \vdots & \vdots & \vdots & \vdots
\end{array}
$$

Similarly, the generalized Pascal's triangle of the fourth order can be constructed. In particular, the *generalized Pascal's triangle of the fourth order* is a number triangle in which at the vertex and on the sides are unities, and each internal element is equal to the sum of the four elements above it: the two neighbors to the left and the two neighbors to the right in the previous row. In the following, five rows of the generalized Pascal's triangle of the fourth order are presented:

$$
\begin{array}{ccccccccccccc}
 & & & & & & 1 & & & & & & \\
 & & & & & 1 & 1 & 1 & 1 & & & & \\
 & & & & 1 & 2 & 3 & 4 & 3 & 2 & 1 & & \\
 & & & 1 & 3 & 6 & 10 & 12 & 12 & 10 & 6 & 3 & 1 \\
 & 1 & 4 & 10 & 20 & 31 & 40 & 44 & 40 & 31 & 20 & 10 & 4 & 1 \\
 & \vdots & \vdots & \vdots & \vdots & \vdots & \vdots & \vdots & \vdots & \vdots & \vdots & \vdots & \vdots
\end{array}
$$

In general, one can construct a generalized Pascal's triangle of the sth order. Denoting the kth element of the nth row of the generalized

Pascal's triangle of the sth order with the symbol $(T_n^k)_s$, $n \geq 0$, $0 \leq k \leq n(s-1)$, we get the recurrent relation

$$(T_{n+1}^k)_s = \sum_{m=0}^{s-1} (T_n^{k-m})_s$$

with the initial conditions $(T_n^k)_s = 0$ for $k < 0$ and $k > n(s-1)$.

It is easy to prove that:

- the element $(T_n^k)_s$ of the generalized Pascal's triangle of the sth order is the coefficient $\binom{n}{k}_s$ of the following expansion:

$$(1 + x + x^2 + \cdots + x^{s-1})^n = \sum_{k=0}^{n(s-1)} \binom{n}{k}_s x^k.$$

□ In fact, for $n = 0$, it holds that

$$(1 + x + x^2 + \cdots + x^{s-1})^0 = 1,$$

i.e., $\binom{0}{0}_s = 1$. For $n = 1$, we have

$$(1 + x + x^2 + \cdots + x^{s-1})^1 = 1 + x + x^2 + \cdots + x^{s-1},$$

i.e., $\binom{1}{0}_s = 1$, $\binom{1}{1}_s = 1$, $\binom{1}{2}_s = 1, \ldots, \binom{1}{s-1}_s = 1$. At last,

$$(1 + x + x^2 + \cdots + x^{s-1})^{n+1}$$
$$= (1 + x + x^2 + \cdots + x^{s-1})^n (1 + x + x^2 + \cdots + x^{s-1})$$
$$= \left(\binom{n}{0}_s + \binom{n}{1}_s x + \binom{n}{2}_s x^2 + \cdots + \binom{n}{n(s-1)}_s x^{n(s-1)} \right)$$
$$\times (1 + x + x^2 + \cdots + x^{s-1})$$
$$= \binom{n}{0}_s + \binom{n}{1}_s x + \binom{n}{2}_s x^2 + \cdots + \binom{n}{n(s-1)}_s x^{n(s-1)}$$
$$+ \binom{n}{0}_s x + \binom{n}{1}_s x^2 + \binom{n}{2}_s x^3 + \cdots + \binom{n}{n}_s x^{n(s-1)+1}$$
$$+ \cdots + \binom{n}{0}_s x^{s-1} + \binom{n}{1}_s x^s + \binom{n}{2}_s x^{s+2}$$
$$+ \cdots + \binom{n}{n(s-1)}_s x^{n(s-1)+(s-1)}$$

122 Catalan Numbers

$$= \binom{n}{0}_s + \left(\binom{n}{1}_s + \binom{n}{0}_s\right) x + \left(\binom{n}{2}_s + \binom{n}{1}_s + \binom{n}{0}_s\right) x^2$$

$$+ \cdots + \left(\binom{n}{s-1}_s + \binom{n}{s-2}_s + \cdots + \binom{n}{0}_s\right) x^{s-1}$$

$$+ \cdots + \left(\binom{n}{n(s-1)}_s + \binom{n}{n(s-1)-1}_s\right)$$

$$+ \cdots + \binom{n}{(n-1)(s-1)}_s\bigg) x^{n(s-1)}$$

$$+ \left(\binom{n}{n(s-1)}_s + \binom{n}{n(s-1)-1}_s\right.$$

$$+ \cdots + \binom{n}{(n-1)(s-1)+1}_s\bigg) x^{n(s-1)+1}$$

$$+ \cdots + \left(\binom{n}{n(s-1)}_s + \binom{n}{n(s-1)-1}_s\right) x^{(n+1)(s-1)-1}$$

$$+ \binom{n}{n(s-1)}_s x^{(n+1)(s-1)}.$$

Noting that, formally, we have the relations $\binom{n}{k}_s = 0$ for any non-negative integer n with $k < 0$ and $k > n(s-1)$, we can state that

$$\binom{n+1}{k}_s = \sum_{k=0}^{s-1} \binom{n}{k}_s, \quad 0 \le k \le n(s-1).$$

Hence, the numbers $\binom{n}{k}_s$ satisfy the same recurrent relation with the same initial conditions as the elements $(T_n^k)_s$ of the generalized Pascal's triangle of the sth order. It follows that, for any non-negative integer n, we have the equality

$$(T_n^k)_s = \binom{n}{k}_s, \quad 0 \le k \le n(s-1). \qquad \square$$

The coefficients $\binom{n}{k}_s$ of the expansion of $(1+x+x^2+\cdots+x^{s-1})^n$ are called the *generalized binomial coefficients of the sth order*. So, we proved the following property:

- The generalized binomial coefficients of the sth order form the generalized Pascal's triangle of the sth order, i.e., they satisfy the recurrent relation

$$\binom{n+1}{k}_s = \sum_{m=0}^{s-1} \binom{n}{k-m}_s,$$

with the initial conditions $\binom{n}{k}_s = 0$ for $k < 0$ and $k > n(s-1)$.

Consider other properties of generalized binomial coefficients of the sth order.

- The generalized binomial coefficient $\binom{n}{k}_s$ is the number of ways to place k different elements in n different boxes so that each box does not contain more than $s-1$ elements.

□ To prove this fact, it is enough to make sure that the number $(C_n^k)_s$ of ways of placing k different balls in n different boxes so that each box contains at most $s-1$ balls satisfies the same recurrent relation with the same initial conditions as the generalized binomial coefficient $\binom{n}{k}_s$.

Indeed, to place k different items in $n+1$ different boxes, let us fix one of the boxes. The fixed box can contain t balls, $0 \le t \le s-1$. If it contains nothing (i.e., zero items), i.e., $t = 0$, then all k items are placed in n remaining boxes in $(C_n^k)_s$ ways. If the fixed box contains exactly one object, i.e., $t = 1$, then the remaining $k-1$ items are placed in the remaining n boxes in $(C_n^{k-1})_s$ ways. Continuing the consideration, we should stop at the case where in the selected box, there are $s-1$ items; in this case, the remaining $k-s+1$ items are placed in the remaining n boxes in $(C_n^{k-s+1})_s$ ways. So, $(C_{n+1}^k)_s = \sum_{m=0}^{s-1}(C_n^{k-m})_s$. The question of obtaining the initial conditions is also not difficult: $(C_n^k)_s = 0$ for $k > n(s-1)$ since this situation, i.e., an arrangement of k items in n boxes so that in each box there are at most $s-1$ items, is impossible. For $k < 0$, we assume that $(C_n^k)_s = 0$ since the corresponding combinatorial situation is also impossible. □

- The generalized Pascal's triangle of the sth order is symmetric with respect to the principal diagonal, i.e., it holds that

$$\binom{n}{k}_s = \binom{n}{(s-1)n-k}_s.$$

☐ This property follows, for example, from the algorithm for building the corresponding generalized Pascal's triangle. ☐

- The sum of the elements of the nth row of the generalized Pascal's triangle of the sth order is s^n:

$$\sum_{k=0}^{(s-1)n} \binom{n}{k}_s = s^n.$$

☐ In fact, it holds that

$$(1+x+x^2+\cdots+x^{s-1})^n = \sum_{k=0}^{(s-1)n} \binom{n}{k}_s x^k;$$

therefore, for $x=1$, we get that

$$s^n = (1+1+\cdots+1)^n = \sum_{k=0}^{(s-1)n} \binom{n}{k}_s 1^k = \sum_{k=0}^{(s-1)n} \binom{n}{k}_s. \quad \square$$

The alternating sum of the elements of a particular row of the generalized Pascal's triangle of the sth order depends on the parity of s.

- The alternating sum of elements of the nth row of the generalized Pascal's triangle of even order is equal to zero, while the alternating sum of the elements of the nth row of the generalized Pascal's triangle of odd order is equal to 1:

$$\sum_{k=0}^{(s-1)n} (-1)^k \binom{n}{k}_s = \begin{cases} 0, & s=2t, \\ 1, & s=2t+1. \end{cases}$$

☐ To prove this property, we again use the equality

$$(1 + x + x^2 + \cdots + x^{s-1})^n = \sum_{k=0}^{(s-1)n} \binom{n}{k}_s x^k.$$

When s is even, i.e., $s = 2t$, we get that

$$0 = (1 - 1 + \cdots + 1 - 1)^n = \sum_{k=0}^{(s-1)n} \binom{n}{k}_s (-1)^k.$$

When s is odd, i.e., $s = 2t + 1$, we get that

$$1 = (1 - 1 + \cdots + 1 - 1 + 1)^n = \sum_{k=0}^{(s-1)n} \binom{n}{k}_s (-1)^k. \qquad \square$$

Based on the recurrent scheme for the construction of the generalized Pascal's triangle, it is not difficult to understand the structure of the descending diagonals of this triangle.

- *The first left- (right-) descending diagonal of the generalized Pascal's triangle of the sth order consists of the consecutive positive integers $1, 2, 3, 4, 5, \ldots$; the second left- (right-) descending diagonal consists of the consecutive triangular numbers $1, 3, 6, 10, 15, \ldots$.*

☐ To prove this fact, it is enough to verify that, for the sequence of elements $\binom{n}{1}_s$, $n = 1, 2, 3, 4, 5, \ldots$, in the first left-descending diagonal, the recurrent relation obtained above takes the form $\binom{n}{1}_s = \binom{n-1}{1}_s + 1$ with the initial condition $\binom{1}{1}_s = 1$, while for the sequence of elements $\binom{n}{2}_s$, $n = 2, 3, 4, \ldots$ in the second left-descending diagonal, this recurrent relation takes the form $\binom{n}{2}_s = \binom{n-1}{1}_s + (n-1)$ with the initial condition $\binom{2}{2}_s = 1$. ☐

As for the ascending diagonals, we can also trace the analogy to the classical case.

- *The sum of the elements of the nth ascending diagonal of the generalized Pascal's triangle of the sth order, $n \geq s$, is the sum of the elements of s previous ascending diagonals of the triangle.*

□ A proof of this fact is carried out in a completely similar way to the proof of the corresponding property of the ordinary Pascal's triangle. □

In particular, for $s = 3$, we get the family of relatives of Fibonacci numbers: the set of *tribonacci numbers* t_n, given by the recurrence relation $t_{n+3} = t_{n+2} + t_{n+1} + t_n$ with initial conditions $t_1 = 0, t_2 = t_3 = 1$.

- In Pascal's triangle of the third order, the sum of the elements of the nth ascending diagonal is equal to the $(n+2)$th tribonacci number.

□ To prove this property, just make sure that the statement is correct for small values of n, and use the previous statement. Indeed, the sum of the elements of the zeroth diagonal and the first and second ascending diagonals are, respectively, $1 = t_2, 1 = t_3$, and $2 = t_4$, and, since the sum X_n of the elements of each ascending diagonal is equal to the sum of the elements of the three previous ascending diagonals, then $X_n = t_{n+2}$ for any non-negative integer n. □

Pascal's pyramid

2.7.11. Let us move on to a three-dimensional analog of Pascal's triangle.

Pascal's pyramid is a number tetrahedron in which the three outer edges consist of unities, each of the side faces is a Pascal's triangle, and any internal element belonging to a given section, parallel to the base of the pyramid, is equal to the sum of three elements located directly above it and belonging to the previous section of the pyramid.

Denote the elements of the nth horizontal section (in fact, a number triangle) of Pascal's pyramid by $P_n^{k,l}$, $n \geq 0$. In the expression $P_n^{k,l}$, if numbering goes from the nth row to the zeroth row, the number n indicates the section's number, the number k is the number of the section's row, and the number l is the number of the corresponding column (diagonal).

The following figure shows the fourth horizontal section of the pyramid:

$$\begin{array}{ccccccc} & & & 1 & & & \\ & & 4 & & 4 & & \\ & 6 & & 12 & & 6 & \\ 4 & & 12 & & 12 & & 4 \\ 1 & 4 & & 6 & & 4 & 1 \end{array}$$

The following illustrates the location of $P_4^{k,l}$ for $n = 4$:

$$\begin{array}{ccccccccc}
 & & & & P_4^{4,0}=1 & & & & \\
 & & & P_4^{3,0}=4 & & P_4^{3,1}=4 & & & \\
 & & P_4^{2,0}=6 & & P_4^{2,1}=12 & & P_4^{2,2}=6 & & \\
 & P_4^{1,0}=4 & & P_4^{1,1}=12 & & P_4^{1,2}=12 & & P_4^{1,3}=4 & \\
P_4^{0,0}=1 & & P_4^{0,1}=4 & & P_4^{0,2}=6 & & P_4^{0,3}=4 & & P_4^{0,4}=1
\end{array}$$

Using the definition, we obtain for numbers $P_{n+1}^{k,l}$ the following recurrent relation:

$$P_{n+1}^{k,l} = P_n^{k-1,l-1} + P_n^{k,l-1} + P_n^{k,l},$$

with the initial conditions $P_0^{0,0} = P_n^{0,0} = P_n^{n,0} = P_n^{0,n} = 1$.

It is easy to see that every element $P_n^{k,l}$ of Pascal's pyramid is a *trinomial coefficient*, that is, the coefficient of the trinomial expansion, or the decomposition of $(1 + x + y)^n$. Denoting the trinomial coefficients by $\binom{n}{k,l}$, where n, l, k are non-negative integers, such that $l + k \leq n$, we can write the decomposition of $(1 + x + y)^n$ in the form

$$(1 + x + y)^n = \sum_{l+k \leq n} \binom{n}{k,l} x^k y^l.$$

As was proved in Chapter 1, the trinomial coefficient $\binom{n}{k,l}$ is the number $P_{(n-k-l,k,l)}$ of permutations with repetitions, calculated using the formula $\binom{n}{k,l} = \frac{n!}{(n-k-l)!k!l!}$.

It is not difficult to check that for trinomial coefficients, the following recurrent relation holds:

$$\binom{n+1}{k,l} = \binom{n}{k,l} + \binom{n}{k-1,l} + \binom{n}{k,l-1},$$

with the initial condition $\binom{0}{0,0} = 1$, $\binom{n}{0,0} = 1$, $\binom{n}{n,0} = \binom{n}{0,n} = 1$.

128 Catalan Numbers

□ In order to prove this fact, we use the formula $\binom{n}{k,l} = \frac{n!}{(n-k-l)!k!l!}$. In fact, we obtain that

$$\binom{n}{k,l} + \binom{n}{k-1,l} + \binom{n}{k,l-1}$$

$$= \frac{n!}{(n-k-l)!k!l!} + \frac{n!}{(n-k-l+1)!(k-1)!l!} + \frac{n!}{(n-k-l+1)!k!(l-1)!}$$

$$= \frac{n!}{(n-k-l)!(k-1)!(l-1)!} \left(\frac{1}{kl} + \frac{1}{(n-k-l+1)k} + \frac{1}{(n-k-l+1)l} \right)$$

$$= \frac{n!}{(n-k-l)!(k-1)!(l-1)!} \cdot \frac{(n-k-l+1)+k+l}{(n-k-l+1)kl}$$

$$= \frac{n!}{(n-k-l)!(k-1)!(l-1)!} \cdot \frac{n+1}{(n-k-l+1)kl}$$

$$= \frac{(n+1)!}{(n-k-l+1)!k!l!} = \frac{(n+1)!}{((n+1)-k-l)!k!l!} = \binom{n+1}{k,l}.$$

Moreover, it holds that $\binom{0}{0,0} = \frac{0!}{0!0!0!} = 1$, $\binom{n}{0,0} = \frac{n!}{n!0!0!} = 1$, $\binom{n}{n,0} = \binom{n}{0,n} = \frac{n!}{0!n!0!} = 1$.

Thus, it is proven that any element $P_n^{k,l}$ of Pascal's pyramid coincides with the trinomial coefficient $\binom{n}{k,l}$. □

- The trinomial coefficients $\binom{n}{k,l}$ form Pascal's pyramid, i.e., they satisfy the recurrent relation

$$\binom{n+1}{k,l} = \binom{n}{k-1,l} + \binom{n}{k,l-1} + \binom{n}{k,l},$$

with the initial conditions $\binom{0}{0,0} = \binom{n}{0,0} = \binom{n}{n,0} = \binom{n}{0,n} = 1$.

- From a combinatorial point of view, the number $\binom{n}{k,l}$ is the number of placements of n different objects into three different boxes so that k objects fall into the first box, l objects fall into the second box, and $n-k-l$ objects fall into the third box.

☐ To prove this fact, we arrange all objects in a row (or just enumerate them), and then attach to each object the number of the box it falls into. The resulting sequence of numbers of boxes forms a permutation with repetitions, consisting of k numbers 1, l numbers 2, and $n-k-l$ numbers 3. Each placement defines such a permutation. Conversely, each such permutation determines an exact placement: the first box contains objects attached with 1; in the second box, we place objects attached with 2; and the third box contains objects attached with 3. This construction establishes a one-to-one correspondence between the permutations with repetitions and the placements of n different objects in three different boxes. Therefore, the number of different placements is $\binom{n}{k,l} = \frac{n!}{(n-k)!(k-l)!l!}$. ☐

- Pascal's pyramid has three axes of symmetry:

$$\binom{n}{k,l} = \binom{n}{l,k} = \binom{n}{n-k-l,l} = \binom{n}{k,n-k-l}.$$

☐ This property can be checked directly using the above explicit formula: $\binom{n}{k,l} = \frac{n!}{(n-k-l)!k!l!}$. ☐

- The sum of all elements of the nth horizontal section of Pascal's pyramid is equal to 3^n:

$$\sum_{k,l \geq 0,\ k+l \leq n} \binom{n}{k,l} = 3^n.$$

☐ In fact, as

$$(1+x+y)^n = \sum_{k,l \geq 0,\ k+l \leq n} \binom{n}{k,l} x^k y^l,$$

then, for $x = 1, y = 1$, we obtain

$$3^n = (1+1+1)^n = \sum_{k,l \geq 0,\ k+l \leq n} \binom{n}{k,l} 1^k 1^l = \sum_{k,l \geq 0,\ k+l \leq n} \binom{n}{k,l}.$$ ☐

We can also get an analog of the alternating sum of the elements of a given row of Pascal's triangle, considering the numbers $(-1)^l \binom{n}{k,l}$:

• *The alternating sum of all the elements of the nth horizontal section of Pascal's pyramid is equal to 1:*

$$\sum_{k=0}^{n}\sum_{l=0}^{n-k}(-1)^l\binom{n}{k,l}=1.$$

☐ Using the decomposition $(1+x+y)^n = \sum_{k=0}^{n}\sum_{l=0}^{n-k}\binom{n}{k,l}x^k y^l$ for $x=1, y=-1$, we get that

$$1=(1+1-1)^n=\sum_{k=0}^{n}\sum_{l=0}^{n-k}\binom{n}{k,l}1^k(-1)^l=\sum_{k=0}^{n}\sum_{l=0}^{n-k}(-1)^l\binom{n}{k,l}. \quad \square$$

2.7.12. Pascal's pyramid, like Pascal's triangle, has several connections with figurate numbers.

• *The number of elements in the nth horizontal section of Pascal's pyramid is the $(n+1)$th triangular number $S_3(n+1)$.*

☐ Indeed, the number of elements in the nth horizontal section of the pyramid is equal to the number of non-negative integer solutions to the equation $k+l+t=n$. In turn, this number can be calculated, presenting each solution as a sequence 111+11111+11111, in which the number of unities is n and the number of characters "+" is two. It is clear that the number of such sequences as well as the number of non-negative integer solutions to the equation $k+l+t=n$ are equal to

$$P_{(n,2)}=\frac{(n+2)!}{n!2!}=\frac{(n+2)(n+1)}{2}=S_3(n+1).$$

However, we can immediately get the result by comparing the recurrent definition of triangular numbers and the recurrent algorithm for the construction of Pascal's pyramid. ☐

Now, it is not difficult to find in Pascal's pyramid the set of tetrahedral numbers.

- *The number of elements in the first n horizontal sections of Pascal's pyramid, from the zeroth section to the $(n-1)$th section, is the nth tetrahedral number $S_3^3(n)$.*

□ This property follows from the fact that the nth tetrahedral number $S_3^3(n) = \frac{n(n+1)(n+2)}{6} = \binom{n+2}{3}$ is, by definition, the sum of the first n triangular numbers. However, we can check this statement by direct summation. ∎

2.7.13. Now, we formulate another property of the pyramid that illustrates the presence of a close relationship between binomial and trinomial coefficients.

- *There exists the following identity:*

$$\binom{n}{k,l} = \binom{n}{n-k}\binom{n-k}{l}.$$

□ In fact, it holds that

$$\binom{n}{k,l} = \frac{n!}{(n-k-l)!\,k!\,l!} = \frac{n!}{k!(n-k)!} \cdot \frac{(n-k)!}{l!(n-k-l)!}$$

$$= \binom{n}{n-k}\binom{n-k}{l}.$$

∎

This relationship allows us to use Pascal's triangle for the construction of the nth horizontal section of Pascal's pyramid for any $n > 0$. Thus, the nth horizontal section of the pyramid can be obtained from the "partial" Pascal's triangle, which contains rows up to the nth one, by multiplying the elements of each represented row by the elements of the last, or nth, row rotated counterclockwise by an angle of $\pi/2$. This rule is illustrated for the fourth section, the arrangement of the elements of which has been analyzed earlier:

```
1                  1                          1
4×              1     1                  4         4
6×    →      1    2    1       →      6     12        6
4×         1   3    3    1          4    12    12    4
1×       1   4    6    4    1     1   4    6    4    1
```

Exercises

1. Prove that $\binom{n}{k}_s = \sum_{m=0}^{\lfloor k/s \rfloor} (-1)^m \binom{n}{m} \binom{n+m-sk-1}{n-1}$.
2. Prove the following recurrent relation: $\binom{n}{k}_{s+1} = \sum_{m=0}^{n} \binom{n}{m} \binom{m}{k-m}_s$, $s \geq 2$.
3. Prove that trinomial coefficients satisfy the relation

$$\sum_{i=0}^{n} \sum_{j=0}^{n-i} \binom{m}{i,j} \binom{n}{k-i, l-j} = \binom{n+m}{k, l},$$

where $\binom{n}{k,l} = 0$, if $\min\{n, k, l, n-k-l\} < 0$.

4. Prove that *polynomial coefficients* $\binom{n}{k_1, k_2, \ldots, k_m}$, i.e., the coefficients of the decomposition

$$(x_1 + x_2 + \cdots + x_m)^n = \sum_{k_1, k_2, \ldots, k_m} \binom{n}{k_1, k_2, \ldots, k_m} x_1^{k_1} x_2^{k_2} \ldots x_m^{k_m},$$

can be calculated using the formula $\binom{n}{k_1, k_2, \ldots, k_m} = \frac{n!}{k_1! k_2! \ldots k_m!}$.

5. Show that the binomial coefficient $\binom{n}{k}$ is a special case of the polynomial coefficient. Show that the trinomial coefficient $\binom{n}{k,l}$ is a special case of the polynomial coefficient.

References

[AbSt72], [Aign82], [Ande03], [BaCo87], [Berg71], [BeSl95], [BiBa70], [Bond93], [Bron01], [Buch09], [Comt74], [CoRo96], [Cofm75], [DeKo13], [Deza17], [Deza18],[Deza21], [Deza23], [Deza24], [DeMo10], [Dick05], [Erus00], [Eule48], [Ficht01], [FlSe09], [Gelf98], [Gelf59], [GGL95Û], [GuJa90], [Hagg04], [Hall70], [GaSa92], [Hogg69], [Hons91], [Jablo01], [Knut97], [KoKo77], [KuAD88], [Land94], [Land02], [LiWi92], [Lips88], [Mark0], [Mazu10], [Mazu10], [Pasc54], [Plou92], [Rior79], [Rior80], [RoTe09], [Robb55], [Roge78], [Rybn82], [Smit84], [Stru87], [Stan97], [Uspe76], [Vile14], [VVV06], [Voro61], [VoKu84], [Wiki24].

Chapter 3
Catalan Numbers

3.1. Construction of Catalan Numbers

A history of the question

3.1.1. In combinatorial mathematics, Catalan numbers are a sequence of natural numbers that occur in various counting problems, often involving recursively defined objects. They are named after the French and Belgian mathematician Eugène Charles Catalan (1814–1894), who worked on continued fractions, descriptive geometry, number theory, and combinatorics.

His notable contributions included discovering a periodic minimal surface in the space $\mathbb{R}^3$; stating the famous *Catalan's conjecture* (*the only positive integer solution to the equation $x^a - y^b = 1$, $a, b > 1$, is $x = 3, a = 2, y = 2, b = 3$*), which was eventually proved in 2002 by Preda Mihăilescu; and introducing the number set, now known as Catalan numbers, to solve combinatorial problems.

E. Ch. Catalan was born in Bruges in 1814. In 1825, he traveled to Paris and learned mathematics at École Polytechnique, where he met Joseph Liouville (1833). In December 183, he was expelled, along with most of the students in his class, for political reasons. He resumed his studies in January 1835, graduated that summer, and went on to teach at Châlons-sur-Marne. In 1841, he obtained his degree in mathematics and went on to Charlemagne College to teach descriptive geometry. The University of Liège appointed him the chair of analysis in 1865. In 1879, still in Belgium, he became a journal editor. In 1883, he worked for the Belgian Academy of Science in the field of number theory. He died in Liège, Belgium, in 1894.

3.1.2. Catalan's sequence was described in the 18th century by Leonhard Euler (1707–1783), who was interested in the number of different ways of dividing a polygon into triangles.

E. Ch. Catalan discovered the connection of this number set to parenthesized expressions during his exploration of the Towers of Hanoi puzzle.

The reflection counting trick (the second proof) for Dyck words was found by Désiré André in 1887.

In 1988, it came to light that Catalan's sequence had been used in China by a Mongolian mathematician Mingantu (circa 1692 – circa 1763) in 1730. That is when he started to write his book *Ge Yuan Mi Lu Jie Fa (The Quick Method for Obtaining the Precise Ratio of Division of a Circle)*, which was completed by his student Chen Jixin in 1774 but published 60 years later.

The name "Catalan numbers" ws coined by an American mathematician, John Francis Riordan (1903–1988) (see [Stan97]).

Catalan numbers are commonly denoted by C_n ([Stan13], [Stan15], [GKP94], [PeSk03], [Deza18], [Deza24], etc.) or $c(n)$ ([GuJa90]), and less commonly by u_n ([LiWi92]).

Recurrent definition of Catalan numbers

3.1.3. A *Catalan number* is a member of the sequence $C_0, C_1, C_2, \ldots, C_n, \ldots$, built according to the following recurrent low:

$$C_0 = 1, \quad \text{and} \quad C_n = C_0 C_{n-1} + C_1 C_{n-2} + \cdots + C_{n-1} C_0 \quad \text{for } n \geq 1.$$

So, we get

$$C_0 = 1, \quad C_1 = C_0C_0 = 1 \cdot 1 = 1,$$
$$C_2 = C_0C_1 + C_1C_0 = 1 \cdot 1 + 1 \cdot 1 = 2,$$
$$C_3 = C_0C_2 + C_1C_1 + C_2C_0 = 1 \cdot 2 + 1 \cdot 1 + 2 \cdot 1 = 5,$$
$$C_4 = C_0C_3 + C_1C_2 + C_2C_1 + C_3C_0 = 1 \cdot 5 + 1 \cdot 2 + 2 \cdot 1 + 5 \cdot 1 = 4,$$
$$C_5 = C_0C_4 + C_1C_3 + C_2C_2 + C_3C_1 + C_4C_0$$
$$= 1 \cdot 14 + 1 \cdot 5 + 2 \cdot 2 + 5 \cdot 1 + 14 \cdot 1 = 42,$$
$$C_6 = C_0C_5 + C_1C_4 + C_2C_3 + C_3C_2 + C_4C_1 + C_5C_0$$
$$= 1 \cdot 42 + 1 \cdot 14 + 2 \cdot 5 + 5 \cdot 2 + 14 \cdot 1 + 42 \cdot 1 = 132.$$

This definition allows us to derive a very simple rule for the consecutive construction of members of the sequence $C_0, C_1, C_2, \ldots, C_n, \ldots$. In fact, this rule was already known to a German-Hungarian scientist and a contemporary of L. Euler, Johann Andreas von Segner (1704–1777). The corresponding algorithm is as follows. *In order to get the next member of Catalan's sequence: write in the usual order the already built members of this sequence; under them, write the same numbers, but in the reverse order; multiply each top number by the number below it; and add up all the obtained elements.* For example, in order to find the number C_7, it is enough to write out one after another the first seven Catalan numbers $C_0, C_1, \ldots, C_6$, and below, write the same numbers in the reverse order:

C_0	C_1	C_2	C_3	C_4	C_5	C_6
1	1	2	5	14	42	132
132	42	14	5	2	1	1

Multiplying each top number by the corresponding bottom number and adding all the obtained products, we get $C_7 = 429$:

$$\begin{array}{c} \quad 1 \quad\; 1 \quad\; 2 \quad\; 5 \quad 14 \quad 42 \quad 132 \\ \times \\ \; 132 \quad 42 \quad 14 \quad 5 \quad 2 \quad 1 \quad 1 \\ \hline 1 \cdot 132 + 1 \cdot 42 + 2 \cdot 14 + 5 \cdot 5 + 14 \cdot 2 + 42 \cdot 1 + 132 \cdot 1 = 429 \end{array}.$$

Using this simple algorithm, it is easy to find the first few terms of the sequence of Catalan numbers: 1, 1, 2, 5, 14, 42, 132, 429, 1430, 4862, ... (sequence A000108 in the OEIS).

Exercises

1. Using the recurrent relation for Catalan numbers, find C_n, $n = 8, 9, 10, 11, 12$.
2. Check that the Catalan numbers C_n, $n = 0, 1, 2, 3, 4, 5, 6, 7$, can be obtained by dividing each element of the principal diagonal of Pascal's triangle by the number $n + 1$. Use this algorithm to calculate C_n, $n = 8, 9, 10, 11, 12$.
3. Check that the Catalan numbers C_n, $n = 1, 2, 3, 4, 5, 6, 7$, can be obtained by the following recurrent relation: $C_n = \frac{4n-2}{n+1} C_{n-1}$, $C_0 = 1$.
4. Check that the Catalan number C_n, $n = 2, 3, 4, 5, 6, 7$, can be obtained by the following formula given by L. Euler: $C_n = \frac{2 \cdot 6 \cdot 10 \cdots (4n-6)}{1 \cdot 2 \cdot 3 \cdots n}$. Use this formula for the calculation of C_n, $n = 8, 9, 10, 11, 12$.
5. Check that the Catalan numbers C_n, $n = 0, 1, 2, 3, 4, 5, 6, 7$, can be obtained by the construction of the following isosceles number triangle: the left side of the triangle is constructed from unities; the last element of the each row coincides with the previous element of the same row; each internal element of the triangle is the sum of its left and right-up neighbors. Use this triangle for the calculation of C_n, $n = 8, 9, 10, 11, 12$.

3.2. Catalan Numbers and Counting Problems

3.2.1. Catalan numbers are interesting primarily because they appear, often quite unexpectedly, in many problems from various fields of mathematics, especially in solving many combinatorial counting problems. We consider some of these examples.

Euler's polygon division problem

3.2.2. *Euler's polygon division problem* is the problem of finding the number of ways a plane convex polygon of n sides can be divided into triangles by non-crossing diagonals. L. Euler first proposed this problem to Ch. Goldbach in 1751.

Let us try to find the number of *Euler's triangulations of a convex $(n+2)$-gon*, i.e., the number of ways to cut a convex $(n+2)$-gon into triangles by non-crossing diagonals of this $(n+2)$-gon.

Note that in each of these cases, the number of diagonals used is $n-1$, and the number of resulting triangles is n.

The answer for $n = 1$, i.e., for a triangle, is trivial. We use no diagonals, and the number of Euler's triangulations is equal to unity.

For $n = 2$, we can use either of two diagonals of a convex quadrangle (square), which gives two possible ways of Euler's triangulation. They are similar: from a vertex of the square comes out a diagonal.

For $n = 3$, all five ways of Euler's triangulation of a convex pentagon are essentially similar, too: from a vertex of the pentagon comes out two diagonals.

For $n = 4$, we get the first non-trivial answer: there are 14 possibilities for Euler's triangulation of a convex hexagon. They form four groups, depending on which (gray) triangle adjoins the left-up side of the hexagon.

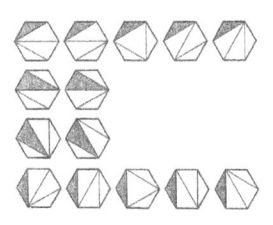

In order not to draw all possible configurations for a convex heptagon, let us fix one of its sides and classify the triangulations of the heptagon depending on which triangle adjoins this side. It gives us five different cases (defining by fixed gray triangles), which we can see in the following figure.

In the first and last cases, the number of triangulations is 14 because after the cutting of the chosen gray triangle. we get a hexagon; for it, we have 14 possibilities of Euler's triangulation. In the second and fourth cases, the deletion of the chosen gray triangle cuts the heptagon into a triangle and a pentagon. The triangle does not need to be cut (i.e., there is exactly one triangulation of the triangle), but the pentagon, as we know, gives fives possibilities for triangulation. In the third case, the deletion of the chosen gray triangle gives two quadrangles. Since each of them can be triangulated in two ways, we get $2 \cdot 2 = 4$ possibilities for a triangulation of the heptagon. So, we have proven that a convex heptagon can be triangulated in $14 + 5 + 2 \cdot 2 + 5 + 14 = 42$ ways.

Considering the convex octagon, we similarly get $42 + 14 + 2 \cdot 5 + 5 \cdot 2 + 14 + 42 = 132$ possibilities. For a convex nonagon we have $132 + 42 + 2 \cdot 14 + 5 \cdot 5 + 14 \cdot 2 + 42 + 132 = 429$ possibilities, and for a convex decagon, we have $429 + 132 + 2 \cdot 42 + 5 \cdot 14 + 14 \cdot 5 + 42 \cdot 2 + 132 + 429 = 1430$ possibilities.

For the convenience of further argumentation, we consider a biangle ("bigon"), which is simply a segment; we assume that the number of triangulations of this biangle is 1.

Now, it is easy to prove the following fact:

- *The number of ways to cut a convex $(n + 2)$-gon, $n \geq 0$, into triangles by non-crossing diagonals of this $(n + 2)$-gon is equal to the nth Catalan number C_n.*

□ In fact, this is true for $n = 0$, $n = 1$, and $n = 2$: as was shown before, the corresponding number of triangulations of a convex $(n + 2)$-gon is equal to 1 (i.e., to C_0), to 1 (i.e., to C_1), and to 2 (i.e., to C_2).

Consider now a positive integer $n \geq 3$. Suppose that the number of Euler's triangulations of any convex $(k+2)$-gon, $0 \leq k \leq n-1$, is equal to C_k, and try to prove, that the number of Euler's triangulations of a convex $(n+2)$-gon is C_n. In order to do this, fix one of the sides of a given convex $(n+2)$-gon, and consider n possibilities, highlighting a triangle adjacent to the fixed side (the illustrations for $n = 3, 4$, and 5 were given earlier). This highlighted triangle divides the $(n+2)$-gon into two polygons that we can now split independently of each other. In this cases, we get:

- a biangle (bigon) and an $(n+1)$-gon (that gives $C_0 C_{n-1}$ triangulations of the $(n+2)$-gon);
- a triangle (3-gon) and an n-gon (that gives $C_1 C_{n-2}$ triangulations of the $(n+2)$-gon);
- a quadrangle (4-gon) and an $(n-1)$-gon (that gives $C_2 C_{n-3}$ triangulations of the $(n+2)$-gon);
$\vdots$
- an $(n+1)$-gon and a biangle (bigon) (that gives $C_{n-1} C_0$ triangulations of the $(n+2)$-gon).

Thus, the total number of Euler's triangulations of a convex $(n+2)$-gon is equal to $C_0 C_{n-1} + C_1 C_{n-2} + \cdots + C_{n-1} C_0$, i.e., the nth Catalan number C_n. □

Catalan's problem

3.2.3. *Catalan's problem* is the problem of finding the number of different ways in which a product of several different ordered factors can be calculated by pairs. In other words, it is the problem of finding the number of *binary bracketings* of several letters. This problem was considered and solved in 1838 by E. Ch. Catalan.

Consider a product formed by $n+1$ factors $a_1, a_2, \ldots, a_{n+1}$. Let us arrange $n-1$ pairs of brackets in these factors. How many such arrangements exist?

In fact, there are exactly two possibilities to place one pair of brackets in three factors: $a_1(a_2 a_3)$ and $(a_1 a_2)a_3$.

For four factors $a_1a_2a_3a_4$ and two pairs of brackets, there are five ways: $((a_1a_2)a_3)a_4$, $(a_1(a_2a_3))a_4$, $a_1((a_2a_3)a_4)$, $a_1(a_2(a_3a_4))$, and $(a_1a_2)(a_3a_4)$.

It is easy to see also that the number of bracket arrangements for two factors is 1 (there are no brackets). Moreover, we can assume that the number of brackets' arrangements for one factor is 1, too.

So, for small numbers of n, we have checked that the numbers of bracket arrangements is given by Catalan numbers. In general, it is easy to prove the following statement.

- *The number of arrangements of $n-1$ pairs of brackets in $n+1$ factors is equal to the nth Catalan number C_n, $n \geq 0$.*

□ This fact is already proven for $n = 0, 1, 2, 3$. In fact, the number of bracket arrangements for one factor (i.e., $n = 0$) is equal to 1 (i.e., C_0): a_1. The number of bracket arrangements for two factors (i.e., $n = 1$) is also equal to 1 (i.e., C_1): a_1a_2. The number of bracket arrangements for three factors (i.e., $n = 2$) is equal to 2 (i.e., C_2): $a_1(a_2a_3)$ and $(a_1a_2)a_3$. The number of bracket arrangements for four factors (i.e., $n = 3$) is equal to five (i.e., C_3): $((a_1a_2)a_3)a_4$, $(a_1(a_2a_3))a_4$, $a_1((a_2a_3)a_4)$, $a_1(a_2(a_3a_4))$, and $(a_1a_2)(a_3a_4)$.

For five factors $a_1a_2a_3a_4a_5$ and three pairs of brackets, we have 14 possibilities. To make sure of this, it is not necessary to write them all out. We can simply note that there are five possibilities of the form $a_1(a_2a_3a_4a_5)$, two possibilities of the form $(a_1a_2)(a_3a_4a_5)$, two possibilities of the form $(a_1a_2a_3)(a_4a_5)$, and five possibilities of the form $(a_1a_2a_3a_4)a_5$.

Consider now a positive integer $n \geq 6$. In order to show that the nth Catalan number C_n can be interpreted as the number of arrangements of $n-1$ pairs of brackets in $n+1$ factors, suppose that the statement is true for any integer k, $0 \leq k \leq n-1$: for $k-1$ pairs of brackets and $k+1$ factors, the number of possible arrangements is C_k, $1 \leq k \leq n-1$; moreover, the number of bracket arrangements for one factor is 1 (i.e., C_0).

Consider now the arrangements of $n-1$ pairs of brackets in $n+1$ factors. Divide all possible arrangements of $n-1$ brackets into the $n+1$ factors $a_1a_2\cdots a_na_{n+1}$ in several groups. The elements of the first group have the form $a_1(a_2 \cdots a_n a_{n+1})$, i.e., in the right-hand side of the arrangement, we have the construction $(a_2 \cdots a_n a_{n+1})$ with the n factors $a_2, \ldots, a_n, a_{n+1}$, while in the left-hand side of the arrangement we have the single element a_1. In this case, one of the pairs of brackets is used to surround these n factors: $a_2, \ldots, a_n, a_{n+1}$, and the remaining $n-2$ pairs are used for arrangements in these n factors: $a_2 \cdots a_n a_{n+1}$. The number of such arrangements is $C_{n-1} = 1 \cdot C_{n-1} = C_0 C_{n-1}$. The elements of the second group have the form $(a_1 a_2)(a_3 a_4 \cdots a_n a_{n+1})$, i.e., in the right-hand side of the arrangement, we have the construction $(a_3 \cdots a_n a_{n+1})$ with the $n-1$ factors $a_3, \ldots, a_n, a_{n+1}$, while in the left-hand side of the arrangement, we have the construction $a_1 a_2$ with two factors. In this case, one pair of brackets is used to surround the left-side construction, $(a_1 a_2)$, and one pair of brackets i used to surround the right-side construction, $(a_3 a_4 \cdots a_n a_{n+1})$. The remaining $n-3$ pairs of brackets are used for arrangements in the $n-1$ factors $a_3 a_4 \cdots a_n a_{n+1}$. The number of such arrangements is $C_{n-2} = 1 \cdot C_{n-2} = C_1 C_{n-2}$. The third group has the form $(a_1 a_2 a_3)(a_4 \cdots a_n a_{n+1})$, i.e., in the right-hand side of the arrangement, we have the construction $(a_4 \cdots a_{n+1})$ with $n-2$ factors $a_4, \ldots, a_n, a_{n+1}$, while in the left-hand side of the arrangement, we have the construction $a_1 a_2 a_3$ with three factors. In this case, as before, one pair of brackets is used to surround the left-side construction, $(a_1 a_2 a_3)$, and another pair of brackets is used to surround the right-side construction, $(a_4 \cdots a_n a_{n+1})$. The remaining $n-3$ pairs of brackets we divide into two groups. In fact, one pair is used now for arrangements in the factors $a_1 a_2 a_3$ of the left-side construction $(a_1 a_2 a_3)$ (C_2 possibilities), while $n-4$ pairs are used for arrangements in the factors $a_4 \cdots a_n a_{n+1}$ of the right-side constriction $(a_4 \cdots a_n a_{n+1})$ (C_{n-3} possibilities). So, for the third group $(a_1 a_2 a_3)(a_4 \cdots a_n a_{n+1})$, the total number of possible arrangements is $C_2 C_{n-3}$. In general, the sth group has the form $(a_1 a_2 \cdots a_s)(a_{s+1} \cdots a_n a_{n+1})$, with the right-side construction $(a_{s+1} \cdots a_n a_{n+1})$, and the left-side construction $(a_1 \cdots a_s)$, $s = 2, 3, \ldots, n-1$. In this case, one pair of brackets is used to surround the elements of the left-side construction, $(a_1 a_2 \cdots a_s)$, and another pair of brackets is used to surround the elements of the right-side construction, $(a_{s+1} \cdots a_n a_{n+1})$. The remaining $n-3$ pairs of brackets are divided into two groups. In fact, $s-2$ pairs are used now for the arrangements in the factors $a_1 \cdots a_s$ of the left-side construction $(a_1 a_2 \cdots a_s)$ (C_{s-1} possibilities), while $n-s-1$ pairs of brackets are used for the arrangements of the elements of the right-side construction $a_{s+1} \cdots a_n a_{n+1}$ (C_{n-s} possibilities). So, for the sth group $(a_1 \cdots a_s)(a_{s+1} \cdots a_n a_{n+1})$, the total number of possible arrangements is $C_{s-1} C_{n-s}$. The last group has the form $(a_1 a_2 \cdots a_n) a_{n+1}$, i.e., in the right-hand side of the arrangement, we have the single element a_{n+1}, while in the left-hand side of the arrangement, we have the construction $(a_1 \cdots a_n)$ with n factors. In this case, one of the pairs of brackets is used to surround the first n factors, and the remaining $n-2$ pairs are used for arrangements in these n factors: $a_1 \cdots a_n$. The number of such arrangements is $C_{n-1} = C_{n-1} \cdot 1 = C_{n-1} C_0$.

As a result, we get that the number of ways to arrange $n-1$ pairs of brackets in $n+1$ factors is

$$C_0 C_{n-1} + C_1 C_{n-2} + C_2 C_{n-3} + \cdots C_{n-3} C_2 + C_{n-2} C_1 + C_{n-1} C_0,$$

i.e., it is equal to the nth Catalan number C_n. □

3.2.4. In 1961, a New Zealand mathematician, Henry George Forder (1889–1981), found a simple way to establish a one-to-one correspondence between Euler's triangulations of an $(n+2)$-gon and the arrangement of $n-1$ pairs of brackets in an alphabetical chain consisting of $n+1$ factors, $a_1, \ldots, a_n, a_{n+1}$ ([Ford61]).

In particular, let us designate all the sides of a given $(n+2)$-gon, except for one (called the *base*), by different letters, $a_1, a_2, \ldots, a_n, a_{n+1}$. With certain Euler's triangulation of this polygon, denote each diagonal forming a triangle with two adjacent sides of the polygon by the letters of these sides, taken in brackets. All remaining diagonals are sequentially denoted by sequences of letters in the same way, combining the designations of the other two sides of the corresponding triangle. The base is indicated last, and the designation for it, having the form of a bracket arrangement with $n-1$ pairs of brackets in the $n+1$ factors

$a_1 a_2 \cdots a_n a_{n+1}$, is uniquely determined by the specified triangulation of this $(n+2)$-gon.

Using this algorithm for a triangle, having as a *base* one of its sides, we get the arrangement $a_1 a_2$. For two squares, we get the arrangements $(a_1 a_2)a_3$ and $a_1(a_2 a_3)$, correspondingly. For five pentagons the corresponding arrangements are $a_1((a_2 a_3)a_4)$, $(a_1(a_2 a_3))a_4$, $a_1(a_2(a_3 a_4))$, and $((a_1 a_2)a_3)a_4$, $(a_1 a_2)(a_3 a_4)$.

3.2.5. There exists another formulation of the same problem. Define *k-power* as a construction, composed of the numbers $k+1, n, \ldots, 3, 2$ (in this specified order) and only using the operation of exponentiation and the corresponding arrangement of brackets. For example, for $k = 2$, we have exactly one 2-power: $3^2 = 9$; for $k = 3$, we have two 3-powers: $(4^3)^2 = 4^6$, and $4^{(3^2)} = 4^9$. For $n = 4$, we have five 4-powers: $((5^4)^3)^2 = 5^{24}$, $5^{(4^{(3^2)})} = 5^{262144}$, $(5^{(4^3)})^2 = 5^{128}$, $(5^4)^{(3^2)} = 5^{36}$, $5^{((4^3)^2)} = 5^{4096}$. It is easy to prove that:

- the number of $(n+1)$-powers is the nth Catalan number C_n, $n \geq 0$.

□ There exists an obvious bijection between the set of arrangements of $n-1$ pairs of brackets in $n+1$ factors and the set of all $(n+1)$-powers: instead of the factors $a_1, a_2, \ldots, a_n, a_{n+1}$, we simply use the numbers $n+1, n, \ldots, 3, 2$, and instead of the sequence $a_1 a_2 \cdots a_{n+1}$, we simply use the power $(n+1)^{n^{\cdots^2}}$. For example, the sequence $(a_1 a_2)a_3$ corresponds to the power $(4^3)^2$, while the sequence $a_1(a_2 a_3)$ corresponds to the power $(4^3)^2$. As the number of arrangements of $n-1$ pairs of brackets in $n+1$ factors is equal to the nth Catalan number C_n, $n \geq 0$, we get a proof of the statement. □

Brackets' problem

3.2.6. The so-called *correctly matched brackets* problem (or *brackets problem*, *correctly matched parentheses problem*) is closely connected to Catalan's problem and also gives as a result the sequence of Catalan numbers.

In fact, if, considering Catalan's problem, we add to the set of $n-1$ pairs of brackets another pair, giving an "external restriction" (unambiguously recoverable when working with specified factors), then we can place brackets without using factors, leaving only the brackets themselves for the construction. Thus, for two factors a_1a_2, instead of the arrangement a_1a_2, we can use the arrangement (a_1a_2) and, therefore, the correctly matched brackets arrangement (). For three factors $a_1a_2a_3$, instead of the arrangements $a_1(a_2a_3)$ and $(a_1a_2)a_3$, we can use the arrangements $(a_1(a_2a_3))$ and $((a_1a_2)a_3)$ and, therefore, the correctly matched brackets arrangements ()() and (()). Placing n pairs of brackets correctly, we get again the nth Catalan number C_n.

Let us consider this problem in detail.

An arrangement of n pairs of brackets is called *correct* if it satisfies the following conditions:

- the number of opening brackets is equal to the number of closing brackets;
- in any initial segment of the arrangement, the number of closing brackets cannot be more than the number of opening brackets.

So, one pair of brackets can be correctly placed in only one way: (); two pairs can be correctly placed in two ways: ()() and (()); three pairs can be correctly placed in five ways: ()()(), ()(()), (())(), (()()), and ((())).

It is easy to prove the following fact:

- *The number of correct arrangements of n pairs of brackets is equal to the nth Catalan number C_n, $n \geq 0$.*

□ It is already checked that, for $n = 0, 1, 2, 3$, the statement is true. In fact, for $n = 1$, we have exactly one possibility, (); it is equal to C_1. For $n = 2$, we have two possibilities: ()() and (()); it is equal to C_2. For $n = 3$, we have five possibilities: ()()(), ()(()), (())(), (()()), and ((())); it is equal to C_3. For $n = 0$, we can also assume that zero pairs of brackets can be correctly arranged in only one way; it is equal to C_0.

Consider now a positive integer $n \geq 4$. Suppose that the number of correct arrangements of k pairs of brackets, $0 \leq k \leq n-1$, is equal to C_k, and try to prove that the number of correct arrangements of n pairs of brackets is C_n. In order to do this, consider the first bracket. It is an opening one. Then, let us find its pair, i.e., the first closing bracket, such that the substring starting with the first bracket and ending in its pair-bracket is correct, or, in particular, has the same number of opening and closing brackets. In this case, all other pairs of brackets are divided into two groups. The first one, having, say, s pairs, is placed between ("in") two chosen brackets. The second one, having $n - s - 1$ pairs, is placed to the right ("out") of the two chosen brackets. Here, s is an integer such that $0 \leq s \leq n - 1$. The first group can be correctly arranged in C_s ways. The second group can be correctly arranged in C_{n-s-1} ways. So, we have in total exactly C_n possibilities:

$$C_0 C_{n-1} + C_1 C_{n-2} + \cdots + C_s C_{n-s-1} + \cdots + C_{n-1} C_0 = C_n. \quad \square$$

For example, if we consider four pairs of brackets, then the case $s = 0$ gives us five (i.e., $C_0 C_3$) arrangements ()(()()()), ()(()(())), ()((())()), ()(((()))), and ()((()))); the case $s = 1$ gives us two (i.e., $C_1 C_2$) arrangements (())(()()) and (())((())); the case $s = 2$ gives us two (i.e., $C_2 C_1$) arrangements (()())(()) and ((()))(()); at last, the case $s = 3$ gives us five (i.e., $C_3 C_0$) arrangements (()()())(), (()(()))(), ((())())(), ((()()))(), and ((()))().

3.2.7. There are many other formulations of the brackets problem.

Given positive integer n, define a *Dyck word of length $2n$* as a sequence consisting of n symbols X and n symbols Y, in which each initial segment contains at least as many X symbols as Y symbols.

For $n = 1$, we have exactly one Dyck word: XY; for $n = 2$, we have two such words: $XXYY$ and $XYXY$; for $n = 3$, five such words exist: $XXXYYY$, $XYXXYY$, $XYXYXY$, $XXYYXY$, and $XXYXYY$. It is also natural to assume that there is exactly one Dyck word of length zero.

- The number of Dyck words of length $2n$ is equal to the nth Catalan number C_n, $n \geq 0$.

$\square$ In order to prove that the number of Dyck words of length $2n$ is C_n, it is enough to replace the symbol X by the symbol (of the opening bracket and the symbol Y by the symbol) of the closing bracket. Such a transform gives a bijection between the set of all Dyck words of length $2n$ and the set of all correct arrangements of n pairs of brackets. □

3.2.8. Given positive integer n, consider the set of sequences consisting of n "1" and n "-1", all partial sums of which are non-negative. For $n = 1$, there is only one such sequence: $1, -1$ (with partial sums $1, 1 + (-1) = 0$); for $n = 2$, there exist two such sequences: $1, 1, -1, -1$ (with partial sums $1, 1+1 = 2, 1+1+(-1) = 1, 1+1+(-1)+(-1) = 0$) and $1, -1, 1, -1$ (with partial sums $1, 1+(-1) = 0, 1+(-1)+1 = 1, 1+(-1)+1+(-1) = 0$).

It is easy to prove that:

- *the number of sequences consisting of n "1" and n "-1", all partial sums of which are non-negative, is equal to the nth Catalan number C_n, $n \geq 0$.*

□ In order to prove this statement, consider a Dyck word of length $2n$. If instead of the symbols X and Y, we use the symbols "1" and "-1", respectively, we get a sequence composed of n "1" and n "-1", all partial sums of which are non-negative. However, we can use for the proof correct bracket structures with n pairs of brackets: we should replace the symbol "(" with the symbol "1" and the symbol ")" with the symbol "-1". □

Similarly, the nth Catalan number gives the answer to the *ballot problem* (see [Weis24] and [Gess92]): in an election, if X and Y are candidates and there are $2n$ voters, n voting for A and n for B, in how many ways can the ballots be counted so that Y is never ahead of X?

3.2.9. For a given positive integer n, define a *mountain with n ascents and n descents* as a broken line composed of n segments of the equal length inclined at an angle of $45°$ to the positive direction of the OX-axis and of n segments of the same length inclined at an angle $135°$ to the positive direction of the OX-axis, such that the resulting "start" and "end" are located on the same level.

- *The number of mountains with n ascents and n descents is equal to the nth Catalan number C_n, $n \geq 0$.*

□ The proof can be obtained using correct bracket structures. In fact, it is enough to build a one-to-one correspondence between the set of all correct arrangements of n pairs of brackets and the set of mountains with n ascents and n descents. The corresponding law is very simple: each "ascending" segment corresponds to an opening bracket; each "descending" segment corresponds to a closing bracket. □

3.2.10. Similarly, C_n gives the number of routes on the upper-right quadrant of the integer lattice $\mathbb{Z}^2$, from the point $(0,0)$ to the point $(2n,0)$, if we can move only right-up and right-down, but it is forbidden to move below the OX-axis. The counting of such paths is equivalent to the counting of Dyck words: X stands for "move right-up," and Y stands for "move right-down." (Also, compare this problem with the mountains problem.)

Equivalently, C_n is the number of monotonic lattice paths along the edges of a grid with $n \times n$ square cells which do not pass above the diagonal. A monotonic path is one which starts at the lowest-left corner, finishes at the upper-right corner, and consists entirely of edges pointing rightward or upward. The counting of such paths is equivalent to the counting of Dyck words: X stands for "move right," and Y stands for "move up."

We consider this situation in detail in Chapter 5.

3.2.11. Note that a bijection between the correct bracket arrangements and the set of all Euler's triangulations of a convex $(n+2)$-gon can also be constructed. For a given convex $(n+2)$-gon, fix one of its sides. Mark it by the symbol "(". Mark all drawn diagonals of the polygon by the symbol "(" and all other sides of the polygon, except the last "right" side neighboring to the initial fixed side, by the symbol ")". Note that for a convex $(n+2)$-gon, there are $n+2$ sides and $n-1$ diagonals; so, we obtain exactly n pairs of brackets. Therefore, we can obtain a correct bracket sequence of length $2n$, going clockwise along the considered polygon, starting from the fixed side and gathering all marks. If we go along the interior of the polygon, we should go back and continue.

Handshake problem

3.2.12. Suppose that there are several pairs of guests at a round table. Each guest shakes hands with one of the others so that no two pairs of joined arms intersect. In how many ways can this be done? It turns out that for n pairs, the answer is C_n, the nth Catalan number.

- *Let $2n$ people be seated around a circular table. Let all of them be simultaneously shaking hands with another person at the table in such a way that none of the arms cross each other. The number of such handshakes is the nth Catalan number C_n, $n \geq 0$.*

For the convenience of consideration, let us reformulate the problem by having on some circle $2n$ points and considering such a division of these points per pairs for which we get disjoint chords. The following figure below shows the cases $n = 0, 1, 2, 3$.

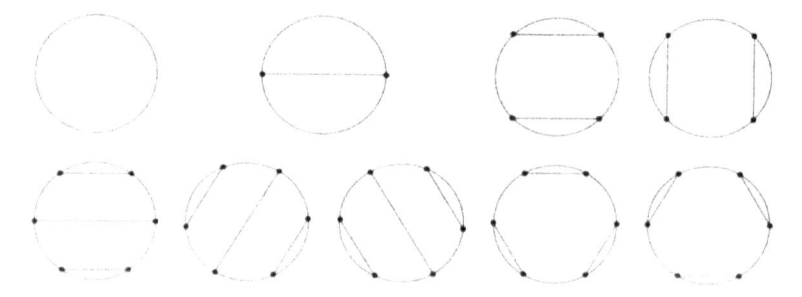

- *The number of partitions of $2n$ points marked on a circle per pairs so that the corresponding chords are not suppressed is equal to the nth Catalan number C_n, $n \geq 0$.*

□ There exists exactly one (i.e., C_0) way to solve the problem for $n = 0$, exactly one (i.e., C_1) way for $n = 1$, exactly two (i.e., C_2) ways for $n = 2$, and exactly five (i.e., C_3) ways for $n = 3$.

Consider now a positive integer $n \geq 4$. Suppose that the number of partitions of $2k$ points marked on a circle per pairs so that the corresponding chords are not suppressed is equal to the kth Catalan number C_k for all $2 \leq k \leq n-1$. Show that the number of partitions of $2n$ points is equal to the nth Catalan number C_n.

In order to do this, fix one of the marked points, and classify all possible configurations depending on which chord comes out from this point. Namely, considering from a given point one or another chord within the framework of the problem in an obvious way splits the remaining $2n - 2$ points into two subsets, each of which contains an even number of points (possibly equal to zero). Holding the rest of the chords is now possible only "inside" each of the derived subsets. If one of the subsets contains $2m$ points, then the second subset contains $2(n - m - 1)$ points. The number of partitions of $2m$ points into pairs so that the corresponding chords do not cross is equal to the mth Catalan number, $0 \leq m \leq n - 1$. The number of partitions of $2(n - m - 1)$ points into pairs so that the corresponding chords do not cross is equal to the $(n - m - 1)$th Catalan number C_{n-m-1}, $0 \leq n - m - 1 \leq n - 1$. Since each series of partitions is carried out independently of each other, the total number of partitions for a given configuration is $C_m C_{n-m-1}$. Summing up the results obtained for each of the possible configurations, i.e., for each $0 \leq m \leq n - 1$, we get that the total number of partitions of $2n$ points marked on a circle per pairs so that the corresponding chords do not cross is equal to the nth Catalan number C_n:

$$C_0 C_{n-1} + C_1 C_{n-2} + \cdots + C_{n-1} C_0 = C_n. \qquad \square$$

Enumeration of trees

3.2.13. It is widely known that Catalan numbers appear quite often when counting the number of some classes of trees.

Recall that a *tree* is a connected graph that does not have cycles ([Ore80], see also Chapter 1). The degree of the vertex of the tree (in general, of any graph) is the number of edges included in it. The vertices of the tree having the degree 1 are called *leaves*. Other vertices of a tree are called *internal vertices*. A *binary tree* is a tree whose internal vertices all have degree 3. A tree is called a *rooted tree* if it has a selected vertex, called the *root*. If the tree is placed on a plane without self-intersecting edges, it is called a *plane*. It is easy to prove that any tree with n vertices has exactly $n-1$ edges ([Ore80], see also Chapter 1).

It is well known (see, for example, [DeMo10]) that:

- there are exactly n^{n-2} trees with n vertices, $n \geq 2$.

Consider the class of all plane binary trees with n internal vertices. A direct check indicates that there is exactly one plane binary tree with one internal vertex, exactly two plane binary trees with two internal vertices, and exactly five plane binary trees with three internal vertices. The plane binary trees with n internal vertices for $n = 1, 2, 3$ are presented in the following figure.

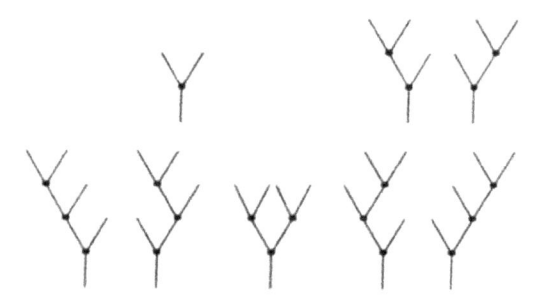

Note that plane binary trees in the figure can be called *planted*, as any such tree has a *trunk*; thus, the graph is drawn to simulate a tree growing up from the ground. The lowest leaf can be considered the root of this planted tree.

The English mathematician Arthur Cayley (1821–1895) proved the following fact:

- *The number of plane binary trees with n internal vertices is equal to the nth Catalan number C_n, $n \geq 0$.*

□ In fact, we have already checked that the statement is true for $n = 1, 2, 3$: there exists exactly one (i.e., C_1) plane binary tree with one internal vertex; there exist exactly two (i.e., C_2) plane binary trees with two internal vertices; there exist exactly five plane binary trees with three internal vertices. Moreover, we can assume that there exists exactly one (i.e., C_0) plane binary tree with zero internal vertices; it is the only tree with two (hanging) vertices.

In order to prove the statement for any non-negative integer n, it is enough to verify that the number of plane binary trees with n internal vertices is equal to the number of partitions of a convex $(n+2)$-gon into triangles (in fact, into n triangles) by diagonals, not intersecting within this $(n+2)$-gon.

To do it, let us build a one-to-one correspondence between plane binary trees with n internal vertices and partitions of a regular $(n+2)$-gon into triangles according to the following law:

- placing one of the hanging vertices (leaves) of the tree "outside" of the internal part of a given convex $(n+2)$-gon, let us cross along the edge of the tree to the nearest side of the $(n+2)$-gon and place the next vertex of the tree in the triangle we fell into;
- draw the next two edges so that they intersect the remaining two sides of a given triangle and "end" either in a new triangle or "outside" of the internal part of the $(n+2)$-gon; in the second case, a vertex that is outside of the $(n+2)$-gon, is no longer used, that is, it remains a hanging vertex (leaf); it corresponds to the intersection of the sides of the $(n+2)$-gon.

Conversely, for each plane binary tree with n internal vertices, we can restore the corresponding partition of a given $(n+2)$-gon into n triangles. To get this correspondence, let us first build the sides of the polygon "crossing" the edges of the tree on which the leaves are suspended; then, sequentially draw the diagonals of the polygon so that each diagonal crosses exactly one edge of the tree; as a result, a set of triangles is obtained such that each triangle contains exactly one of the internal vertices of the tree (see the diagonals a, b, c obtained in the following figure).

Thus, we have obtained a bijection between the set of all plane binary trees with n internal vertices and the set of all Euler's triangulations of a given convex $(n+2)$-gon. Since the number of such triangulations is equal to the nth Catalan number C_n, the number of plane binary trees with n internal vertices is also equal to the nth Catalan number C_n.

We can prove the statement using a one-to-one correspondence between the set of plane binary trees with n internal vertices and the set of correct arrangements of n pairs of brackets. This bijection was proposed by the Polish mathematician Jan Lukasiewicz (1878–1956). In fact, mark all leaves of a (planted) plane binary tree, except for the last right and the lowest, by closing brackets, and all internal vertices by opening brackets; imagine a worm crawling up the trunk and around the entire tree and collecting the available brackets on the way. In this case, any plane binary three with n internal vertices (and $n+2$ leaves) corresponds to some correct arrangements of n pairs of brackets.

However, the proof can be obtained by direct consideration. In fact, the statement was checked for $n = 0, 1, 2, 3$: there exist exactly C_n plane binary trees with n internal vertices if $n = 0, 1, 2, 3$. Consider now a positive integer $n \geq 4$. Suppose that the statement is true for any non-negative integer k, $0 \leq k \leq n-1$, and prove it for n. In fact, consider any plane binary tree with n internal vertices, $n \geq 4$. Fix a hanging vertex of this graph, and represent the edge, going from this vertex, as a trunk; in other words, present the considered tree as a planted tree. Consider now the lowest internal vertex of the obtained planted tree. We have exactly two edges going up from this vertex. So, we get two (planted) plane binary subtrees of our binary tree, with these two edges as their trunks. Consider the left such tree. It has s internal vertices, where s is a non-negative integer, such that $0 \leq s \leq n-1$. Then, the right subtree has $n-s-1$ internal vertices, $0 \leq n-s-1$. By induction, we can construct exactly C_s "left" subtrees and exactly C_{n-s-1} "right" subtrees, altogether $C_s C_{n-s-1}$ possibilities. So, for $s = 0$, we get $C_0 C_{n-1}$ possible plane binary trees with n internal vertices (in this case, the "left" subtree is just the tree with two vertices); for $s = 1$, we get $C_1 C_{n-2}$ possible plane binary trees with n internal vertices; ...; for $s = n - 2$ we get $C_{n-2}C_1$ possible plane binary trees with n internal vertices; for $s = n - 1$, we get $C_{n-1}C_0$ possible plane binary trees with n internal vertices (in this case, the "right" subtree is just the tree with two vertices). As a result, the total number of plane binary trees with n internal vertices is equal to the nth Catalan number:

$$C_0 C_{n-1} + C_1 C_{n-2} + \cdots + C_{n-2} C_1 + C_{n-1} C_0 = C_n. \qquad \square$$

3.2.14. For a given plane binary tree, if we know the number of its internal vertices, it is easy to find the number of its hanging vertices (leaves) and the number of its edges.

- *A plane binary tree with n internal vertices has exactly $n+2$ hanging vertices (leaves).*

□ Consider a plane binary tree with n internal vertices. Let us take its planted representation and count all vertices of this tree, starting from the lowest internal vertex and moving up. In fact, starting from a given internal vertex, we can move up by one of the two edges "leaving" this vertex. On this way, we get at the next step either a new internal vertex or a hanging vertex (leave). Since moving from the lowest internal vertex we come across another $n - 1$ internal vertices, then in total we have to use $2n$ "leaving" edges, two edges from each internal vertex. On $n - 1$ of these edges will be "suspended" some internal vertices. On the remaining $2n - (n - 1) = n + 1$ edges will be "suspended" hanging vertices (leaves). Adding to them the lowest hanging vertex (the root), we get the result: a plane binary tree with n internal vertices has exactly $n + 2$ hanging vertices (leaves).

On the other hand, the same result can be obtained by general consideration. In fact, *the degree sum formula* states that, for a given undirected graph G, it holds that

$$\sum_{v \in V} d(v) = 2|E|,$$

where V is the set of vertices of the graph G and E is the set of edges of the graph G. That is, the sum of the vertex degrees equals twice the number of edges. Moreover, the number of edges of a tree with k vertices is equal to $k-1$, $k \geq 1$. If a plane binary tree has n internal vertices and m leaves, then it has exactly $n+m-1$ edges. For any internal vertex, its degree is 3, while for any leaf, its degree is 1. So, we get the equality

$$3n + m = \sum_{v \in V} d(v) = 2|E| = 2(n+m-1),$$

and $m = n + 2$. □

Thus, we have proved the following result:

- The number of plane binary trees with $n+2$ leaves is equal to the nth Catalan number C_n, $n \geq 0$.

As any plane binary tree with n internal vertices has exactly $n+2$ hanging vertices (leaves), we obtain a tree with $2n+2$ vertices; it has exactly $2n+1$ edges. The following statement is a simple consequence of this fact:

- The number of plane binary trees with $2n+1$ edges is equal to the nth Catalan number C_n, $n \geq 0$.

3.2.15. In terms of computer science, a *binary tree* is a tree data structure in which each node has at most two children, referred to as the left child and the right child. A *rooted binary tree* has a *root node*, and every node has at most two children. A *full* (or *proper, strict*) *binary tree* is a rooted binary tree in which every node has either zero or two children.

It is easy to see that any plane planted binary tree with n internal vertices corresponds to a full binary tree with n internal nodes (one of which, the root, has the degree 2, and all others have the degree 3). In fact, it is enough to choose as the root the lowest internal vertex of the planted binary tree, and delete the lowest ("planted") hanging vertex. So, as a consequence, we obtain that:

- the number of all full binary trees with n internal nodes is equal to the nth Catalan number C_n, $n \geq 0$.

Obviously, any full binary tree with n internal nodes has exactly $n+1$ leaves. It follows from the constructed bijection between the set of all plane planted binary trees with n internal vertices and the set of all full binary trees with n internal nodes. Any plane planted binary tree with n internal vertices has $n+2$ leaves. As we delete exactly one (lowest) leaf, the resulting full binary tree has exactly $n+1$ leaves. So, we can state that:

- *the number of all full binary trees with $n+1$ leaves is equal to the nth Catalan number C_n.*

3.2.16. Consider now the set of all plane rooted trees. It is easy to check that there is exactly one plane rooted tree with zero edges, exactly one plane rooted tree with one edge, exactly two plane rooted trees with two edges, exactly five plane rooted trees with three edges, and exactly 14 plane rooted trees with four edges.

Note that any plane rooted tree can be presented as a *planted tree*, i.e., it can be drawn to simulate a tree growing up from the root in the ground. In general, we can prove the following result:

- *The number of plane rooted trees with n edges is equal to the nth Catalan number C_n, $n \geq 0$.*

□ This statement for small n is already verified. The number of plane rooted trees with n edges is equal to 1 (i.e., C_0) for $n=0$, to 1 (i.e., C_1) for $n=1$, to 2 (i.e., C_2) for $n=2$, to 5 (i.e., C_3) for $n=3$, and to 14 (i.e., C_4) for $n=4$.

The simplest way to prove the general statement is to build a one-to-one correspondence between the set of all plane binary trees with n internal vertices and the set of all plane rooted trees with n edges. This correspondence was proposed by Frank R. Bernhart.

Let us imagine that a given plane binary tree is depicted so that the edges, going from each internal vertex, are directed up, down, and to the right. Imagine that each horizontal edge shrinks to a point and disappears. If on the right end of the deleted horizontal edge is the vertex of degree 3, then it is carried to the left and merges with the point at its left. All the vertical edges become different. This simple transformation translates any plane binary tree with n internal vertices (and, hence, with $n+2$ leaves and $2n+1$ edges) to a rooted tree with n edges when discarding "superfluous" lowest vertex. (See the algorithm for this transformation for the case of $n = 1$ in the following figure.)

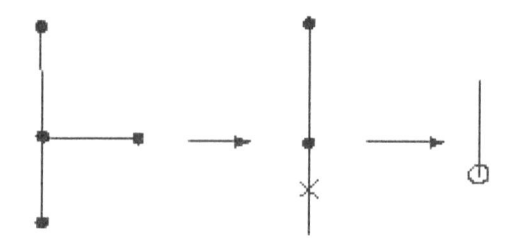

As we have exactly C_n plane binary trees with n internal vertices, we have exactly C_n plane rooted trees with n edges.

On the other hand, we can use for a proof the correct arrangements of brackets. In fact, there exists a simple one-to-one-correspondence between the set of all plane rooted trees with n edges and the set of all correct bracket arrangements with n pair of brackets. For a given (planted) plane rooted tree with n edges, let us mark each of its edge by one pair of brackets, putting an opening bracket to the left and a closing bracket to the right. Now, starting from the root and moving from the left to the right along the edges of the tree, we can collect all n pairs of brackets, obtaining a corresponding correct arrangement.

However, the proof can be obtained by direct consideration. In fact, the statement was checked for $n = 0, 1, 2, 3, 4$: there exist exactly C_n plane rooted trees with n edges if $n = 0, 1, 2, 3, 4$. Consider now a positive integer $n \geq 5$. Suppose that the statement is true for any non-negative integer k, $0 \leq k \leq n - 1$, and prove it for n. In fact, consider a planted representation of any plane rooted tree with n edges, $n \geq 5$. Fix the left edge, going from the root, and delete it. This operation gives us two plane rooted subtrees of the initial tree. The "right" subtree has the root of the initial tree as its root; the "left" subtree has as its root the vertex that is adjacent to the root of the initial tree. Consider the "left" subtree. It has s edges, where s is a non-negative integer, such that $0 \leq s \leq n - 1$. Then, the "right" subtree has $n - s - 1$ edges, $0 \leq n - s - 1$. By induction, we can construct exactly C_s "left" subtrees and exactly

C_{n-s-1} "right" subtrees, altogether with $C_s C_{n-s-1}$ possibilities. So, restoring the removed edge, for $s = 0$, we get $C_0 C_{n-1}$ possible plane rooted trees with n edges (in this case, the "left" subtree is just the graph with one vertex); for $s = 1$, we get $C_1 C_{n-2}$ possible plane binary trees with n internal vertices; ...; for $s = n-2$, we get $C_{n-2} C_1$ possible plane binary trees with n internal vertices; for $s = n-1$, we get $C_{n-1} C_0$ possible plane binary trees with n internal vertices (in this case, the "right" subtree is just the graph with one vertex). As a result, the total number of plane rooted trees with n edges is equal to the nth Catalan number:

$$C_0 C_{n-1} + C_1 C_{n-2} + \cdots + C_{n-2} C_1 + C_{n-1} C_0 = C_n. \qquad \square$$

Since each tree with n edges has $n + 1$ vertices, we also get the following result:

- *The number of plane rooted trees with $n+1$ vertices is equal to the nth Catalan number C_n.*

Enumeration of partitions

3.2.17. A *partition* of a set is a grouping of its elements into non-empty subsets in such a way that every element is included in exactly one subset, i.e., X is a disjoint union of the subsets.

Formally, a family of sets S is a *partition* of X if the following conditions hold:

- $\emptyset \notin S$, the family S does not contain the empty set;
- $\bigcup_{A \in P} A = X$, the union of the sets in S is equal to X;
- $(\forall A, B \in P)\ A \neq B \implies A \cap B = \emptyset$, the intersection of any two distinct sets in S is empty.

The sets in S are called the *parts* (or *blocks*, *cells*) of the partition.

Starting with $n = 1$, we get exactly one partition of a 1-set (say, the set $\{1\}$): $\{\{1\}\}$. For the 2-set $\{1, 2\}$, we have two partitions: $\{\{1, 2\}\}$ and $\{\{1\}, \{2\}\}$. The 3-set $\{1, 2, 3\}$ has five partitions: $\{\{1\}, \{2\}, \{3\}\}$, $\{\{1, 2\}, \{3\}\}$, $\{\{1, 3\}, \{2\}\}$, $\{\{1\}, \{2, 3\}\}$, and $\{\{1, 2, 3\}\}$. The number of partitions of a given n-set is, by definition, the nth Bell number $B(n)$. So, taking into account $B(0) = 1$, we obtain that the sequence of Bell numbers starts with the numbers $1, 1, 2, 5, 15, 52, 203, 877, 4140, \ldots$ (sequence A000110 in the OEIS).

3.2.18. Consider now the so-called *non-crossing partitions*. In fact, consider a finite set that is linearly ordered, or (equivalently, for the purpose of this definition) arranged in a cyclic order like the vertices of a regular n-gon. No generality is lost by taking this set to be $V_n = \{1, \ldots, n\}$. So, let us arrange the set $\{1, 2, 3, \ldots, n\}$ in a cyclic order (for example, placing the numbers $1, 2, 3, \ldots, n$ in this specified order on a circle). A partition of the set $\{1, 2, 3, \ldots, n\}$ is called *non-crossing* if no two partition blocks "cross" each other, i.e., if a and b belong to one partition block and x and y to another, they are not arranged in the order $axby$.

In other words, if we draw an arc based at a and b and another arc based at x and y, then the two arcs cross each other if the order is $axby$ but not if it is $axyb$ or $abxy$. So, a non-crossing partition is a partition without crossing arcs. For example, the partition $\{\{1, 2\}, \{3, 4\}\}$ of the set $\{1, 2, 3, 4\}$ is non-crossing, while the partition $\{\{1, 3\}, \{2, 4\}\}$ is not.

For $n = 1$, the only partition of the set $\{1\}$ is non-crossing: $\{\{1\}\}$. For $n = 2$, we have two partitions of the set $\{1, 2\}$: $\{\{1, 2\}\}$ and $\{\{1\}, \{2\}\}$. Both partitions are non-crossing. All five partitions of the set $\{1, 2, 3\}$ are also non-crossing: $\{\{1\}, \{2\}, \{3\}\}$, $\{\{1, 2\}, \{3\}\}$, $\{\{1, 3\}, \{2\}\}$, $\{\{1\}, \{2, 3\}\}$, and $\{\{1, 2, 3\}\}$. If $n = 4$, we obtain 15 partitions of the set $\{1, 2, 3, 4\}$. The only one partition in the set that is crossing is $\{\{1, 3\}, \{2, 4\}\}$.

Consider a partition of the set $V_n = \{1, 2, 3, \ldots, n\}$: $V_n = S_1 \cup \cdots \cup S_q$. Instead of each element k of the set $V_n = \{1, 2, \ldots, n\}$, let us write a symbol of the class S_i, into which the element k falls in this partition $V_n = S_1 \cup \cdots \cup S_q$: symbol a for S_1, symbol b for S_2, etc. So, any partition of V_n can be represented by a sequence of length n, consisting of the elements $a, b, \ldots$. For example, the partition $V_4 = \{1, 2\} \cup \{3, 4\}$ corresponds to the sequence $aabb$, the partition $V_4 = \{1, 2\} \cup \{3, 4\}$ to the sequence $abab$, and the partition $V_4 = \{1, 2\} \cup \{3\} \cup \{4\}$ to the sequence $aabc$.

Connecting by the arcs the same sequence's elements, we see that in some sequences, the arcs intersect (in our example, only for the sequence *abab*), and in some sequences, they do not. By the construction, a sequence has no intersecting arcs if and only if the corresponding partition is non-crossing.

- *The number of non-crossing partitions of an n-set is equal to the nth Catalan number C_n.*

☐ We can prove this statement directly. In fact, for a partition of a given n-set (say, the set $V_n = \{1, 2, \ldots, n\}$), there are at most n partition blocks; therefore, in order to represent this partition by a sequence, we should use at least one symbol a_1 and at most n symbols $a_1, \ldots, a_n$. It is easy to see that the statement is true for $n = 0, 1, 2, 3, 4$. We have already shown that there is exactly one (non-crossing) partition for the empty set, there is exactly one (non-crossing) partition for a 1-set, there are exactly two (non-crossing) partitions for a 2-set, there are exactly five (non-crossing) partitions for a 3-set, and there are exactly 14 non-crossing partitions for a 4-set. Consider a positive integer $n \geq 5$, and suppose that the statement is true for all non-negative k, $0 \leq k \leq n-1$. Consider the set of all sequences, corresponding to partitions of an n-set. Any such sequence starts with the symbol a_1. If there is the only one symbol a_1, fix it; if there are several symbols a_1, fix the last of these symbols. In any case, before this fixed symbol a_1, we have s places, $s = 0, 1, 2, \ldots, n-1$, and after this fixed symbol, we have $n - s - 1$ places, $0 \leq n - s - 1 \leq n - 1$. In order to obtain the sequence, corresponding to a partition of the considered n-set, we can use for the first group of places any sequence corresponding to a partition of an s-set (represented by the symbols $a_1, \ldots, a_s$), and for the second group of places, we use any sequence corresponding to a partition of an $(n-s-1)$-set (represented by the symbols $a_2, \ldots, a_{n-1}$, not used in the first group). By induction, the number of possible constructions is $C_s C_{n-s-1}$. So, summing all these values for $s = 0, 1, 2, \ldots, n-1$, we obtain that the number of all sequences, corresponding to non-crossing partitions of an n-set, is equal to the nth Catalan number C_n:

$$C_0 C_{n-1} + C_1 C_{n-2} + \cdots + C_{n-2} C_1 + c_{n-1} C_0 = C_n.$$

Obviously, the correspondence between the non-crossing partitions of an n-set and sequences used in the proof is a bijection. So, the number of non-crossing partitions of an n-set is also equal to the nth Catalan number C_n, $n \geq 0$. □

3.2.19. A similar result can be obtained for a subclass of partitions of a $2n$-set. Let us consider partitions of a $2n$-set (say, the set $\{1, 2, \ldots, 2n\}$), in which every block is of size 2. For $n = 1$, we have exactly one such partition of the set $\{1, 2\}$: $\{\{1, 2\}\}$. It is non-crossing. For $n = 2$, we have three such partitions of the set $\{1, 2, 3, 4\}$: $\{\{1, 2\}, \{3, 4\}\}$, $\{\{1, 4\}, \{2, 3\}\}$, $\{\{1, 3\}, \{2, 4\}\}$. The only one partition from this list is crossing: $\{\{1, 3\}, \{2, 4\}\}$. For $n = 3$, there are 15 such partitions; exactly five of them are non-crossing.

Consider a partition of the set $V_{2n} = \{1, 2, 3, \ldots, 2n\}$ with all blocks of size 2: $V_n = S_1 \cup \cdots \cup S_n$, $|S_1| = |S_2| = \cdots |S_n| = 2$. Instead of each element k of the set $V_{2n} = \{1, 2, \ldots, 2n\}$, let us write a symbol of the class S_i into which the element k falls in this partition: symbol a_1 for S_1, symbol a_2 for S_2,..., symbol a_n for S_n. So, any such partition of V_{2n} can be represented by a sequence of length $2n$, consisting of two elements a_1, two elements $a_2, \ldots$, two elements a_n. For example, the partition $V_4 = \{1, 2\} \cup \{3, 4\}$ corresponds to the sequence $a_1 a_1 a_2 a_2$, and the partition $V_4 = \{1, 2\} \cup \{3, 4\}$ corresponds to the sequence $a_1 a_2 a_1 a_2$.

Connecting by arcs the same sequence's elements, we see that, by construction, a sequence has no intersecting arcs if and only if the corresponding partition is non-crossing.

- *The number of non-crossing partitions of the set $\{1, \ldots, 2n\}$ in which every block is of size 2 is equal to the nth Catalan number C_n.*

□ In this case, the proof is quite simple. In fact, it is enough to construct a one-to-one correspondence between the set of all non-crossing partitions of the set $\{1, \ldots, 2n\}$ in which every block is of size 2 and the set of all correct arrangements of n pairs of brackets. The rule of correspondence is as follows: replace the symbol a_i, $i = 1, 2, \ldots, n$ with an opening bracket if it is the first symbol of the pair $a_i a_i$ used in the considered sequence; replace the symbol a_i, $i = 1, 2, \ldots, n$

with a closing bracket if it is the second symbol of the pair $a_i a_i$ used in the considered sequence. As the number of correct arrangements of n pairs of brackets is equal to the nth Catalan number C_n, the statement is proven. □

Enumeration of permutations

3.2.20. A *permutation* of a finite set S is defined as a bijection from S to itself.

In mathematical texts, it is customary to denote permutations using lowercase Greek letters. Commonly, either α and β or σ, τ, and π are used. As the properties of permutations do not depend on the nature of the set elements, we can consider the permutations of the set $V_n = \{1, 2, \ldots, n\}$.

In Cauchy's *two-line notation*, we list the elements of S in the first row, and for each one, we list its image below it in the second row. For instance, a particular permutation of the set $S = \{1, 2, 3, 4, 5\}$ can be written as $\sigma = \begin{pmatrix} 1 & 2 & 3 & 4 & 5 \\ 2 & 5 & 4 & 3 & 1 \end{pmatrix}$. If there is a natural order for the elements of S, we can omit the first row and write the permutation in *one-line notation* as

$$\sigma = (\sigma(x_1) \; \sigma(x_2) \; \sigma(x_3) \; \ldots \; \sigma(x_n)),$$

that is, as an ordered arrangement of the elements of S. In mathematical literature, a common usage is to omit parentheses for one-line notation. The one-line notation is also called the *word representation* of a permutation. The above example would then be $\sigma = 2, 5, 4, 3, 1$. It is also typical to use commas to separate these entries only if some have two or more digits. So, we obtain that $\sigma = 25431$. This form is more compact and is common in elementary combinatorics and computer science.

- The number P_n of all permutations of a given n-set is equal to $n!$.

□ This simple fact can be proven by induction. There exist only one permutation for the set $\{1\}$: $\sigma = 1$. There exit exactly two permutations for the set $\{1, 2\}$: $\sigma_1 = 12$ and $\sigma_2 = 21$. So, $P_1 = 1!$ and

$P_2 = 2!$. Suppose that $P_{n-1} = (n-1)!$, and consider the set of all permutations of the set $\{1, 2, \ldots, n\}$. As for a given permutation of the set $\{1, 2, \ldots, n-1\}$, we have exactly n places for the new element n (before, between, and after the elements of the considered permutation of the set $\{1, 2, \ldots, n-1\}$), we obtain that $P_n = n \cdot P_{n-1}$. So, $P_n = n \cdot (n-1)! = n!$. □

3.2.21. Consider now the notion of *pattern-avoiding permutations*.

A *pattern* is a smaller permutation embedded in a larger permutation. More precisely, a permutation σ of an n-set contains a permutation α of a k-set, $k \leq n$, as a pattern if we can find k digits of σ that appear in the same relative order as the digits of α. Consider, for example, the permutation $\sigma = 18274653$. If, instead of looking at all eight digits, we focus on the digits 8, 4, and 6, we find that the largest of these three digits comes first, the smallest of these digits comes second, and the middle of these three digits comes last, just as in the permutation $\alpha = 312$. So, the digits 846 are called a 312-*pattern* inside σ. It is easy to see that the permutation σ also contains the patterns 123, 132, 231, and 321. On the other hand, it does not contain the pattern 213. In this case, we say that the permutation σ *avoids* the pattern 213.

Pattern-avoiding permutations have provided many counting problems. In particular:

- the number of permutations of an n-set that avoid the pattern 213 is equal to the nth Catalan number C_n, $n \geq 0$.

□ In fact, the statement is true for $n = 0, 1, 2$. There is one permutation of the empty set, there is one permutation for a 1-set, there are two permutations of a 2-set; all of them are 213-avoiding. For $n = 3$, there are six permutations, five of them are 213-avoiding: 123, 132, 231, 312, 321. For $n = 4$, there are 24 permutations, 14 of them are 213-avoiding: 1234, 1243, 1342, 1423, 1432, 4123, 4132, 3412, 4312, 2341, 2431, 3421, 4231, 4321.

Consider a positive integer $n \geq 5$. Suppose that the statement is true for any s, $0 \leq s \leq n-1$. For a given 213-avoiding permutation of length n, find and fix the number 1. It is placed on some position i,

$1 \leq i \leq n$. Then, the numbers placed on $i-1$ positions before 1 must be all larger than the numbers placed on $n-i$ positions after 1; otherwise, we have a 213-pattern using the fixed number 1 as the smallest number in the pattern. So, we can obtain a 213-avoiding permutation of length n simultaneously arranging $i-1$ numbers before 1 in one of the 213-avoiding permutations of length $i-1$ and $n-1$ numbers after 1 in one of the 213-avoiding permutations of length $n-i$. By induction, for a fixed i, we obtain $C_{i-1}C_{n-i}$ possibilities. In total, the number of 213-avoiding permutations of length n is equal to the nth Catalan number C_n:

$$C_0 C_{n-1} + C_1 C_{n-2} + \cdots + C_{n-1} C_0 = C_n.$$ □

3.2.22. In general, no matter which permutation α of length 3 is chosen as the pattern, it turns out that

- *the number of α-avoiding permutations of length n is given by the nth Catalan number C_n for any pattern $\alpha \in \{123, 132, 213, 231, 312, 321\}$.*

□ The arguments for avoiding 132, 231, or 312 are quite similar to our 213-avoiding argument above, while the arguments for avoiding 123 and 321 are more complicated. However, there are several one-to-one constructions that prove this statement ([ClKi08] and [Pudw24]).
□

In particular, as any permutation with no three-term increasing subsequence is just an 123-avoiding permutation, we obtain the following fact:

- *The number of permutations of length n with no three-term increasing subsequence is equal to the nth Catalan number C_n, $n \geq 0$.*

3.2.23. In mathematics and computer science, a *stack-sortable permutation* (or *tree permutation*) is a permutation whose elements may be sorted by an algorithm whose internal storage is limited to a single-stack data structure (see, for example, [Knot77]).

For a formal definition, consider a permutation σ of the set $\{1, 2, \ldots, n\}$. Write σ as $\sigma = unv$, where n is the largest element in σ and u and v are shorter sequences. Set $S(\sigma) = S(u)S(v)n$, with S being the identity for one-element sequences. In these conditions, a permutation σ of the set $\{1, 2, \ldots, n\}$ is called *stack-sortable* if $S(\sigma) = 12 \cdots (n-1)n$.

Thus, the only permutation of the empty set is stack-sortable, as is the only permutation of the set $\{1\}$. There are two permutations of the set $\{1, 2\}$; both are stack-sortable: $S(12) = S(1)S(2) = 12$; $S(21) = S(1)S(2) = 12$. For $n = 3$, there are six permutations, five of which are stack-sortable: 123, 132, 213, 312, 321. For example, $S(231) = S(2)S(1)3 = 213$, while $S(321) = S(21)3 = S(1)23 = 123$. For $n = 4$, there are 24 permutations, 14 of which are stack-sortable: 4123, 4132, 4213, 4312, 4321, 1234, 1324, 2134, 3124, 3214, 1423, 1432, 1243, 2143. For example,

$$S(2134) = S(213)4 = S(21)34 = S(1)234 = 1234,$$

while

$$S(3421) = S(3)S(21)4 = 3S(1)24 = 3124.$$

It is easy to prove by induction that the stack-sortable permutations are exactly the permutations that do not contain the permutation pattern 231. So, they are counted by Catalan numbers and may be placed in a bijection with many other combinatorial objects, including Dyck paths and binary trees.

- the number of stack-sortable permutations of an n-set is equal to the nth Catalan number C_n, $n \geq 0$.

Exercises

1. Check that the number of Euler's triangulations of a convex $(n+2)$-gon is equal to the nth Catalan number C_n if $n = 5, 6, 7, 8, 9, 10$.
2. Check that the nth Catalan number C_n can be interpreted as the number of arrangements of $n-1$ pairs of brackets for $n+1$ factors if $n = 5, 6, 7, 8, 9, 10$.
3. Use Forder's method to build a one-to-one correspondence between Euler's triangulations of a convex $(n+2)$-gon and the

arrangements of $n-1$ pairs of brackets for $n+1$ factors if $n = 2, 3, 4, 5, 6, 7, 8, 9, 10$.
4. Build illustrations for the handshake problem for n pairs of guests if $n = 5, 6, 7, 8, 9, 10$. Make sure that in these cases, the handshake problem is equivalent to the problem of splitting of $2n$ points on a circle into pairs so that the corresponding chords are not crossed.
5. Build a tree with n vertices, $n = 1, 2, 3, 4, 5, 6, 7, 8, 9, 10$. Make sure that the number of edges of a tree with n vertices is equal to $n-1$. Prove this statement. Make sure that the constructed trees are plane. Prove that any tree can be represented on the plane as a plane graph, i.e., without self-intersection of its edges. Build all plane binary trees with n internal vertices (leaves) for $n = 0, 1, 2, 3, 4, 5, 6, 7, 8, 9, 10$. Represent each of these trees as a planted tree. Check that each of these trees has exactly $n+2$ hanging vertices (leaves) and $2n+1$ edges. Make sure that for each n, the number of such trees is equal to the nth Catalan number C_n.
6. Build all plane rooted trees with n edges, $n = 0, 1, 2, 3, 4, 5, 6, 7, 8, 9, 10$. Represent each of these trees as a planted tree. Make sure that each of these trees has $n+1$ vertices. Make sure that for each n, the number of such trees is the nth Catalan number C_n.
7. Prove that Catalan numbers list *oriented trees* on n vertices, i.e., rooted trees that have a order for *children* vertices with a common *parent*.
8. Prove that the number of correct arrangements of n pairs of brackets is equal to the nth Catalan number C_n using a one-to-one correspondence between the set of correct arrangements of n pairs of brackets and the set of plane binary trees with n internal vertices, proposed by J. Lukasiewicz. Prove that this algorithm provides a correct bracket arrangement. Give the corresponding illustrations for $n = 2, 3, 4, 5, 6, 7, 8, 9, 10$.
9. Prove that Catalan numbers represent the number of ways to linearize the Cartesian product of two linear ordered sets, one of which consists of two elements and the second one consists of n elements.

10. Prove that the nth Catalan number C_n expresses the number of polyominoes with perimeter $2k + 2$.

3.3. Closed Formula for Catalan Numbers

3.3.1. Other definitions of Catalan numbers can be found in the literature. For example, in the widely known work of Conway and Guy, *The Book of Numbers* ([CoGu96]), the elements $C_0, C_1, C_2, \ldots, C_n, \ldots$ of the sequence of Catalan numbers are defined using *Pascal's triangle*:

$$C_n = \frac{1}{n+1} \cdot \binom{2n}{n},$$

where $\binom{2n}{n} = \frac{(2n)!}{n!n!}$, $n \geq 0$, is a *central binomial coefficient*.

Thus, we can obtain the sequence $1, 1, 2, 5, 14, \ldots$ of Catalan numbers by writing out the central elements $1, 2, 6, 20, 70, \ldots$ of Pascal's triangle and dividing them by the numbers $1, 2, 3, 4, 5, \ldots$, respectively:

```
                  1
                 1 1
                1 2 1
               1 3 3 1
              1 4 6 4 1
            1 5 10 10 5 1
           1 6 15 20 15 6 1
         1 7 21 35 35 21 7 1
        1 8 28 56 70 56 28 8 1
         ⋮ ⋮ ⋮ ⋮ ⋮ ⋮ ⋮ ⋮ ⋮
```

3.3.2. The equivalence of the two specified definitions of Catalan numbers can be proven in many ways. In the following section, we use the generating function of the sequence of Catalan numbers for this purpose. Now, let us consider the simplest combinatorial method for a proof (see [Wiki24] and [PWZ96]).

- There exists the following relation:
$$(n+1)C_n = (4n-2)C_{n-1}, \quad C_0 = 1.$$

□ In fact, let us use Euler's triangulation of convex polygons to build the relationship between the numbers C_{n-1} and C_n.

For a given $(n + 1)$-gon, let us highlight one of its sides as the basic one. Let us triangulate this $(n + 1)$-gon. It is easy to see that, regardless of the type of the obtained triangulation, we construct $n-2$ diagonals and therefore receive, in total, exactly $2n - 1$ edges. Select one of the specified edges, and assign to it one or another orientation. Since for each triangulation we can do this in $2 \cdot (2n-1) = 4n-2$ ways and the number of possible triangulations is C_{n-1}, there is exactly $(4n - 2)C_{n-1}$ "decorated" triangulations of a given $(n+1)$-gon (with a basic side and an oriented edge).

Now, consider an $(n + 2)$-gon and again highlight one of its sides as the basic one. After triangulation of this $(n + 2)$-gon, select one of its sides other than the basic one. Since for each triangulation we can do it in $n + 1$ ways and the number of triangulations of $(n + 2)$-gon is C_n, there is exactly $(n+1)C_n$ such "decorated" triangulations of a given $(n + 2)$-gon (with a basic side and another highlighted side).

Now, we can build a one-to-one correspondence between "decorated" triangulations obtained in the first and second cases.

In particular, in order to obtain a "decorated" triangulation of a given $(n + 1)$-gon (with a basic side and an oriented edge) from some "decorated" triangulation of a given $(n + 2)$-gon (with a basic side and another highlighted side), we should compress to an edge the triangle in a "decorated" triangulation of the $(n + 2)$-gon, one of the sides of which turn out to be selected. In this case, the selected side is compressed to the point; we should orient the obtained edge in the direction "to the compressed side" (equivalently, to the point obtained after the compression).

On the other hand, in order to obtain a "decorated" triangulation of a given $(n+2)$-gon (with a basic side and another highlighted side) from some "decorated" triangulation of a given $(n + 1)$-gon (with a basic side and an oriented edge), we should turn the oriented edge in a "decorated" triangulation of the considered $(n + 2)$-gon into a triangle by splitting it "in the end" and highlight the new side of the obtained triangle.

Thus, we have proved that
$$(4n-2)C_{n-1} = (n+1)C_n.$$
In other words, we got for the Catalan numbers the following recurrent relation:
$$C_n = \frac{4n-2}{n+1}C_{n-1}, \quad C_0 = 1. \qquad \square$$

Now, we can easily obtain the closed formula for the Catalan numbers.

- *For the nth Catalan number, it holds that*
$$C_n = \frac{1}{n+1}\binom{2n}{n}.$$

☐ In fact, it is easy to check that the sequence $f(n) = \frac{(2n)!}{n!(n+1)!}$, $n = 0, 1, 2, \ldots$, satisfies the same recurrent relation with the same initial condition as the sequence C_n, $n = 0, 1, 2, 3, \ldots$:
$$f(n) = \frac{4n-2}{n+1}f(n-1), \quad f(0) = 1.$$
So, it follows that $f(n) = C_n$ for any $n = 0, 1, 2, 3, \ldots$. In other words, it holds that
$$C_n = \frac{(2n)!}{n!(n+1)!} = \frac{1}{n+1}\frac{(2n)!}{n!n!} = \frac{1}{n+1}\binom{2n}{n}. \qquad \square$$

3.3.3. However, there exist many other proofs of the closed formula for C_n. The following bijective proof provides a natural explanation for the term $n+1$ appearing in the denominator of the formula for C_n.

In order to realize this proof, let us consider all monotonic lattice paths along the edges of a grid with $n \times n$ square cells. A monotonic path is one which starts at the lowest left corner, finishes at the upper right corner, and consists entirely of edges pointing rightward or upward. As any such path has exactly n edges pointing rightward and exactly n edges pointing upward, there are in total $\binom{2n}{n}$ such paths.

Given a monotonic path, the *exceedanse* of the path is defined as the number of vertical edges above the diagonal.

We are interested in monotonic paths which do not pass above the diagonal, i.e., in the monotonic paths with zero exceedance.

Given a monotonic path whose exceedance is not zero, we apply the following algorithm to construct a new path whose exceedance is 1 less than the one we started with:

- Starting from the bottom left, follow the path until it first travels above the diagonal.
- Continue to follow the path until it touches the diagonal again.
- Denote by X the first such edge that is reached.
- Swap the portion of the path occurring before X with the portion occurring after X.

The algorithm causes the exceedance to decrease by 1 for any path that we feed it because the first vertical step starting on the diagonal is the only vertical edge that changes from being above the diagonal to being below it when we apply the algorithm. All the other vertical edges stay on the same side of the diagonal.

This implies that the number of paths with exceedance n is equal to the number of paths with exceedance $n-1$, which is equal to the number of paths with exceedance $n-2$, and so on, down to zero. In other words, we have split up the set of all monotonic paths into $n+1$ equally sized classes, corresponding to the possible exceedances between 0 and n. Each class contains exactly one monotonic path with exceedance 0. Since there are in total $\binom{2n}{n}$ monotonic paths, there are $\frac{1}{n+1}\binom{2n}{n}$ monotonic paths with exceedance 0, and we obtain the result $C_n = \frac{1}{n+1}\binom{2n}{n}$.

3.3.4. We can construct this proof in terms of Dyck words. In fact, consider any sequence consisting of n symbols X and n symbols Y. There are exactly $\binom{2n}{n}$ such sequences.

Given a sequence of length $2n$ consisting of n symbols X, and n symbols Y, the *error* of the sequence is defined as the number of symbols Y that violate the requirements for Dyck words. So, any Dyck word is a sequence with zero error.

Starting with one of the sequences consisting of n symbols X and n symbols Y, let **X** be the first X that brings an initial subsequence to the equality "number of Y = number of X," and configure the sequence as $F\mathbf{X}L$. Obtain the new sequence of the form $L\mathbf{X}F$. The algorithm causes the number of errors to decrease by 1.

This implies that the number of sequences with error n is equal to the number of sequences with error $n - 1$, which is equal to the number of sequences with error $n - 2$, and so on, down to zero. In other words, we have split up the set of all sequences into $n + 1$ equally sized classes, corresponding to the possible errors between 0 and n. Each class contains exactly one Dyck word. Since there are in total $\binom{2n}{n}$ sequences, there are $\frac{1}{n+1}\binom{2n}{n}$ Dyck words, and we obtain the result $C_n = \frac{1}{n+1}\binom{2n}{n}$.

For example, for $n = 2$, we have six sequences consisting of 2 symbols X and 2 symbols Y: $YYXX, YXYX, YXXY, XYYX, XYXY, XXYY$. They form two classes under the considered transform:

$$YYXX \to XYYX \to XXYY; \quad YXYX \to YXXY \to XYXY.$$

For $n = 3$, we have 20 sequences consisting of 3 symbols X and 3 symbols Y. They form five classes under the above transform:

$$YYYXXX \to XYYYXX \to XXYYYX \to XXXYYY;$$
$$YYXYXX \to XYYXYX \to YXXXYY \to XXYYXY;$$
$$YXYYXX \to YYXXXY \to XYXYYX \to XXYXYY;$$
$$YYXXYX \to YXXYYX \to XYYXXY \to XYXXYY;$$
$$YXYXYX \to YXXYXY \to YXXYXY \to XYXYXY.$$

Exercises

1. Calculate the first 10 Catalan numbers by constructing the corresponding number of rows of Pascal's triangle and transforming the numbers lying on its principal diagonal.
2. Calculate the first 10 Catalan numbers using the formula $C_{n+1} = \frac{(2n)!}{n!(n+1)!}$. Compare the complexity of the calculations with that of the above.

3. Calculate the first 10 Catalan numbers using the recurrent relation $C_n = \frac{4n-2}{n+1} C_{n-1}$, $C_0 = 1$. Compare the complexity of calculations when using each of the three algorithms reviewed.
4. Consider the function $f(n) = \frac{(2n)!}{n!(n+1)!}$. Check that $f(0) = 1$. Prove that $f(n) = \frac{4n-2}{n+1} f(n-1)$. Check that $f(n) = C_n$ for $n = 0, 1, 2, 3, 4, 5, 6, 7, 8, 9, 10$.
5. Prove that $C_n = \frac{1}{n+1}((C_n^0)^2 + (C_n^1)^2 + (C_n^2)^2 + \cdots + (C_n^n)^2)$.

3.4. Generating Function of the Sequence of Catalan Numbers

3.4.1. The easiest way to prove the equivalence of the two previously discussed classical definitions of the Catalan numbers is to use the *generating function* of the sequence of Catalan numbers.

- The generating function $f(x)$ of the sequence $1, 1, 2, 5, 14, \ldots$ of Catalan numbers has the form $f(x) = \frac{1-\sqrt{1-4x}}{2}$, i.e., it holds that

$$\frac{1-\sqrt{1-4x}}{2} = C_0 \cdot x + C_1 \cdot x^2 + \cdots + C_n \cdot x^{n+1} + \cdots, \quad |x| < \frac{1}{4}.$$

☐ Consider the formal sum

$$f(x) = C_0 \cdot x + C_1 \cdot x^2 + \cdots + C_n \cdot x^{n+1} + \cdots.$$

Then, using the definition of the Catalan numbers, we can obtain the following result:

$$\begin{aligned}
f(x) \cdot f(x) &= (C_1 x + C_2 \cdot x^2 + \cdots + C_n \cdot x^n + \cdots) \\
&\quad \times (C_1 \cdot x + C_2 \cdot x^2 + \cdots + C_n \cdot x^n + \cdots) \\
&= (C_1 C_1) x^2 + (C_1 C_2 + C_2 C_1) x^3 \\
&\quad + (C_1 C_3 + C_2 C_2 + C_3 C_1) x^4 + \cdots \\
&= C_2 x^2 + C_3 x^3 + C_4 x^4 + \cdots + C_n x^n + \cdots.
\end{aligned}$$

So, it holds that

$$\begin{aligned}
f^2(x) &= C_2 x^2 + C_3 x^3 + C_4 x^4 + \cdots + C_n x^n + \cdots \\
&= (C_1 x + C_2 \cdot x^2 + \cdots + C_n \cdot x^n + \cdots) - C_1 x = f(x) - x,
\end{aligned}$$

i.e., we obtain that
$$f^2(x) = f(x) - x.$$

Solving the obtained quadratic equation $f^2(x) - f(x) + x = 0$ relative to $f(x)$, we get that $f_1(x) = \frac{1+\sqrt{1-4x}}{2}$ and $f_2(x) = \frac{1-\sqrt{1-4x}}{2}$. In order to properly choose the sign of the square root, we should note that the free term of the decomposition $f(x) = C_0 \cdot x + C_1 \cdot x^2 + \cdots + C_n \cdot x^{n+1} + \cdots$ is zero, i.e., $f(0) = 0$. So, it is necessary use the "minus" sign. Therefore, we get the following result:

$$f(x) = \frac{1 - \sqrt{1-4x}}{2}.$$

In order to find the radius of convergence of the series $f(x) = C_0 \cdot x + C_1 \cdot x^2 + \cdots + C_n \cdot x^{n+1} + \cdots$, let us use the following well-known decomposition (see Chapter 1):

$$(1+z)^\alpha = 1 + \alpha z + \frac{\alpha(\alpha-1)}{2!} z^2 + \frac{\alpha(\alpha-1)(\alpha-2)}{3!} z^3 + \cdots$$
$$\cdots + \frac{\alpha(\alpha-1)(\alpha-2)\cdots(\alpha-n+1)}{n!} z^n + \cdots, \quad |z| < 1, \alpha \in \mathbb{R}.$$

In our case, $\alpha = \frac{1}{2}$ and $z = -4x$, which implies that the radius of convergence of the series $C_0 \cdot x + C_1 \cdot x^2 + \cdots + C_n \cdot x^{n+1} + \cdots$ is equal to $\frac{1}{4}$. □

3.4.2. Now, we can prove that the sequence of Catalan numbers can be obtained using the principal diagonal of Pascal's triangle. In fact, it holds the following property:

- The nth Catalan number C_n can be obtained by dividing the element $\binom{2n}{n}$ of the principal diagonal of Pascal's triangle by $\frac{1}{n+1}$:

$$C_n = \frac{1}{n+1} \binom{2n}{n}.$$

□ First of all, consider the central elements $1, 2, 6, 20, 70, \ldots$ of Pascal's triangle, i.e., the elements of the form $\binom{2n}{n}$, $n \geq 0$. It is easy to show that each of these elements is divisible by $n+1$. In fact, it holds that

$$\binom{2n}{n} \cdot \frac{1}{n+1} = \frac{(2n)!}{n!n!(n+1)} = \frac{(2n+1)!}{n!(n+1)!} \cdot \frac{1}{2n+1} = \frac{1}{2n+1}\binom{2n+1}{n},$$

or

$$\binom{2n+1}{n}(n+1) = \binom{2n}{n}(2n+1).$$

Since
$2n + 1 = (n + 1) \cdot 1 + n,$
$n + 1 = n \cdot 1 + 1,$ and
$n = 1 \cdot n + 0,$
$\gcd(2n + 1, n + 1) = 1$, and as $n + 1$ divides $\binom{2n}{n}(2n+1)$, it follows that $n+1$ divides $\binom{2n}{n}$.

As was proved above, the generating function for the sequence of Catalan numbers has the form $f(x) = \frac{1-\sqrt{1-4x}}{2}$. Let us now to go back to the decomposition

$$(1+z)^\alpha = 1 + \alpha z + \frac{\alpha(\alpha-1)}{2!}z^2 + \frac{\alpha(\alpha-1)(\alpha-2)}{3!}z^3 + \cdots$$

$$\cdots + \frac{\alpha(\alpha-1)(\alpha-2)\cdots(\alpha-n+1)}{n!}z^n + \cdots, \quad |z| < 1, \alpha \in \mathbb{R}.$$

For $\alpha = \frac{1}{2}$ and $z = -4x$, we get now the equality

$$\sqrt{1-4x}$$
$$= 1 - \frac{\frac{1}{2}}{1!} \cdot 4x + \frac{\frac{1}{2}(\frac{1}{2}-1)}{2!}(4x)^2 - \frac{\frac{1}{2}(\frac{1}{2}-1)(\frac{1}{2}-2)}{3!}(4x)^3$$
$$+ \cdots + (-1)^n \frac{\frac{1}{2}(\frac{1}{2}-1)\cdots(\frac{1}{2}-n+1)}{n!}(4x)^n + \cdots, \quad |x| < \frac{1}{4}.$$

Then, we have that

$$\sqrt{1-4x}$$
$$= 1 - \frac{2^{2\cdot 1}\frac{1}{2}}{1!} \cdot x + \frac{2^{2\cdot 2}\frac{1}{2}(\frac{1}{2}-1)}{2!}x^2 - \frac{2^{2\cdot 3}\frac{1}{2}(\frac{1}{2}-1)(\frac{1}{2}-2)}{3!}x^3$$
$$+ \cdots + (-1)^n \frac{2^{2\cdot n}\frac{1}{2}(\frac{1}{2}-1)\cdots(\frac{1}{2}-n+1)}{n!}x^n + \cdots, \quad |x| < \frac{1}{4}.$$

174 Catalan Numbers

So, the elementary transformations allow us to obtain that

$$\frac{1-\sqrt{1-4x}}{2}$$
$$= \frac{2^{2\cdot 1-1}\frac{1}{2}}{1!}\cdot x - \frac{2^{2\cdot 2-1}\frac{1}{2}(\frac{1}{2}-1)}{2!}x^2 + \frac{2^{2\cdot 3-1}\frac{1}{2}(\frac{1}{2}-1)(\frac{1}{2}-2)}{3!}x^3$$
$$- \cdots + (-1)^{n-1}\frac{2^{2n-1}\frac{1}{2}(\frac{1}{2}-1)\cdots(\frac{1}{2}-n+1)}{n!}x^n$$
$$+ \cdots = \sum_{n=1}^{\infty}(-1)^{n-1}\frac{2^{2n-1}\frac{1}{2}(\frac{1}{2}-1)\cdots(\frac{1}{2}-n+1)}{n!}x^n, \quad |x|<\frac{1}{4}.$$

As, on the one hand, it holds that

$$\frac{1-\sqrt{1-4x}}{2} = \sum_{n=1}^{\infty} C_{n-1}x^n$$

and, on the other hand, it holds that

$$\frac{1-\sqrt{1-4x}}{2} = \sum_{n=1}^{\infty}(-1)^{n-1}\frac{2^{2n-1}\frac{1}{2}(\frac{1}{2}-1)\cdots(\frac{1}{2}-n+1)}{n!}x^n,$$

we obtain that

$$C_{n-1} = (-1)^{n-1}\frac{2^{2n-1}\frac{1}{2}\left(\frac{1}{2}-1\right)\cdots\left(\frac{1}{2}-n+1\right)}{n!}.$$

It is easy to get that

$$(-1)^{n-1}2^n\frac{1}{2}\left(\frac{1}{2}-1\right)\cdots\left(\frac{1}{2}-n+1\right)$$
$$= (-1)^{n-1}\cdot 1\cdot(1-2)(1-4)\cdots(1-2n+2) = 1\cdot 3\cdot 5\cdot(2n-3)$$
$$= \frac{1\cdot 2\cdot 3\cdot(2n-4)(2n-3)(2n-2)}{2\cdot 4\cdot(2n-4)(2n-2)}$$
$$= \frac{1\cdot 2\cdot 3\cdot(2n-4)(2n-3)(2n-2)}{2^{n-1}\cdot 1\cdot 2\cdot(n-2)(n-1)}$$
$$= \frac{(2n-2)!}{2^{n-1}(n-1)!}.$$

Therefore, it holds that

$$(-1)^{n-1}\frac{2^{2n-1}\frac{1}{2}\left(\frac{1}{2}-1\right)\cdots\left(\frac{1}{2}-n+1\right)}{n!}$$

$$= \frac{(2n-2)!}{n!(n-1)!} = \frac{1}{n}\cdot\frac{(2n-2)!}{(n-1)!(n-1)!} = \frac{1}{n}\cdot\binom{2n-2}{n-1}.$$

So, we have proven that $C_{n-1} = \frac{1}{n}\cdot\binom{2n-2}{n-1}$ and, therefore, $C_n = \frac{1}{n+1}\cdot\binom{2n}{n}$. □

Exercises

1. Prove that the generating function $g(x) = \sum_{n=0}^{\infty} C_n x^n$ of the sequence of Catalan numbers has the form $\frac{1-\sqrt{1-4x}}{2x}$.
2. Prove that the generating function $g(x) = \sum_{n=0}^{\infty} C_n x^n$ of the sequence of Catalan numbers can be represented as $\frac{2}{1+\sqrt{1-4x}}$.
3. Prove that $C_0 + \frac{C_1}{5} + \frac{C_2}{25} + \cdots + \frac{C_n}{5^n} + \cdots = \frac{5-\sqrt{5}}{2}$.
4. Prove that $C_0 + \frac{C_1}{8} + \frac{C_2}{64} + \cdots + \frac{C_n}{8^n} + \cdots = 4 - 2\sqrt{2}$.
5. Let $x^{\underline{n}} = x(x-1)\cdots(x-n+1)$ be a *falling factorial*. Find the values of the expression $(\frac{1}{2})^{\underline{k}}$ for $k = 1, 2, \ldots, 10$. Find the value of the expression $2^k \cdot (\frac{1}{2})^{\underline{k}}$ for $k = 1, 2, \ldots, 10$. Make sure that $(-1)^{k-1} 2^k \cdot (\frac{1}{2})^{\underline{k}} = 1 \cdot 3 \cdot \cdots \cdot (2k-3)$.
6. Prove that the nth Catalan number can be obtained using the formula $C_n = (-1)^n \frac{2^{2n+1}(\frac{1}{2})^{\underline{n+1}}}{(n+1)!}$. Calculate the nth Catalan number using this formula if $n = 1, 2, \ldots, 10$.

3.5. Properties of Catalan Numbers

Simplest properties of Catalan numbers

3.5.1. There are many other formulas connecting Catalan numbers with binomial coefficients:

- *The nth Catalan number can be obtained by subtracting from the corresponding element $\binom{2n}{n}$ of the principal diagonal of Pascal's* triangle the preceding element $\binom{2n}{n-1}$ of the same row:

$$C_n = \binom{2n}{n} - \binom{2n}{n-1}.$$

☐ The proof can be obtained by direct verification:

$$\binom{2n}{n} - \binom{2n}{n-1} = \frac{(2n)!}{(n!)^2} - \frac{(2n)!}{(n-1)!(n+1)!}$$
$$= \frac{(2n)!}{(n!)^2}\left(1 - \frac{n}{n+1}\right) = \frac{1}{n+1}\binom{2n}{n} = C_n.$$

However, we can obtain a proof of this fact in terms of Dyck words. In fact, starting with a "non-Dyck" sequence consisting from n symbols X and n symbols Y, interchange all X's and Y's after the first Y that violates Dyck's condition. In the initial subsequence, ending with the first Y that violates Dyck's condition, there is exactly one more Y than there are X's. Therefore, the remaining part of the "non-Dyck" sequence has one more X than Y's. After the interchange, this part of the "non-Dyck" sequence will have one more Y than X's. Since there are still $2n$ symbols in the resulting sequence, there are now $n+1$ symbols Y and $n-1$ symbols X. Because the interchange process is reversible, we have a bijection between the set of "non-Dyck words" of length $2n$ and the set of sequences of length $2n$ consisting of $n-1$ symbols X and $n+1$ symbols Y. The number of "non-Dyck words" is therefore

$$\binom{2n}{n-1} = \binom{2n}{n+1}.$$

So, the number of Dyck words is obtained by removing the number $\binom{2n}{n-1} = \binom{2n}{n+1}$ of "non-Dyck words" from the total number $\binom{2n}{n}$ of sequences consisting of n symbols X and n symbols Y. As the number of Dyck words of length $2n$ is equal to C_n, we obtain the result

$$C_n = \binom{2n}{n} - \binom{2n}{n-1} = \binom{2n}{n} - \binom{2n}{n+1}.$$

☐

3.5.2. Another formula of this kind has the following form:
- *There exists an equality*
$$C_n = \frac{1}{2n+1}\binom{2n+1}{n}.$$

☐ The proof is similar to the previous one:
$$C_n = \frac{1}{n+1}\binom{2n}{n} = \frac{(2n)!}{n!n!(n+1)}$$
$$= \frac{1}{2n+1} \cdot \frac{(2n+1)!}{n!(n+1)!} = \frac{1}{2n+1}\binom{2n+1}{n}. \quad \square$$

Moreover:
- *there exists an equality*
$$C_n = \frac{1}{n}\binom{2n}{n-1}.$$

☐ The proof is also similar:
$$C_n = \frac{1}{n+1}\binom{2n}{n} = \frac{(2n)!}{n!n!(n+1)} = \frac{1}{n} \cdot \frac{(2n)!}{(n-1)!(n+1)!}$$
$$= \frac{1}{n} \cdot \binom{2n}{n-1} = \frac{1}{n} \cdot \binom{2n}{n+1}. \quad \square$$

3.5.3. The following property gives the already known recurrent relation for the elements of the sequence of Catalan numbers.
- *For Catalan numbers, the following recurrent relation holds*:
$$C_n = \frac{2(2n-1)}{n+1}C_{n-1}, \quad C_0 = 1.$$

☐ Indeed, simple transformations make it possible to check that
$$\frac{C_n}{C_{n-1}} = \frac{(2n)!}{(n+1)n!n!} \cdot \frac{n(n-1)!(n-1)!}{(2n-2)!} = \frac{n(2n-1)2n}{(n+1)n^2} = \frac{2(2n-1)}{n+1}. \quad \square$$

Based on this statement, we can get formulas for the calculation of Catalan numbers using *multifactorials*.

The first such formula was proposed by L. Euler.

- *There exists the formula*
$$C_n = \frac{2 \cdot 6 \cdot 10 \cdots (4n-2)}{1 \cdot 2 \cdot 3 \cdots (n+1)} = \frac{(4n-2)!!!!}{(n+1)!}, \quad n \geq 1,$$
where $n!!!! = n(n-4)(n-8) \cdots$ is the *quadruple factorial*.

☐ Let us prove this formula using induction by n and the previous property. For $n = 1$, it holds that $C_1 = 1 = \frac{2}{(1+1)!}$. Going from n to $n+1$, we get
$$C_{n+1} = \frac{2(2n+1)}{n+2} \cdot C_n = \frac{4n+2}{n+2} \cdot \frac{2 \cdot 6 \cdot 10 \cdot (4n-2)}{(n+1)!}$$
$$= \frac{2 \cdot 6 \cdot 10 \cdots (4n-2)(4n+2)}{(n+2)!}$$
$$= \frac{2 \cdot 6 \cdot 10 \cdots (4n-2)(4(n+1)-2)}{((n+1)+1)!} = \frac{(4(n+1)-2)!!!!}{((n+1)+1)!}.$$
☐

Simple transformations allow us to rewrite the previous formula to another form.

- *There exists the formula*
$$C_n = 2^n \cdot \frac{1 \cdot 3 \cdots (2n-1)}{1 \cdot 2 \cdots (n+1)} = 2^n \cdot \frac{(2n-1)!!}{(n+1)!}, \quad n \geq 1,$$
where $n!! = n(n-2)(n-4) \cdots$ is the *double factorial*.

☐ In fact,
$$C_n = \frac{2 \cdot 6 \cdot 10 \cdots (4n-2)}{(n+1)!} = \frac{2^n \cdot 1 \cdot 3 \cdots (2n-1)}{(n+1)!}$$
$$= \frac{2^n (2n-1)!!}{(n+1)!}.$$
☐

Considerations in the previous section allow us to calculate the nth Catalan number using the generalized binomial coefficients $\binom{n}{\alpha}$.

- *The following equality holds:*
$$C_n = (-1)^n \cdot 2^{2n+1} \cdot \binom{n+1}{\frac{1}{2}},$$
where $\binom{n}{\alpha} = \frac{\alpha(\alpha-1)(\alpha-2)\cdots(\alpha-n+1)}{1 \cdot 2 \cdots n}$, $n \in \mathbb{N}$, $\alpha \in \mathbb{R}$.

☐ A proof has the following form:

$$C_n = \frac{2^n \cdot 1 \cdot 3 \cdot \dots \cdot (2n-1)}{(n+1)!} = \frac{2^{2n+1} \cdot \frac{1}{2} \cdot \frac{1}{2} \cdot \frac{3}{2} \cdot \dots \cdot \frac{2n-1}{2}}{(n+1)!}$$

$$= \frac{2^{2n+1} \cdot (-1)^n \cdot \frac{1}{2} \cdot (\frac{1}{2} - 1) \cdot \dots \cdot (\frac{1}{2} - n)}{1 \cdot 2 \cdot \dots \cdot (n+1)}$$

$$= 2^{2n+1} \cdot (-1)^n \binom{n+1}{\frac{1}{2}}. \qquad \square$$

Using the definition of the *falling factorial*, we can rewrite the previous formula in the following form:

- The following equality holds:

$$C_n = \frac{(-1)^n \cdot 2^{2n+1} \cdot (\frac{1}{2})^{\underline{n+1}}}{(n+1)!},$$

where $\alpha^{\underline{n}} = \alpha(\alpha-1)(\alpha-2) \cdot \dots \cdot (\alpha-n+1)$, $n \in \mathbb{N}$, $\alpha \in \mathbb{R}$.

☐ In fact, this equality coincides with the previous one; we only use different notations. ☐

3.5.4. There exists a formula which allows us to calculate the nth Catalan number C_n using the previous elements of the sequence and binomial coefficients.

- The following equality holds:

$$C_n = \sum_{k \geq 0} \binom{n-1}{2k} C_k 2^{n-2k-1}.$$

☐ For example, $C_3 = 5 = 1 \cdot 1 \cdot 2^2 + 1 \cdot 1 \cdot 2^0 = \binom{2}{0}C_0 2^2 + \binom{2}{2}C_1 2^0$. The general result can be obtained using a combinatorial approach, in particular, using the rhyme scheme model for the calculation of Catalan numbers. ☐

Catalan's triangles

3.5.5. *Catalan's triangle* is a number triangle in which the mth element of the nth row, $n \geq 0$, $0 \leq m \leq n$, is defined by the formula

$C_{nm} = \frac{(n+m)!(n-m+1)}{m!(n+1)!}$. So,

$$C_{00} = \frac{0! \cdot 1}{0! \cdot 1!} = 1,$$

$$C_{10} = \frac{1! \cdot 2}{0! \cdot 2!} = 1, \quad C_{11} = \frac{2! \cdot 1}{1! \cdot 2!} = 1,$$

$$C_{20} = \frac{2! \cdot 3}{0! \cdot 3!} = 1, \quad C_{21} = \frac{3! \cdot 2}{1! \cdot 3!} = 2, \quad C_{22} = \frac{4! \cdot 1}{2! \cdot 3!} = 2,$$

and so on.

The first few rows of Catalan's triangle are presented as follows:

				1				
			1		1			
		1		2		2		
	1		3		5		5	
1		4		9		14		14
1	5		14		28		42		42
1	6	20		48		90		132		132
⋮	⋮	⋮	⋮	⋮	⋮	⋮

However, there is a much simpler way to build Catalan's triangle. To get it, note that:

- the left-hand side of Catalan's triangle consists of unities.

☐ In fact, it holds that

$$C_{n0} = \frac{n!(n+1)}{0!(n+1)!} = 1. \qquad \blacksquare$$

In addition:

- the right-hand side of Catalan's triangle consists of the Catalan numbers $C_0, C_1, C_2, \ldots$.

☐ In fact, we have

$$C_{nn} = \frac{(2n)! \cdot 1}{n!(n+1)!} = \frac{1}{n+1} \cdot \frac{(2n)!}{n!n!} = \frac{1}{n+1} \binom{2n}{n} = C_n. \qquad \blacksquare$$

Finally, it is easy to make sure that:

- *any internal element $C_{(n+1)m}$ of Catalan's triangle can be obtained as the sum of its left and top-right neighbors.*

☐ In fact,
$$C_{nm} + C_{(n+1)(m-1)}$$
$$= \frac{(n+m)!(n-m+1)}{m!(n+1)!} + \frac{(n+m)!(n-m+3)}{(m-1)!(n+2)!}$$
$$= \frac{(n+m)!}{(n+1)!(m-1)!} \left(\frac{n-m+1}{m} + \frac{n-m+3}{n+2} \right)$$
$$= \frac{(n+m)!}{(n+1)!(m-1)!} \cdot \frac{(n+m+1)(n+2) + m(n-m+3)}{m(n+2)}$$
$$= \frac{(n+m)(n^2 - nm + n + 2n - 2m + 2 + nm - m^2 + 3m)}{(n+1)!(m-1)!(n+2)m}$$
$$= \frac{(n+m)!(n+m+1)(n-m+2)}{m!(n+2)!}$$
$$= \frac{(n+1+m)!(n-m+2)}{m!(n+2)!} = C_{(n+1)m}. \qquad \square$$

Therefore, the construction of Catalan's triangle can be carried out recursively:

- the leftmost element of the nth triangle's row is equal to unity: $C_{n0} = 1$, $n \geq 0$;
- the rightmost element of the nth triangle's row is equal to the nth Catalan number: $C_{nn} = C_n$, $n \geq 0$;
- any internal element of the nth triangle's row can be obtained as the sum of its left and top-right neighbors: $C_{nm} = C_{n(m-1)} + C_{(n-1)m}$, $n \geq 0$.

In fact, we can simply start with the top unity $C_{00} = 1$ and construct the triangle using only the above rule of summation, considering any "empty" place as that containing zero: any element of the triangle can be obtained as the sum of its left and top-right neighbors. In this way, we obtain $C_{10} = 0 + 1 = 1$, $C_{11} = 1 + 0 = 1$, $C_{20} = 0 + 1$, $C_{21} = 1 + 1 = 2$, $C_{22} = 2 + 0 = 2$, $C_{30} = 0 + 1$, $C_{31} = 1 + 2 = 3$, $C_{32} = 3 + 2 = 5$, $C_{33} = 5 + 0 = 5$, etc.

3.5.6. This number triangle has a number of other interesting properties:

- *The two last elements of the nth row of Catalan's triangle coincide with the nth Catalan number:*

$$C_{n(n-1)} = C_{nn} = C_n, \quad n \geq 1.$$

□ We can check this fact by direct computation:

$$C_{n(n-1)} = \frac{(n+n-1)! \cdot 2}{(n-1)!(n+1)!} = \frac{(2n-1)! \cdot 2n}{(n-1)! \cdot n \cdot (n+1)!}$$

$$= \frac{(2n)!}{n!(n+1)!} = C_{nn} = C_n.$$

On the other hand, this property follows from the recurrent construction: $C_n = C_{nn} = C_{n(n-1)} + 0 = C_{n(n-1)}$. □

- *The sum of all elements of any row of Catalan's triangle is equal to the last element of the next row and, therefore, to the $(n+1)$th Catalan number C_{n+1}:*

$$C_{n0} + C_{n1} + \cdots + C_{nn} = C_{(n+1)(n+1)} = C_{n+1}.$$

□ In fact, it holds that

$$C_{n0} + C_{n1} + \cdots + C_{nn}$$

$$= \frac{n!(n+1)}{0!(n+1)!} + \frac{(n+1)!n}{1!(n+1)!} + \cdots + \frac{(2n)! \cdot 1}{n!(n+1)!}$$

$$= \frac{1}{n+1}\left(\binom{n}{0}(n+1) + \binom{n+1}{1} \cdot n + \cdots + \binom{n+n}{n} \cdot 1\right)$$

$$= \frac{1}{n+1}\binom{2n+2}{n+2} = \frac{(2n+2)!}{(n+1)n!(n+2)!} = \frac{(2n+2)!}{((n+1)!)^2(n+2)}$$

$$= \frac{1}{n+2}\binom{2(n+1)}{n+1} = C_{n+1}.$$

On the other hand, this fact can be easy obtained by induction using the recurrent nature of Catalan's triangle. In fact, the statement is true for $n = 0$: $C_{00} = 1 = C_{11} = C_1$. Similarly, it is true for $n = 1$:

$C_{10} + C_{11} = 1 + 1 = 2 = C_{22} = C_2$. Consider a positive integer $n \geq 2$. Suppose that the statement is true for $n-1$, i.e., it holds that

$$C_{(n-1)0} + C_{(n-1)1} + \cdots + C_{(n-1)(n-1)} = C_{nn} = C_n.$$

Since by the recurrent construction of the triangle we have that

$$C_{(n+1)(n+1)} = C_{(n+1)n} + 0 = C_{(n+1)n}, \quad C_{(n+1)n} = C_{(n+1)(n-1)} + C_{nn},$$
$$C_{(n+1)(n-1)} = C_{(n+1)(n-2)} + C_{n(n-1)}, \ldots, C_{(n+1)2} = C_{(n+1)1} + C_{n2},$$
$$C_{(n+1)1} = C_{(n+1)0} + C_{n1}, \quad C_{(n+1)0} = 0 + C_{n0} = C_{n0} = 1,$$

we obtain that

$$\begin{aligned} C_{(n+1)(n+1)} &= C_{(n+1)n} = C_{(n+1)(n-1)} + C_{nn} \\ &= C_{(n+1)(n-2)} + C_{n(n-1)} + C_{nn} = \cdots = C_{(n+1)1} \\ &+ C_{n2} + \cdots + C_{n(n-1)} + C_{nn} = C_{(n+1)0} \\ &+ C_{n1} + C_{n2} + \cdots + C_{n(n-1)} + C_{nn} \\ &= C_{n0} + C_{n1} + C_{n2} + \cdots + C_{n(n-1)} + C_{nn}. \quad \square \end{aligned}$$

It can be proven that:

- the number C_{nm}, i.e., the mth element of the nth row of Catalan's triangle, expresses the number of sequences consisting of n "1" and m "−1", all partial sums of which are non-negative.

□ For example, using three "1" and one "−1", we can construct exactly three such sequences:

$$1, 1, 1, -1 \text{ (partial sums } 1, 2 = 1 + 1, 3 = 1 + 1 + 1,$$
$$2 = 1 + 1 + 1 + (-1)),$$
$$1, 1, -1, 1 \text{ (partial sums } 1,$$
$$2 = 1 + 1, 1 = 1 + 1 + (-1), 2 = 1 + 1 + (-1) + 1),$$
$$1, -1, 1, 1 \text{ (partial sums } 1,$$
$$0 = 1 + (-1), 1 = 1 + (-1) + 1, 2 = 1 + (-1) + 1 + 1).$$

It corresponds to the number $C_{31} = 3$.

It is easy to see that there is exactly one sequence consisting of n "1" and zero "−1": it is $1, 1, \ldots, 1$. Obviously, all partial sums of such a sequence are non-negative. Assuming that there exists exactly one sequence consisting of zero "1" and zero "−1", we get that the statement is true for the elements C_{n0}, $n \geq 0$. Moreover, there exists exactly one such sequence consisting of one "1" and one "−1" (it is $1, -1$); there exist exactly two such sequences consisting of two "1" and one "−1" (they are $1, 1, -1$, and $1, -1, 1$); there exist exactly two such sequences consisting of two "1" and two "−1" (they are $1, 1, -1, -1$ and $1, -1, 1, -1$). So, the statement is true for $C_{11} = 1$, $C_{21} = 2$, and $C_{22} = 2$.

In general, consider a positive integer $n \geq 3$. Suppose that the statement is true for any sequence consisting of $n - 1$ "1" and m "−1", as well as for any sequence consisting of n "1" and $m - 1$ "−1", where $0 \leq m \leq n - 1$. Consider the set of all sequences consisting of n "1" and m "−1", all partial sums of which are non-negative. Any such sequence has as the last term or "1" or "−1". If the last term is "−1", its deletion gives us a sequence consisting of n "1" and $m - 1$ "−1", all partial sums of which are non-negative. The number of such sequences is $C_{n(m-1)}$. If the last term is "1", its deletion gives us a sequence consisting of $n - 1$ "1" and m "−1", all partial sums of which are non-negative. The number of such sequences is $C_{(n-1)m}$. Therefore, the number of sequences consisting of n "1" and m "−1", all partial sums of which are non-negative, is C_{nm}:

$$C_{(n-1)m} + C_{n(m-1)} = C_{nm}.$$

□

Note that we obtain an additional proof of the following fact:

- *The number of sequences consisting of n "1" and n "−1", all partial sums of which are non-negative, is equal to the nth Catalan number C_n.*

Moreover, we can state that:

- *the number of sequences consisting of n "1" and $n - 1$ "−1", all partial sums of which are non-negative, is equal to the nth Catalan number C_n.*

3.5.7. There are other number triangles associated with Catalan numbers. Let us build one of them using the design of Pascal's triangle: starting with extreme unities, we act according to Pascal's rule, "each next number is equal to the sum of the numbers, located above them on the right and on the left."

But instead of Pascal's isosceles triangle, we try to construct using this rule the rectangular triangle by drawing a vertical line to the left of which we cannot move. Filling the right-hand side of the triangle by unities, we get all the remaining elements of the construction as the sum of the numbers located above them on the right and left. Because for elements in the left vertical the upper-left neighbor is not defined, we consider them equal to the upper-right neighbor, treating the corresponding sum as degenerate:

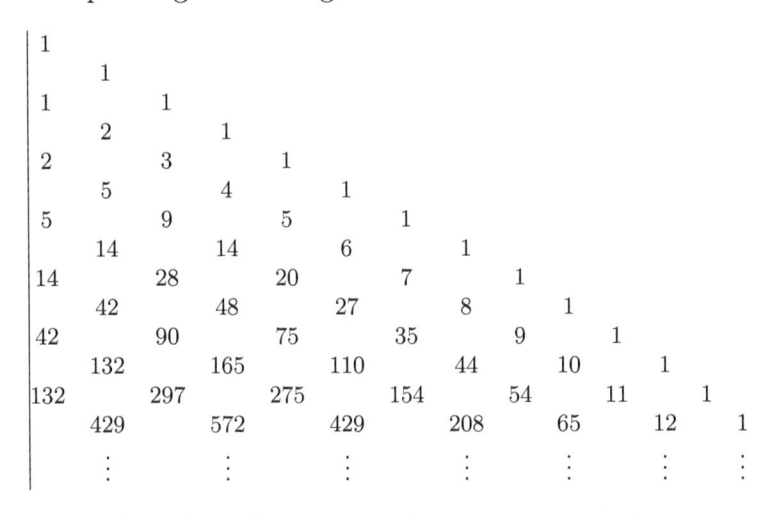

```
| 1
|       1
| 1            1
|       2            1
| 2            3            1
|       5            4            1
| 5            9            5            1
|      14           14            6            1
|14           28           20            7            1
|      42           48           27            8            1
|42           90           75           35            9            1
|     132          165          110           44           10            1
|132          297          275          154           54           11            1
|     429          572          429          208           65           12            1
|      :            :            :            :            :            :            :
```

Having carefully looked at the resulting structure, it is easy to note that:

- *the numbers appearing on the left vertical of the constructed triangle are Catalan numbers.*

☐ To determine the cause of this phenomenon, consider in the constructed triangle the *correct paths*, that is, the paths which begin at the vertex of the triangle, use only valid right-down and left-down movements, and end at the left vertical of the triangle. It follows from this construction that any number located on the left vertical is equal to the number of correct paths, coming to this point from the vertex. Note that any correct path, going to the nth point of the left vertical, $n \geq 0$, has exactly $2n$ steps (movements): n right-down steps and n left-down steps; moreover, each such path starts with the right-down step and ends at the left-down step.

It is easy to see that *the number of correct paths with $2n$ steps is equal to the number of root plane trees with n edges*: placing the root of the tree at the top of the triangle, consider a right-down movement as building a new edge and a left-down movement as returning back along the already constructed edges.

However, we can construct a bijection between the set of correct paths of length $2n$ and the set of Dyck words of length $2n$: just put instead of a right-down step the symbol X and instead of a left-down step the symbol Y.

Since the number of rooted plane trees with n edges (equivalently, the number of Dyck words of length $2n$) is equal to the nth Catalan number C_n, the number of correct paths with $2n$ edges is equal to C_n, which implies that each element on the left vertical of the constructed triangle is indeed a Catalan number.

Moreover, we can prove the statement by direct consideration. In fact, the statement is true for $n = 0, 1, 2, 3$. Consider a positive integer $n \geq 4$. Suppose that the statement is true for any k, $0 \leq k \leq n-1$. Consider any correct path with $2n$ steps. It always starts from the right-down step. Find the left-down step corresponding to the first right-down step, i.e., the first left-down step of the considered path, leading to a point on the left vertical. Fix both chosen steps. In this case, our path is divided into two parts. The first part contains two fixed steps. "Between" these two fixed steps, there are s pairs of steps, $0 \leq s \leq n-1$. Then, in the second part, located "out" of the two fixed steps, there are $n-s-1$ pairs of steps, $0 \leq n-s-1 \leq n-1$. By induction, the number of possible constructions for the first part is equal to C_s, while the number of possible constructions for the second part is equal to C_{n-s-1}. So, for a fixed s, we have $C_s C_{n-s-1}$ possibilities. In total, going through all possible values, $0, 1, \ldots, n-1$, of s, we obtain that the number of correct paths with $2n$ steps is equal to the nth Catalan number C_n:

$$C_0C_{n-1} + C_1C_{n-2} + \cdots + C_{n-2}C_1 + C_{n-1}C_0 = C_n. \qquad \square$$

Number-theoretic properties of Catalan numbers

3.5.8. It turns out that:

- for $n \geq 4$, the Catalan number C_n is composite, that is, the only Catalan's primes are the numbers $C_2 = 2$ and $C_2 = 5$.

$\square$ Indeed, from the formula

$$C_n = \frac{1}{n+1}\binom{2n}{n} = \frac{1}{n+1} \cdot \frac{(2n)!}{n!n!},$$

it follows that if a prime number p divides C_n, then $p \leq 2n$. On the other hand, it is easy to show that

$$C_n > 2n + 1, \quad \text{if } n \geq 4.$$

In fact, for $n = 4$, there is an inequality $C_4 = 14 > 2 \cdot 4 + 1 = 9$. Consider a positive integer $n \geq 5$, and suppose that the statement is true for $n - 1$, i.e., $C_{n-1} > 2(n-1) + 1 = 2n - 1$. Going from $n - 1$ to n, note that $\frac{2(2n-1)}{n+1} \geq 2$ for all $n \geq 2$ and $4n - 2 > 2n + 1$ for all $n \geq 4$. So, it follows that

$$C_n = \frac{2(2n-1)}{n+1}C_{n-1} \geq 2(2n-1) = 4n - 2 > 2n + 1, \quad \text{if } n \geq 5.$$

Thus, we have proven that, for $n \geq 4$, the nth Catalan number C_n cannot be a prime. As any such number is greater than unity, it should be composite. $\square$

3.5.9. Moreover, it is not difficult to show that in most cases, Catalan numbers are even.

- The Catalan number C_n is odd if and only if $n = 2^k - 1$, $k = 0, 1, 2, 3, \ldots$.

$\square$ For small values of n, it is obviously true: $C_0 = 1$, and $0 = 2^0 - 1$; $C_1 = 1$, and $1 = 2^1 - 1$; $C_2 = 2$ is even; $C_3 = 5$, and $3 = 2^2 - 1$;

$C_4 = 14$ is even. In general, the statement is true, as the sum $C_0C_{n-1} + C_1C_{n-2} + \cdots + C_{n-1}C_0$, giving the value of C_n, reads equally from the left to the right and from the right to the left; therefore, all the numbers C_{2n}, $n > 0$, are even:

$$C_{2n} = C_0C_{2n-1} + C_1C_{2n-2} + \cdots + C_{n-1}C_n$$
$$+ C_nC_{n-1} + \cdots + C_{2n-2}C_1 + C_{2n-1}C_0$$
$$= 2(C_0C_{2n-1} + C_1C_{2n-2} + \cdots + C_{n-1}C_n).$$

By the same reason, the number C_{2n+1} is even (odd) if and only if the number C_n is even (odd):

$$C_{2n+1} = C_0C_{2n} + C_1C_{2n-1} + \cdots + C_{n-1}C_{n+1} + C_nC_n$$
$$+ C_{n+1}C_{n-1} + \cdots + C_{2n-2}C_1 + C_{2n-1}C_0$$
$$= 2(C_0C_{2n-1} + C_1C_{2n-2} + \cdots + C_{n-1}C_n) + C_nC_n.$$

Since $C_0 = 1$ is odd, then the Catalan number C_1 is odd; since C_1 is odd, the Catalan number C_3 is odd; since C_3 is odd, the Catalan number C_7 is odd, etc. So, we have proven that the numbers $C_0, C_1, C_3, C_7, C_{15}, C_{31}, \ldots$ are odd. As for any $n = 2^k - 1$, $k = 0, 1, 2, \ldots$, it holds that $2n + 1 = 2^{k+1} - 1$, we have proven that all Catalan numbers C_n with index $n = 2^k - 1$, $k = 0, 1, 2, 3, \ldots$, are odd.

On the other hand, for a positive integer n, if the number $2n+1 \neq 2^k - 1$ for some $k \geq 1$, then $n \neq 2^{k-1} - 1$ for some $k \geq 1$. So, for a given positive integer n, let us suppose that for all $0 \leq s \leq n - 1$, the statement is true, i.e., the Catalan number C_s is odd if and only if $n = 2^t - 1$, $t = 0, 1, 2, 3, \ldots$. If the number $2n + 1 \neq 2^k - 1$ for some $k \geq 1$, then, as we noted above, the number $n \neq 2^t - 1$ for some $t \geq 0$. By induction, the number C_n is even, and hence, the number

$$C_{2n+1} = 2(C_0C_{2n-1} + C_1C_{2n-2} + \cdots + C_{n-1}C_n) + C_nC_n$$

is even. □

The first few odd Catalan numbers are therefore 1, 5, 429, 9694845, 14544636039226909, ... (sequence A038003 in the OEIS).

3.5.10. The representation of Catalan numbers by Euler's triangulations of convex polygons allow us to prove the following statement:

- For a given positive integer $n > 1$, if the number $n + 2$ is the natural power of a prime number, $n + 2 = p^k$, $p \in P$, $k \in \mathbb{N}$, then the Catalan number C_n is divisible by p:

$$C_n \equiv 0 (\mathrm{mod}\ p).$$

□ For example, $2 + 2 = 2^2$, and $C_2 = 2 \equiv 0 (mod\ 2)$; $5 + 2 = 7$, and $C_5 = 42 \equiv 0 (mod\ 7)$; $7 + 2 = 3^2$, and $C_7 = 429 \equiv 0 (mod\ 3)$.

The proof is based on certain algebraic considerations. The group $\mathbb{Z}_{n+2}$ of residue classes modulo $n+2$ acts on a set of triangulations of a regular $(n+2)$-gon by rotations. If $n > 1$, this action does not have fixed points. Therefore, for $n + 2 = p^k$, $p \in P$, $k \in \mathbb{N}$, the length of each of its orbits is divisible by p; therefore, the number C_n of Euler's triangulations of a regular $(n + 2)$-gon is divisible by p. □

3.5.11. The representation of Catalan numbers by regular bracket structures gives another number-theoretic property of these numbers.

- If the number n is the natural power of a prime, $n = p^k$, $p \in P$, $k \in \mathbb{N}$, then $C_n \equiv 2 (mod\ p)$.

□ For example, since $2 = 2^1$, $C_2 = 2 \equiv 2 \equiv 0 (mod\ 2)$; since $3 = 3^1$, $C_3 = 5 \equiv 2 (mod\ 3)$; since $4 = 2^2$, $C_4 = 14 \equiv 2 \equiv 0 (mod\ 2)$; since $5 = 5^1$, $C_5 = 42 \equiv 2 (mod\ 5)$, etc.

The proof is based on the fact that the group $\mathbb{Z}_{2n}$ of residue classes modulo $2n$ acts on the set of correct arrangements of n pairs of brackets according to the following rule: the generator of this group is represented by a cyclic shift by unity. This shift is as follows:

- the first left (opening) bracket is erased;
- instead of it, we add the last right (closing) bracket to the structure;
- the closing bracket, paired with the first left (opening) bracket erased, is replaced with the opening bracket;
- all other brackets of the considered correct arrangement do not change.

For $n = 2$, we have two possible arrangements of two pairs of brackets: ()() and (()). They are transformed from one to another, i.e., they form one orbit:

$$()() \to (()) \to ()().$$

For $n = 3$, we have five possible arrangements of three pairs of brackets: ()()(), ()(()), (())(), (()()), and ((())). Under the transformation, they form two orbits:

$$()()() \to (()()) \to ()()(); \quad ()(()) \to ((())) \to (())() \to ()(()).$$

For $n = 4$, we have 14 possible arrangements of three pairs of brackets: ()()()(), ()()(()), ()(())(), ()(()()), ()((())), (())()(), (())(()), (()())(), ((()))(), (()()()), (()(())), ((())()), ((()())), and ((((())))).

Under the transformation, they form three orbits:

$$()()()() \to (()()()) \to ()()()();$$

$$()()(()) \to (()(())) \to ()(())() \to ((())()) \to (())()() \to$$

$$\to ()((())) \to (((()))) \to (()())() \to ()()(());$$

$$()((())) \to ((())) \to ((())) \to (()(())) \to ()(()) \to ()((())).$$

If $n > 1$, this action does not have fixed points. Exactly one orbit of this action has a length of 2. It consists of the parenthesis structures $()() \cdots ()$ and $(() \cdots ())$.

The lengths of the remaining orbits are divisible by p if $n = p^k$. It means that p divides the number $C_n - 2$. □

- For a given odd prime p, the kth Catalan number C_k is divisible by p for any $\frac{p-1}{2} < k < p - 1$:

$$C_k \equiv 0 \pmod{p}, \quad \text{where } p \in P, \quad p \geq 3, \quad \frac{p-1}{2} < k < p - 1.$$

☐ By the divisibility property of binomial coefficients, it holds that $C_k = \frac{(2k)!}{k!(k+1)!} \equiv 0 \pmod{p}$. For example, if $p = 5$ and $k = 3$, then $C_3 = 5 \equiv 0 \pmod 5$; if $p = 11$ and $k = 6, 7, 8, 9$, then $C_k \equiv 0 \pmod{11}$: the numbers 132, 429, 1430, and 4862 are divisible by 11. □

Asymptotic behavior of Catalan numbers

3.5.12. Consider the behavior of the sequence of Catalan numbers on infinity.

- For large positive integers n, it holds that

$$C_n \sim \frac{4^n}{\sqrt{\pi}n^{3/2}}, \quad i.e., \quad \lim_{n\to\infty} \frac{C_n}{\frac{4^n}{\sqrt{\pi}n^{3/2}}} = 1.$$

□ In order to determine the principal term of the asymptotic of the nth Catalan number C_n, let us use the Stirling formula (see Chapter 1):

$$n! \sim \sqrt{2\pi n}\left(\frac{n}{e}\right)^n.$$

Now, it is easy to check that

$$C_n = \frac{(2n)!}{(n!)^2(n+1)} \sim \frac{\sqrt{4\pi n}\left(\frac{2n}{e}\right)^{2n}}{2\pi n \left(\frac{n}{e}\right)^{2n}(n+1)}$$

$$= \frac{4^n}{\sqrt{\pi n}(n+1)} = \frac{4^n}{\sqrt{\pi}(n^{3/2}+n)}.$$

As $n^{3/2} + n \sim n^{3/2}$, $C_n \sim \frac{4^n}{\sqrt{\pi}n^{3/2}}$. □

A more accurate asymptotic analysis shows that Catalan numbers are approximated by the fourth-order approximation

$$C_n \sim \frac{4^{n-5}(8n(16n(8n-9)+145)-1155)}{\sqrt{\pi}n^{9/2}}.$$

The numbers of decimal digits in C_{10^n} for $n = 0, 1, 2, \ldots$ are $1, 5, 57, 598, 6015, 60199, 602051, 6020590, 60205987, 602059978, \ldots$ (sequence A114466 in the OEIS). The digits converge to the digits in the decimal expansion of $\log_{10} 4 = 0.602059991\cdots$ (sequence A114493 in the OEIS).

Exercises

1. Calculate the first 10 Catalan numbers, subtracting from the corresponding element of the principal diagonal of Pascal's triangle the previous element of the same triangle's row: $C_n = \binom{2n}{n} - \binom{2n}{n-1}$.
2. Calculate the first 10 Catalan numbers using the formula $C_n = \frac{2 \cdot 6 \cdot 10 \cdots (4n-2)}{1 \cdot 2 \cdot 3 \cdots (n+1)}$, obtained by L. Euler.
3. Calculate the first 10 Catalan numbers using the formula $C_n = 2^n \cdot \frac{1 \cdot 3 \cdots (2n-1)}{1 \cdot 2 \cdots (n+1)}$. Compare the complexity of the three algorithms used.
4. Prove that $C_n = \prod_{k=2}^{n} \frac{n+k}{k}$, $n \geq 2$. Calculate the first 10 Catalan numbers using this formula.
5. Build the first 10 rows of Catalan's triangle using the closed formula for the elements C_{nm} and using the recurrent algorithm for the triangle's construction. Compare the complexity of these algorithms.
6. Build the first 20 rows of the "second" Catalan's triangle using Pascal's "truncated" triangle.
7. Among the first twenty Catalan numbers, find all the numbers divisible by 2; divisible by 3; divisible by 5; divisible by 7; divisible by 11. Check that C_n is divisible by p, if $n = p^k - 2 > 1$, $p \in P$, $k \in \mathbb{N}$. Consider the opposite statement: "if C_n is divisible by a prime p, then $n = p^k - 2$." Is this statement true? Give several examples.
8. Among the first twenty Catalan numbers, find all the numbers that give the remainder 2 when divided by 3; when divided by 5; when divided by 7; when divided by 11. Check that $C_n \equiv 2 (mod\ p)$, if $n = p^k$, $p \in P$, $k \in \mathbb{N}$. Consider the opposite statement: "if $C_n \equiv 2 (mod\ p)$, then $n = p^k$, $p \in P$, $k \in \mathbb{N}$." Is this statement true? Give several examples.
9. Using the information on the behavior of Catalan numbers C_n for large n, prove that C_n as a function of n increases with the growth of n faster than any exponential function.

3.6. Polynomials Related to Catalan Numbers

Catalan x-polynomials

3.6.1. Consider the sequence of *Catalan x-polynomials* $C_n(x)$, $n = 0, 1, 2, \ldots$, defined by the recurrent relation

$$C_{n+1}(x) = \sum_{k=0}^{n} x^k C_k(x) C_{n-k}(x), \quad C_0(x) \equiv 1.$$

It is easy to check that

$$C_1(x) \equiv 1, \quad C_2(x) = 1 + x, \quad C_3(x) = 1 + 2x + x^2 + x^3,$$
$$C_4(x) = 1 + 3x + 3x^2 + 3x^3 + 2x^4 + x^5 + x^6.$$

In general, $C_n(x)$ is a polynomial with positive integer coefficients. It follows from the definition that:

- the value of the polynomial $C_n(x)$ at $x = 1$ is the nth Catalan number C_n:

$$C_n(1) = C_n.$$

□ In fact, $C_0(1) = 1$, i.e., $C_0(1) = C_0$. Then, $C_1(1) = C_0(1)C_0(1) = C_0 \cdot C_0 = C_1$, $C_2(1) = C_0(1)C_1(1) + C_1(1)C_0(1) = C_0 C_1 + C_1 C_0 = C_2$, etc. In general,

$$C_n(1) = C_0(1)C_{n-1}(1) + C_1(1)C_{n-2}(1)$$
$$+ \cdots C_{n-2}(1)C_1(1) + C_{n-1}(1)C_0(1)$$
$$= C_0 C_{n-1} + C_1 C_{n-2} + \cdots C_{n-2} C_1 + C_{n-1} C_0 = C_n. \quad \square$$

3.6.2. The Catalan x-polynomials represent a special case of the *Catalan (x, y)-polynomials* $C_n(x, y)$. The recurrent definition of these polynomials requires the introduction of some additional notations, which is beyond the scope of this book. However, for small values of n, these polynomials are

$$C_0(x, y) \equiv 1, \quad C_1(x, y) \equiv 1, \quad C_2(x, y) = x + y,$$
$$C_3(x, y) = xy + y^3 + xy^2 + x^2 y + x^3,$$

$$C_4(x,y) = xy^3 + x^2y^2 + x^3y + xy^4 + x^2y^3 + x^3y^2$$
$$+ x^4y + y^6 + xy^5 + x^2y^4 + x^3y^3 + x^4y^2 + x^5y + x^6,$$

which makes it possible to illustrate the general relations:
$$C_n(x,1) = C_n(x), \quad C_n(1,1) = C_n.$$

Polynomials characterizing a graph's embeddings

3.6.3. Plane trees, enumerated by Catalan numbers, are an example of embedding a graph on a plane (equivalently, on a sphere), that is, its image on the plane without self-intersecting of edges. Not every graph allows for a plane embedding: so, it is well known that the *complete graph* on five vertices K_5 and the *complete bipartite graph* $K_{3,3}$ cannot be embedded on a plane ([Ore80]). However, these two graphs can be placed without self-intersections of the edges on a torus.

This leads to the question of embedding of graphs on an arbitrary two-dimensional surface ([Land02]). Without going into the formal definition of two-dimensional surfaces, we assume that any *closed, oriented two-dimensional surface* is a sphere to which a finite number of handles are glued. A two-dimensional sphere with g glued handles is called a *surface of genus g*: the surface of genus 0 is simply a sphere, whereas a surface of genus 1 is a torus.

An *embedding* of a connected graph on the surface M is such an image of the graph on the surface that:

- each vertex of the graph is mapped onto a point on the surface M, with different vertices mapped onto different points;
- each edge of the graph is matched by a non-self-intersecting segment of a continuous curve on the surface M, with the ends of this segment matching the vertices connected by this edge; the lines corresponding to different edges do not intersect;
- the supplement to the graph's image on M is a disjoint union of two-dimensional regions homeomorphic to a circle (and called the *faces* of the embedding).

The first two requirements in the definition are the same as those in the definition of a graph's embedding on a plane (or on a sphere). The third requirement is new; it states that there are no supplements to the graph's image on M that have handles; in other words, the image of the graph on the surface M should cut all of its handles.

If a graph with V vertices and E edges is embedded on a surface M of genus g and F is the number of resulting faces, then the numbers V, E, F, and g are related by the famous *Euler's formula*:

$$V - E + F = 2 - 2g.$$

The value $2 - 2g$ is called the *Euler characteristic* of the surface M.

3.6.4. Closed, orientable surfaces can be made from polygons by gluing their sides together in pairs. For example, gluing the opposite sides of a square, we get the torus, while gluing the neighboring sides of a square gives a sphere. Consider a regular $2n$-gon and break its sides into pairs in all possible ways. For each such division into pairs, glue the sides belonging to one pair (maintaining the orientability of the surface). The result is a closed, orientable surface. The question we are interested in is: how many ways of gluing give us a surface of genus g, $g = 0, 1, 2, \ldots$.

Let us start with examples. For $n = 2$, we study the gluing of a square. The sides of the square can be divided into pairs in three ways. If we denote the sides of the one pair by the letter a and the sides of the other pair by the letter b, then, starting with a fixed side of the square and moving clockwise around the square, we obtain the sequences *aabb*, *abba*, and *abab*. The first two partitions give a sphere; the third partition gives a torus. For $n = 3$, there are 15 ways to pair the sides of the hexagon. Using a similar model with the letters a, b, and c, we get the sequences *aabbcc*, *accbba*, *accabb*, *ccabba*, *cabbac*, *bacbac*, *abcacb*, *bcacba*, *cacbab*, *ababcc*, *babcca*, *abccab*, *bccaba*, *ccabab*, and *cababc*. It is not difficult to see that the first five partitions give a sphere after gluing, and the last 10 partitions give a torus.

It turns out that it is not difficult to determine the genus of a surface by studying the corresponding gluing. The image of the boundary of a polygon gives an embedded graph on the surface. In such a model, there is exactly one face: the interior of the polygon. We also know the number of edges: it is equal to half the number of sides of the $2n$-gon, that is, n. Now, we can determine the number of vertices of this graph. To do this, we can, say, mark with the same numbers the vertices of the polygon, which go to one vertex of the resulting graph. The number of different markings will be equal to the number of vertices of the resulting graph. So, for example, if we realize the gluing of the opposite sides of the square, we get the marking $1, 1, 1, 1$, that is, the resulting graph will contain exactly one vertex, and Euler's formula will take the form

$$2 - 2g = V - E + F = 1 - 2 + 1 = 0,$$

that is, $g = 1$, and the surface, as was previously obtained directly, is a torus. If we realize the gluing of the neighboring sides of the square, we get the marking $1, 2, 3, 2$, that is, the resulting graph will contain exactly three vertices, and Euler's formula will take the form

$$2 - 2g = V - E + F = 3 - 2 + 1 = 2,$$

that is, $g = 0$, and the surface is a sphere. For a regular hexagon, the resulting graph on a sphere has four vertices, while the resulting graph on a torus has two vertices.

Thus, the genus of the resulting surface is determined by the number of vertices in a graph, glued from the sides of a polygon. Our problem is to list all the possibilities of gluing for a regular $2n$-gon, leading to a surface of genus g.

Match with each number n a polynomial $Q_n(x)$; it is the generating polynomial for the number of ways of gluing the sides of a regular $2n$-gon. The coefficient of x^k in the polynomial $Q_n(x)$ is equal to the number of ways of gluing the sides of a regular $2n$-gon for which the glued graph has exactly k vertices. The genus g of the surface obtained during gluing is completely determined from these data.

Let us write out the first few polynomials $Q_n(x)$. We assume, for convenience, that

$$Q_0(x) = x.$$

Next, it is easy to check that

$$Q_1(x) = x^2.$$

In fact, the only possible gluing of a "regular bigon" gives a graph with two vertices and one edge on the sphere. The next two polynomials were already calculated:

$$Q_2(x) = 2x^3 + x, \quad Q_3(x) = 5x^4 + 10x^2.$$

Writing out all the possible glues of regular octagons and decagons, two more polynomials can be found:

$$Q_4(x) = 14x^5 + 70x^3 + 21x, \quad Q_5(x) = 42x^6 + 420x^4 + 483x^2.$$

However, these calculations are already becoming extremely labor-intensive.

It is not difficult to see that:

- the sum of all the coefficients of the polynomial $Q_n(x)$, that is, its value at $x = 1$, is $(2n-1)!! = 1 \cdot 3 \cdot 5 \cdot \ldots \cdot (2n-1)$.

□ In fact, the sum of all the coefficients of the polynomial $Q_n(x)$ is equal to the number of all the side partitions of a regular $2n$-gon into pairs. To count them, let us fix a side of the polygon. Its side's pair can be chosen in $2n-1$ different ways. Now, let us fix any of the remaining $2n-2$ sides; then, the paired side can be chosen in $2n-3$ different ways, etc.

However, there exists a general formula for the number of partitions of mn different elements into n groups of cardinality m. This number is equal to $\frac{P_{(m,m,m,\ldots,m)}}{n!}$, where $P_{(n_1,n_2,\ldots,n_k)} = \frac{(n_1+n_2+\cdots+n_k)!}{n_1!n_2!\cdots n_k!}$. Thus, for the partitions of $2n$ different elements into n pairs, we have $\frac{P_{(2,2,2,\ldots,2)}}{n!}$ possibilities. However,

$$\begin{aligned}\frac{P_{(2,2,2,\ldots,2)}}{n!} &= \frac{(2n)!}{2!2!\cdots 2!n!} \\ &= \frac{2n(2n-1)(2n-2)(2n-3)(2n-4)\cdots\cdot 2}{2^n \cdot n!} \\ &= (2n-1)(2n-3)\cdots\cdot 3 \cdot 1,\end{aligned}$$

and the statement is proven. □

Let us pay attention to the coefficient of the biggest degree of the polynomial $Q_n(x)$, $n = 0, 1, 2, \ldots$. The sequence of these coefficients begins with the elements $1, 1, 2, 5, 14, \ldots$, and we have reason to assume that it is the sequence of Catalan numbers. Indeed:

- *the degree of the polynomial $Q_n(x)$ is $n+1$, and the coefficient of x^{n+1} in the polynomial $Q_n(x)$ is equal to the nth Catalan number C_n.*

□ The proof of this fact is not difficult. By Euler's formula, the number of vertices in the graph on the glued surface of genus g is determined from the equality

$$2 - 2g = V - E + F = V - n + 1.$$

Therefore, $V = n + 1 - 2g \leq n + 1$, as the genus g is non-negative. The last inequality turns into an equality if and only if $g = 0$, i.e., if and only if the glued surface is a sphere. We call two pairs of sides of the polygon *interleaved* if between two sides from the first pair there is a side from the second pair, whichever direction we go around the boundary of the polygon. In other words, two pairs of sides are interleaved if the segments connecting the midpoints of each pair of sides intersect. If in a partition of the sides into pairs there are interleaved pairs, then their gluing gives a handle, and the resulting surface cannot be a sphere.

On the other hand, if there are no interleaved pairs in a partition of sides into pairs, then the resulting surface is a sphere. Indeed, in such a partition, we can select a pair of adjacent sides glued to each other. Their gluing gives a polygon with fewer number of sides, which are broken into non-interleaved pairs, and we can use induction.

Furthermore, the set of partitions of the sides of a regular $2n$-gon into pairs that do not have interleaved pairs is in one-to-one correspondence with the set of regular bracket structures with n pairs of brackets. Indeed, we fix any initial vertex of the polygon and move from it along the polygon boundary clockwise. Each side of the polygon to be traversed matches the left (opening) or the right (closing) bracket by the following rule: if the passing side is the first side of a new pair, then we compare it to the left bracket; otherwise, we compare it to the right bracket. As a result, we obtain a bracket structure which is correct. Moreover, any correct bracket structure can be obtained exactly once. Since the number of correct bracket structures for n pairs of brackets is C_n, then the statement is completely proven.

On the other hand, we can use for a proof the well-known handshake problem. In fact, consider the set of midpoints of the sides of a regular $2n$-gon. Then, we have $2n$ points lying on a circle. In order to obtain a partition of the sides of this $2n$-gon into non-interleaved pairs, we should just split the obtained $2n$ points into n pairs such that the segments connecting the elements of each pair are non-crossing. It was proven earlier that the number of such splits is equal to the nth Catalan number C_n. □

Exercises

1. Find the polynomials $C_n(x)$ for $n = 5, 6, 7, 8, 9, 10$. Check that $C_n(1) = C_n$. Prove this statement by induction.
2. What are the degrees of the polynomial $C_n(x)$, $n = 0, 1, 2, \ldots, 10$? Which connection exists between the index and the degree of the polynomial $C_n(x)$?
3. Build polynomials $Q_n(x)$ if $n = 6, 7, 8$.
4. Check that $Q_n(1) = (2n-1)!!$ if $n = 1, 2, \ldots, 8$.
5. Find $Q_n(1)$ if $n = 9, 10$.

3.7. Catalan Numbers in the Family of Special Numbers

Catalan numbers and elements of Pascal's triangle

3.7.1. As was previously shown, Catalan numbers are closely related to the elements of Pascal's triangle, i.e., to binomial coefficients. In fact, we have proven at least three formulas that allow us to calculate the nth Catalan number using the elements of Pascal's triangle:

$$C_n = \frac{1}{n+1} \cdot \binom{2n}{n} = \binom{2n}{n} - \binom{2n}{n-1} = \frac{1}{2n+1}\binom{2n+1}{n}.$$

Moreover, we considered a construction that gives the sequence of Catalan numbers as the elements of the left vertical of a number triangle, closely related to Pascal's triangle (see "Catalan's triangles" in Section 3.5).

Finally, we recall that Catalan numbers are closely connected with *generalized binomial coefficients*: $\binom{n}{\alpha} = \frac{\alpha(\alpha-1)(\alpha-2)\cdots(\alpha-n+1)}{1\cdot 2\cdots n}$, $n \in \mathbb{N}$, $\alpha \in \mathbb{R}$. In fact, it holds that

$$C_n = (-1)^n \cdot 2^{2n+1} \cdot \binom{n+1}{\frac{1}{2}}.$$

Catalan numbers and factorial numbers

3.7.2. Any binomial coefficient $\binom{n}{m}$ can be calculated from the classical formula using factorials:

$$\binom{n}{m} = \frac{n!}{m!(n-m)!}.$$

Therefore, Catalan numbers are closely related to the *factorial numbers*.

Recall that the *factorial* of a positive integer n, denoted by $n!$, is the product of all positive integers less than or equal to n:

$$n! = 1 \cdot 2 \cdots (n-1) \cdot n, \quad n \in \mathbb{N}.$$

The value of $0!$ is 1, according to the convention for an empty product.

The sequence of factorials starts with the elements 1, 1, 2, 6, 24, 120, 720, 5040, 40320, 362880, ... (sequence A000142 in the OEIS).

As $C_n = \frac{1}{n+1} \cdot \binom{2n}{n}$, we can state that:

- any Catalan number can be calculated using factorials:

$$C_n = \frac{(2n)!}{(n+1)!n!}, \quad n \geq 0.$$

In particular, $C_0 = \frac{0!}{1!0!} = 1$; $C_1 = \frac{2!}{2!1!} = 1$; $C_2 = \frac{4!}{3!2!} = 2$, $C_3 = 5$.

3.7.3. Moreover, we have at least two formulas connecting Catalan numbers with *multifactorials*.

First, define the *double factorial* (or *semifactorial*) $n!!$ of a positive number n as the product of all the integers from 1 up to n that have the same parity (odd or even) as n (see [Call09]). That is,

$$n!! = \prod_{k=0}^{\lceil \frac{n}{2} \rceil - 1} (n - 2k) = n(n-2)(n-4) \cdots.$$

For even n, the double factorial is $n!! = \prod_{k=1}^{\frac{n}{2}} 2k = n(n-2)(n-4) \cdots 4 \cdot 2$, and for odd n, it is $n!! = \prod_{k=1}^{\frac{n+1}{2}} (2k-1) = n(n-2)(n-4) \cdots 3 \cdot 1$. The double factorial $0!! = 1$ is an empty product.

The sequence of double factorials for even $n = 0, 2, 4, 6, 8, \ldots$ starts as $1, 2, 8, 48, 384, 3840, 46080, 645120, \ldots$ (sequence A000165 in the OEIS). The sequence of double factorials for odd $n = 1, 3, 5, 7, 9, \ldots$ starts as $1, 3, 15, 105, 945, 10395, 135135, \ldots$ (sequence A001147 in the OEIS).

As for any positive integer n, it holds that $C_n = 2^n \cdot \frac{1 \cdot 3 \cdots (2n-1)}{1 \cdot 2 \cdots (n+1)}$, we can state that

- *the nth Catalan numbers can be obtained using factorials and double factorials*:

$$C_n = 2^n \cdot \frac{(2n-1)!!}{(n+1)!}, \quad n \geq 1.$$

In particular, $C_1 = 2^1 \frac{1!!}{2!} = 1$; $C_2 = 2^2 \frac{3!!}{3!} = 2^2 \frac{1 \cdot 3}{1 \cdot 2 \cdot 3} = 2$, $C_3 = 2^3 \frac{5!!}{4!} = 2^3 \frac{1 \cdot 3 \cdot 5}{1 \cdot 2 \cdot 3 \cdot 4} = 5$.

Furthermore, for a given $a \in \mathbb{N}$, the *multifactorial* (more exactly, the α-*factorial* $n!_\alpha$ of a positive integer n is the product of all the integers from 1 up to n that have the same reminder as n after the integer division by α (see [Call09]). In other words, $n!_\alpha$ is the product of positive integers less than or equal to n, which belong to the residue class $\overline{n}_\alpha$ modulo α. As always, $0!_a = 1$ is an empty product.

For $\alpha = 1$, we obtain the classical factorial numbers: $n!_1 = n!$.

For $\alpha = 2$, we obtain the double factorial numbers: $n!_2 = n!!$.

For $\alpha = 3$, the *triple factorial numbers* $n!_3 = n!!!$ can be calculated using the following rule:

$$n!!! = \begin{cases} 3k(3k-3) \cdots 6 \cdot 3, & n = 3k, \\ (3k+1)(3k-2) \cdots 5 \cdot 2, & n = 3k+1, \\ (3k+2)(3k-1) \cdots 4 \cdot 1, & n = 3k+2. \end{cases}$$

The sequence of triple factorial numbers $(3k+1)!!!$, $k = 0, 1, 2, \ldots$, starts with the elements 1, 4, 28, 280, 3640, 58240, 1106560, 24344320, 608608000, 17041024000, ... (see the sequence A007559 in the OEIS). The sequence of triple factorial numbers $(3k+2)!!!$, $k = 0, 1, 2, \ldots$, starts with the elements 2, 10, 80, 880, 12320, 209440, 4188800, 96342400, 2504902400, 72642169600, ... (see the sequence A008544 in the OEIS). The sequence of triple factorial numbers $(3k)!!! = 3^n \cdot n!$, $k = 0, 1, 2, \ldots$, starts with the elements 1, 3, 18, 162, 1944, 29160, 524880, 11022480, 264539520, 7142567040, ... (sequence A008544 in the OEIS).

For $\alpha = 4$, the *quadruple* (or *quartic, 4-fold*) *factorial numbers* $n!_4 = n!!!!$ can be calculated using the following rule:

$$n!!! = \begin{cases} 4k(4k-4)\cdots 8 \cdot 4, & n = 4k, \\ (4k+1)(4k-3)\cdots\cdots 7 \cdot 3, & n = 4k+1, \\ (4k+2)(4k-2)\cdots\cdots 6 \cdot 2, & n = 4k+2, \\ (4k+3)(4k-1)\cdots\cdots 5 \cdot 1, & n = 4k+3. \end{cases}$$

The sequence of quadruple factorial numbers $(4k+1)!!!!$, $k = 0, 1, 2, \ldots$, starts with the elements 1, 5, 45, 585, 9945, 208845, 5221125, 151412625, 4996616625, 184874815125, ... (see the sequence A007696 in the OEIS). The sequence of quadruple factorial numbers $(4k+2)!!!! = \frac{(2n)!}{n!}$, $k = 0, 1, 2, \ldots$, starts with the elements 2, 12, 120, 1680, 30240, 665280, 17297280, 518918400, 17643225600, 670442572800, ... (see the sequence A001813 in the OEIS). The sequence of quadruple factorial numbers $(4k+3)!!!!$, $k = 0, 1, 2, \ldots$, starts with the elements 3, 21, 231, 3465, 65835, 1514205, 40883535, 1267389585, 44358635475, 1729986783525, ... (see the sequence A008545 in the OEIS). The sequence of quadruple factorial numbers $(4k)!!!! = 4^n \cdot n!$, $k = 0, 1, 2, \ldots$, starts with the elements 1, 4, 32, 384, 6144, 122880, 2949120, 82575360, 2642411520, 95126814720, ... (sequence A047053 in the OEIS).

We have proven earlier the following result derived by L. Euler: $C_n = \frac{2 \cdot 6 \cdot 10 \cdots (4n-2)}{1 \cdot 2 \cdot 3 \cdots (n+1)}$, $n \geq 1$. In other word, we can state that:

- *Catalan numbers can be calculated using factorials and quadruple factorials*:

$$C_n = \frac{(4n-2)!!!!}{(n+1)!}, \quad n \geq 1.$$

In particular, $C_1 = \frac{2!!}{2!} = 1$; $C_2 = \frac{6!!}{3!} = \frac{2 \cdot 6}{1 \cdot 2 \cdot 3} = 2$, $C_3 = \frac{10!!}{4!} = \frac{2 \cdot 6 \cdot 10}{1 \cdot 2 \cdot 3 \cdot 4} = 5$.

It is not difficult to detect natural connections between Catalan's sequence and other well-known special numbers in combinatorics: the Stirling numbers of the first and second kind, Bell numbers, Euler zigzag numbers, etc.

Catalan numbers and Mersenne numbers

3.7.4. A *Mersenne number* is a positive integer of the form

$$\mathcal{M}_n = 2^n - 1, \quad n \in \mathbb{N}.$$

The first few Mersenne numbers are 1, 3, 7, 15, 31, 63, 127, 255, 511, 1023, 2047, ... (sequence A000225 in OEIS).

For $2^n - 1$ to be prime, it is necessary that n itself be prime. By this reasoning, many authors require that the exponent n in the definition above be a prime.

Mersenne numbers $2^p - 1$ with prime p form the sequence 3, 7, 31, 127, 2047, 8191, 131071, 524287, 8388607, 536870911, ... (sequence A001348 in OEIS).

In fact, the exponent n which give Mersenne primes takes the values 2, 3, 5, 7, 13, 17, 19, 31, ... (sequence A000043 in the OEIS), and the resulting Mersenne primes are 3, 7, 31, 127, 8191, 131071, 524287, 2147483647, ... (sequence A000668 in the OEIS).

In order to find a connection between Mersenne and Catalan numbers, it is enough to remember the parity property of Catalan number. As only odd Catalan numbers are Catalan numbers with index $2^k - 1$, we can state the following elegant property:

- *The nth Catalan number C_n is odd if and only if its index n is a Mersenne number $2^k - 1$.*

Catalan numbers and Bell numbers

3.7.5. When solving combinatorial problems, considering the cases of small values of n, it is easy to confuse the sequence of Catalan numbers 1, 1, 2, 5, 14, 42, 132, 429, 1430, 4862, ... (sequence A000108 in the OEIS) with the sequence 1, 1, 2, 5, 15, 52, 203, 877, 4140, 21147, ... (sequence A000110 in the OEIS). In a note to his bibliography, which also includes a list of references to the above sequence, Henry W. Gould ([Goul85]) notes that if the objects being counted are sufficiently complex, then at $n = 4$, you can easily skip the 15th object and assume that you are dealing with Catalan's sequence.

The elements of the sequence 1, 1, 2, 5, 15, 52, 203, 877, 4140, 21147, ... are called *Bell numbers* and enumerate the *partitions* of the set of n elements, n = 0, 1, 2, 3, 4, 5, 6, ...

Therefore: for V_1, there is exactly one partition, $\{1\}$; for V_2, there are two partitions, $\{1, 2\}$ and $\{1\} \cup \{2\}$; for V_3, there are five partitions, $\{1, 2, 3\}$, $\{1\} \cup \{2, 3\}$, $\{1, 2\} \cup \{3\}$, $\{1, 3\} \cup \{2\}$, and $\{1\} \cup \{2\} \cup \{3\}$; etc. Let us also assume that the empty set has one partition. Then, denoting the nth Bell number by B_n, $n \geq 0$, we get that $B(0) = 1, B(1) = 1, B(2) = 2, B(3) = 5, B(4) = 15$, etc.

Bell numbers can be constructed in different ways. So, for a poet, the Bell number $B(n)$ is the number of all kinds of rhyme schemes of an n-line poem. For example, a four-line poem has exactly 15 possible rhyme schemes, one of which is the absence of any rhyme: *aaaa, aaab, aaba, aabb, aabc, abaa, abab, abac, abba, abbb, abba, abca, abcb, abcc, abcd*.

Such a transition from partitions to rhymes is easily possible if, instead of each element k of the set $V_n = \{1, 2, \ldots, n\}$, we write a symbol of the class S_i into which the element k falls in a given partition $V_n = S_1 \cup \cdots \cup S_q$: symbol a for S_1, symbol b for S_2, etc. Thus, the partition $V_4 = \{1, 2\} \cup \{3, 4\}$ corresponds to the rhyme *aabb*, and the partition $V_4 = \{1, 2\} \cup \{3\} \cup \{4\}$ corresponds to the rhyme *aabc*.

Connecting by the same arcs the rhyme's elements, we see that in some rhymes, the arcs intersect, and in some rhymes, they do not. Let us call rhymes in which arcs do not intersect *plane rhymes* (Joanna Groumi, 1970). It turns out that:

- *the number of plane rhymes of an n-line poem is equal to the nth Catalan number C_n.*

☐ To prove this fact, we use the correct bracket structures.

It was already shown that *the number of correct arrangements of n pairs of brackets is equal to the nth Catalan number C_n*.

Now, we can build a one-to-one correspondence between the set of all plane rhymes for an n-line poem and the set of all correct arrangements of n pairs of brackets. For a given plane rhyme, the correspondence is constructed according to the following law:

- put opening bracket before each new symbol of the rhyme;
- close the bracket (the required number of times) after the last symbol of the rhyme, i.e., in the case when the set of symbols of this type is completely exhausted.

So, for $n = 1$, the only rhyme a corresponds to the bracket arrangement (). For $n = 2$, the rhyme aa corresponds to the arrangement (()), and the rhyme ab corresponds to the arrangement ()(). For $n = 3$, the rhyme aaa corresponds to the arrangement ((())), the rhyme aab corresponds to the arrangement (())(), the rhyme aba corresponds to the arrangement (()()), the rhyme abb corresponds to the arrangement ()(()), and the rhyme abc corresponds to the arrangement ()()(). For $n = 4$, the rhyme $aaaa$ corresponds to the arrangement ((())), the rhyme $abba$ corresponds to the arrangement ((())()), etc.

At the same time, the unambiguous recovery of a rhyme according to the given arrangement is possible only if the plane rhymes are involved in the construction. So, the rhyme $abab$ gives the same arrangement of brackets, i.e., the arrangement ((())()) for the rhyme $abba$.

Thus, we proved that the nth Catalan number C_n is equal to the number of plane rhymes of an n-line poem.

However, we can prove this statement directly. In fact, for an n-line poem, there are at most n rhyme schemes; therefore, in order to represent them, we should use at least one symbol a_1 and at most n symbols $a_1, \ldots, a_n$. It is easy to see that the statement is true for $n = 0, 1, 2, 3, 4$. We have already shown that there is exactly 1 (plane) rhyme for a 0-line poem, there is exactly 1 (plane) rhyme for a 1-line poem, there are exactly 2 (plane) rhymes for a 2-line poem, there are exactly 5 (plane) rhymes for a 3-line poem, and there are exactly 14 plane rhymes for a 4-line poem. Consider a positive integer $n \geq 5$, and suppose that the statement is true for all non-negative k, $0 \leq k \leq n-1$. Consider the set of all rhyme schemes for an n-line poem. Any such scheme starts with the symbol a_1. If there is only the symbol a_1, fix it; if there are several symbols a_1, fix the last of these symbols. In any case, before this fixed symbol a_1, we have s places, $s = 0, 1, 2, \ldots, n-1$, and after this fixed symbol, we have $n-s-1$ places, $0 \leq n-s-1 \leq n-1$. In order to obtain a rhyme scheme for the n-line poem from this construction, we can use for the first group of places any rhyme scheme for an s-line poem (represented by the symbols $a_1, \ldots, a_s$), and for the second group of places, we can use any rhyme scheme for an $n-s-1$-line poem (represented by the symbols $a_2, \ldots, a_{n-1}$, not used in the first group). By induction, the number of possible constructions is $C_s C_{n-s-1}$. So, summing all these values for $s = 0, 1, 2, \ldots, n-1$, we obtain that the number of all the plane rhymes of an n-line poem is equal to the nth Catalan number C_n:

$$C_0 C_{n-1} + C_1 C_{n-2} + \cdots + C_{n-2} C_1 + C_{n-1} C_0 = C_n. \qquad \square$$

Catalan numbers and Euler zigzag numbers

3.7.6. Recall that an *alternating permutation* (or *zigzag permutation*) of the set $\{1, 2, 3, \ldots, n\}$ is an arrangement of those numbers so that each entry is alternately greater or lesser than the previous entry.

The numbers A_n of *up-down* permutations on $\{1, 2, \ldots, n\}$ are known as *Euler zigzag numbers* (or *up-down numbers*). The first few

values of A_n are 1, 1, 1, 2, 5, 16, 61, 272, 1385, 7936, 50521,...
(sequence A000111 in the OEIS).

The numbers Z_n count the permutations of $\{1,\ldots,n\}$ that are either up-down or down-up (or both, for $n < 2$). So, $Z_n = 2A_n$ for $n \geq 2$; the first few values of Z_n are 1, 1, 2, 4, 10, 32, 122, 544, 2770, 15872, 101042,... (sequence A001250 in the OEIS).

These numbers satisfy a simple recurrent relation, similar to that of Catalan numbers: by splitting the set of alternating permutations (both down-up and up-down) of the set $\{1, 2, \ldots, n, n+1\}$ according to the position k of the largest entry $n+1$, one can show that

$$Z_n = 2A_{n+1} = \sum_{k=0}^{n} \binom{n}{k} A_k A_{n-k}, \quad n \geq 1.$$

This recurrent relation allows us to obtain the exponential generating function $A(x) = \sum_{n=0}^{\infty} A_n \frac{x^n}{n!}$ for the sequence A_n, $n = 0, 1, 2, \ldots$. In fact, it holds that

$$A(x) = \tan\left(\frac{\pi}{4} + \frac{x}{2}\right) = \sec x + \tan x,$$

i.e., $A(x)$ is the sum of the secant and tangent functions. This result is known as *André's theorem* (see [Stan10]).

3.7.7. Since the secant function is even and tangent is odd, we obtain two additional classes of numbers.

The numbers A_{2n} with even indices are called *secant numbers* (or *zig numbers*); they are the numerators in the Maclaurin series of $\sec x$. The first few values are 1, 1, 5, 61, 1385, 50521, 2702765, 199360981, 19391512145, 2404879675441,... (sequence A000364 in the OEIS). Secant numbers are related to the *Euler numbers* E_n, which are defined by the Taylor series expansion

$$\frac{1}{\cosh t} = \frac{2}{e^t + e^{-t}} = \sum_{n=0}^{\infty} \frac{E_n}{n!} \cdot t^n.$$

The sequence E_n, $n = 0, 1, 2, \ldots$, starts with the numbers $1, 0, -1, 0, 5, 0, -61, 0, 1385, 0, \ldots$ (sequence A122045 in the OEIS), and it is easy to prove that Euler numbers with odd indices are equal to zero, while $E_{2n} = (-1)^n A_{2n}$. So, it holds that

$$A_{2n} = |E_{2n}|.$$

The numbers A_{2n+1} with odd indices are called *tangent numbers* (or *zag numbers*). The first few values are 1, 2, 16, 272, 7936, 353792, 22368256, 1903757312, 209865342976, 29088885112832, ... (sequence A000182 in the OEIS). The tangent numbers are closely related to *Bernoulli numbers*. The relation is given by the formula

$$B_{2n} = (-1)^{n-1} \frac{2n}{4^{2n} - 2^{2n}} A_{2n-1}, \quad n \geq 1.$$

Generalizations of Catalan numbers

3.7.8. In combinatorial mathematics, the *Lobb number* $L_{m,n}$, $n \geq m \geq 0$, counts the number of ways that $n+m$ opening brackets and $n-m$ closing brackets can be arranged to form the start of a valid sequence of balanced parentheses. They are named after Andrew Lobb, who used them to give a simple inductive proof of the formula for the n-th Catalan number.

Lobb numbers form a natural generalization of the Catalan numbers, which count the number of complete strings of balanced parentheses of a given length. Thus, the nth Catalan number is equal to the Lobb number $L_{0,n}$: $C_n = L_{0,n}$.

The (m,n)th Lobb number $L_{m,n}$ can be given in terms of binomial coefficients by the formula

$$L_{m,n} = \frac{2m+1}{m+n+1} \binom{2n}{m+n}, \quad n \geq m \geq 0.$$

An alternative expression for the Lobb number $L_{m,n}$ is

$$L_{m,n} = \binom{2n}{m+n} - \binom{2n}{m+n+1}.$$

The triangle of these numbers starts as

```
           1
         1   1
        2   3   1
       5   9   5   1
     14  28  20   7   1
    42  90  75  35   9   1
```

(sequence A039599 in the OEIS), where the right-hand side is formed from unities, $L_{n,n} = 1$, and the left-hand side is formed from the Catalan Numbers, $L_{0,n} = \frac{1}{1+n}\binom{2n}{n} = C_n$.

Apart from counting sequences of parentheses, the Lobb numbers also count the number of ways in which $n + m$ copies of the value 1 and $n - m$ copies of the value -1 may be arranged into a sequence such that all of the partial sums of the sequence are non-negative.

3.7.9. In 1874, Catalan ([Cata74]) observed that the numbers

$$C^S(n, m) = \frac{(2m)!(2n)!}{(m+n)!m!n!}$$

are positive integers for the integer non-negative parameters n and m. Subsequently, these numbers were called *Super Catalan numbers*.

For $m = 1$, these numbers form a sequence of double Catalan numbers ([Gess92]):

$$C^S(n, 1) = 2 \cdot \frac{(2n)!}{(n+1)!n!n!} = 2C_n, \quad n = 0, 1, 2, \ldots.$$

For $m = 0$ and $m = n$, these numbers coincide with central binomial coefficients:

$$C^S(n, 0) = \frac{(2n)!}{n!n!} = \binom{2n}{n}, \quad C^S(n, n) = \frac{(2n)!(2n)!}{(2n)!n!n!}$$

$$= \frac{(2n)!}{n!n!} = \binom{2n}{n}, \quad n = 0, 1, 2, \ldots.$$

Thus, for $m = 0$, $m = 1$, and $m = n$, the numbers $C^S(n, m)$ have a simple combinatorial interpretation. In fact, the central binomial coefficient $\binom{2n}{n}$ counts the number of paths from the point $(0, 0)$ to the point (n, n) on the square lattice $\mathbb{Z}^2$ if the only "up" and "right" movements are allowed (see Chapter 5). In general, the central binomial coefficient $\binom{2n}{n}$ is the number of arrangements such that there are an equal number of two types of objects. So, there are $\binom{2n}{n}$ arrangements of n copies of A and n copies of B. The same central binomial coefficient $\binom{2n}{n}$ is also the number of words of length $2n$ made up of A and B within which, as one reads from left to right, there are never more B's than A's at any point. For example, if $n = 2$, there are six words of length 4 in which each prefix has at least as many copies of A as of B: $AAAA$, $AAAB$, $AABA$, $AABB$, $ABAA$, $ABAB$.

As for Catalan numbers, the collection of related combinatorial problems is enormous.

Other combinatorial interpretations of $C^S(n,m)$ are known only for $m = 2$ and $m = 3$ ([GeXi05]). Thus, the numbers $C^S(n,2)$ count cubic trees with n interior vertices (or the number of hexagonal trees with n nodes). The question of finding such an interpretation for the general case remains an open problem.

3.7.10. *Fuss–Catalan numbers*, named after Nicolas Fuss (1755–1826) and E. Ch. Catalan, are defined as follows:

$$A_m(p,r) = \frac{r}{mp+r}\binom{mp+r}{m}.$$

The number $A_m(p,r)$ gives the number of (oriented) forests containing r p-ary trees with m internal vertices (of out-degree p).

In some publications, numbers of this form are called *two-parameter Fuss–Catalan numbers* (or *Raney numbers*), while ordinary Fuss–Catalan numbers (*one-parameter Fuss–Catalan numbers*) are considered the numbers $A_m(p,r)$ for $r = 1$:

$$A_m(p,1) = \frac{1}{mp+1}\binom{mp+1}{m}.$$

These numbers count the number of p-ary trees with m internal vertices.

For $p = 2$ and $r = 1$, we get the usual Catalan numbers:

$$A_m(2,1) = \frac{1}{2m+1}\binom{2m+1}{m} = C_m, \quad m = 0,1,2,\ldots .$$

However, there are many other interesting special cases of Fuss–Catalan numbers. So, for $p = 0$, we get the ordinary binomial coefficients:

$$A_m(0,r) = \frac{r}{r}\binom{r}{m} = \binom{r}{m}.$$

In other words, it turns out that the numbers $A_m(0,r)$ form Pascal's triangle.

At $p = 1$, again, quite unexpectedly, we get Pascal's triangle, "enumerated" along descending diagonals. Thus,

$$A_m(1,1) = \frac{1}{m+1}\binom{m+1}{m} = 1,$$

$$A_m(1,2) = \frac{2}{m+2}\binom{m+2}{m} = m+1,$$

$$A_m(1,3) = \frac{3}{m+3}\binom{m+3}{m} = \frac{(m+1)(m+2)}{2},$$

$$A_m(1,4) = \frac{4}{m+4}\binom{m+4}{m} = \frac{(m+1)(m+2)(m+3)}{6},\ldots.$$

Thus, it can be stated that, for $p = 1$ and $r = 1$, the sequence $A_m(1,1)$, $m = 0, 1, 2, \ldots$ coincides with the sequence of unities, 1, 1, 1,...; for $p = 1$ and $r = 2$, the sequence $A_m(1,2)$ coincides with the sequence of positive integers, $1, 2, 3, \ldots$; for $p = 1$ and $r = 3$, the sequence $A_m(1,3)$, $m = 0, 1, 2, \ldots$ coincides with the sequence of triangular numbers; for $p = 1$ and $r = 4$, the sequence $A_m(1,4)$, $m = 0, 1, 2, \ldots$ coincides with the sequence of tetrahedral numbers, and, in general, for $p = 1$ and a fixed positive integer r, the sequence $A_m(1,r)$, $m = 0, 1, 2, \ldots$ coincides with the sequence of simplicial numbers of dimension r.

3.7.11. Another generalization of the Catalan numbers is defined ([Klar70] and [HiPe91]) by

$$d_k^p = \frac{1}{k}\binom{pk}{k-1} = \frac{1}{(p-1)k+1}\binom{pk}{k},\quad k \geq 1.$$

The value d_k^p gives ([HiPe91]):

- the number of p-ary trees with k source-nodes;
- the number of ways of dividing a convex polygon into k disjoint $(p+1)$-gons with non-intersecting polygon diagonals;
- the number of p-good paths from $(0, -1)$ to $(k, (p-1)k-1)$ (a lattice path from one point to another is p-good if it lies completely below the line $y = (p-1)x$).

The usual Catalan numbers are a special case with $p = 2$: $C_k = d_k^2$.

A further generalization can be obtained as follows. Let $p > 1$ be a positive integer and $q \leq p-1$. Then, define $d_{q0}^p = 1$, and let d_{qk}^p be the number of p-good paths from $(1, q-1)$ to $(k, (p-1)k-1)$, $k \geq 0$ ([HiPe91]). Formulas for d_{qi}^p include the generalized Jonah formula

$$\binom{n-q}{k-1} = \sum_{i=1}^{k} d_{qi}^p \binom{n-pi}{k-i},$$

and the closed formula

$$d_{qk}^p = \frac{p-q}{pk-q}\binom{pk-q}{k-1}.$$

A recurrent relation is given by

$$d_{qk}^p = \sum_{i,j} d_{p-r,i}^p d_{q+r,j}^p, \quad i,j,r \geq 1, k \geq 1, q < p-r, \quad \text{and} \quad i+j = k+1.$$

Exercises

1. Calculate the values of the Super Catalan numbers $C^S(n,m)$ for $n,m \in \{0,1,2,3,\ldots,10\}$. What properties does the corresponding number array possess? Find in it the sequence of double Catalan numbers. Check that $C^S(n,m) = \sum_k (-1)^k \binom{2n}{n-k}\binom{2m}{m+k}$.
2. Prove that $A_m(p,r) = \frac{r}{m!}\prod_{i=1}^{m-1}(mp+r-i) = r \cdot \frac{\Gamma(mp+r)}{\Gamma(1+m)\Gamma(m(p-1)+r+1)}$, where $\Gamma(z)$ is the gamma function. Check that $A_0(p,r) = 1$; $A_1(p,r) = r$; $A_2(p,1) = p$.
3. Prove that there is a recurrent relation $A_m(p,r) = A_m(p,r-1) + A_{m-1}(p,p+r-1)$ with the initial conditions $A_m(p,0) = 0$ and $A_0(p,r) = 1$.
4. Prove that there is a recurrent relation $A_m(p,s+r) = \sum_{k=0}^{m} A_k(p,r)A_{m-k}(p,s)$.
5. Prove that $A_m(p,r) = \frac{r}{m(p-1)+r}\binom{mp+r-1}{m} = \frac{r}{m}\binom{mp+r-1}{m-1}$.

References

[AbSt72], [Abra74], [Alte71], [AlKu73], [Apos86], [BaCo87], [Bern99], [BeSl95], [BoRe14], [Bron01], [BoBa03], [Brua97],

[Camp84], [Cata44], [Cata74], [ChMo75], [Chu87], [CoGu96], [Comt74], [ChRe14], [CYY20], [Dave47], [Dede63], [DeZa80], [DeDe12], [Deza17], [Deza18], [Deza21], [Deza24], [Dick05], [Dorr65], [EgGa88], [Ficht01], [FlSe09], [Gard61], [Gard76], [Gard88], [GoJa83], [GKP94], [Goul85], [Guy58], [Hara03], [HaPa77], [Hons73], [Hons85], [Knut97], [Knut76], [Kord95], [KoSa06], [Kost82], [MaWo00], [Mazu10], [Ore80], [Pasc54], [Poly56], [Roge78], [Rose18], [Sand78], [Sier64], [Sing78], [Sloa24], [SlPl95], [Stan97], [Stan13], [Stan15], [Stru87], [Uspe76], [Weis24], [Wiki24].

Chapter 4
Relatives of Catalan Numbers

4.1. Motzkin Numbers

A history of the question

4.1.1. In combinatorial mathematics, *Motzkin numbers* are a sequence of positive integers that occur in various counting problems. They are named after the Israeli-American mathematician Theodore Motzkin (1908–1970).

T. Motzkin grew up in Berlin and started studying mathematics at an early age, entering university when he was only 15. He received his PhD in 1934 from the University of Basel under the supervision of Alexander Ostrowski.

In 1935, T. Motzkin was appointed to the Hebrew University in Jerusalem. In 1936, he was an invited speaker at the International Congress of Mathematicians in Oslo. During World War II, he worked as a cryptographer for the British government.

In 1948, T. Motzkin moved to the United States. After two years at Harvard and Boston College, he was appointed at UCLA in 1950, becoming a professor in 1960. He worked there until his retirement.

T. Motzkin was a mathematician of great erudition, versatility, and ingenuity. The range of his work included beautiful and important contributions to the theory of linear inequalities and programming, approximation theory, convexity, combinatorics, algebraic geometry, number theory, algebra, function theory, and numerical analysis.

He first developed the "double description" algorithm of polyhedral combinatorics and computational geometry. He was the first to prove the existence of principal ideal domains that are not Euclidean domains, with $\mathbb{Z}\left[\frac{1+\sqrt{-19}}{2}\right]$ being the first example given by him.

He found the first explicit example of a non-negative polynomial which is not a sum of squares, known as the *Motzkin polynomial*: $x^4y^2 + x^2y^4 - 3x^2y^2 + 1$. The quote "complete disorder is impossible," describing Ramsey theory, is attributed to him.

The *Motzkin transposition theorem*, *Motzkin–Taussky theorem*, *Fourier–Motzkin elimination*, and *Motzkin numbers* ([Motz48]) are named after him.

4.1.2. The combinatorial and algebraic properties, applications, and generalizations of Motzkin numbers have been widely studied in several papers. The reader can find an extensive survey in [Bern99], [CWZ20], [DoSh77], [OsJe15], [WaZh15], [ZhQi17], and [Sloa24]. One of the first articles on Motzkin numbers is the work of Donaghey and Shapiro [DoSh77]. They presented some combinatorial interpretations of this numerical set. Oste and Van der Jeugt [OsJe15] provided a good summary of the different variations based on the generalizations of Motzkin paths, which are one of the combinatorial interpretations of Motzkin numbers. Wang and Zhang [WaZh15] examined a type of generalized Motzkin sequence, which is based on the extension of Catalan numbers. Zhao and Qi [ZhQi17] also studied this sort of generalization. Recently, Chen *et al.* [CWZ20] presented a new Motzkin triangle.

Recurrent construction of Motzkin numbers

4.1.3. Initially, *Motzkin numbers* are introduced as number of ways of drawing any number of non-intersecting chords joining n

(labeled) points on a circle. Such definition we can find in [Wiki24], [Sloa24], etc.

However, according to our structure of consideration, we will define the sequence of Motzkin numbers recursively.

A *Motzkin number* is a member of the sequence M_0, M_1, $M_2, \ldots, M_n, \ldots$, built according to the following recurrent low:

$$M_0 = 1, \quad \text{and} \quad M_{n+1} = M_n + M_0 M_{n-1} + M_1 M_{n-2}$$
$$+ M_2 M_{n-3} + \cdots + M_{n-1} M_0 \quad \text{for } n \geq 0.$$

So, we get

$$M_0 = 1, \quad M_1 = M_0 = 1, \quad M_2 = M_1 + M_0 M_0 = 1 + 1 \cdot 1 \cdot 1 = 2,$$
$$M_3 = M_2 + M_0 M_1 + M_1 M_0 = 2 + 1 \cdot 1 + 1 \cdot 1 = 4,$$
$$M_4 = M_3 + M_0 M_2 + M_1 M_2 + M_2 C_0 = 4 + 1 \cdot 2 + 1 \cdot 1 + 2 \cdot 1 = 9,$$
$$M_5 = M_4 + M_0 M_3 + M_1 M_2 + M_2 M_1 + M 3_3$$
$$M_0 = 9 + 1 \cdot 4 + 1 \cdot 2 + 2 \cdot 1 + 4 \cdot 1 = 21,$$
$$M_6 = M_5 + M_0 M_4 + M_1 M_3 + M_2 M_2 + M_3 M_1 + M_4$$
$$M_0 = 21 + 1 \cdot 9 + 1 \cdot 4 + 2 \cdot 2 + 4 \cdot 1 + 9 \cdot 1 = 51, \quad \text{etc.}$$

This definition allows us to get a simple rule for consecutive construction of members of the sequence $M_0, M_1, M_2, \ldots, M_n, \ldots$. The corresponding algorithm is thus: *in order to get the next member of the Motzkin's sequence, write in the usual order already built members of this sequence, excluding the last one; under them write the same numbers, but in the reverse order; multiply each top number by the number below it, and add up all the obtained elements; add the result with the last of already built members of this sequence.* For example, in order to find the number M_7, it is enough to consider the numbers $M_0 = 1, M_1 = 1, \ldots, M_5 = 21, M_6 = 51$, write out one after another the first six Motzkin numbers $M_0, M_1, \ldots, M_5$, and below write the same numbers in the reverse order.

M_0	M_1	M_2	M_3	M_4	M_5
1	1	2	4	9	21
21	9	4	2	1	1

Multiplying each top number by the corresponding bottom number, adding all obtained products and the number M_6, we get $M_7 = 127$:

$$\begin{array}{cccccc} 1 & 1 & 2 & 4 & 9 & 21 \\ \times & & & & & \\ 21 & 9 & 4 & 2 & 1 & 1 \\ \hline 1\cdot 21 \;+\; 1\cdot 9 \;+\; 2\cdot 4 \;+\; 4\cdot 2 \;+\; 9\cdot 1 \;+\; 21\cdot 1 \;+\; 51 \;=\; 127 \end{array}.$$

Using this simple algorithm, it is easy to check that the sequence M_n, $n = 0, 1, 2, 3, \ldots$, starts with elements $1, 1, 2, 4, 9, 21, 51, 127, 323, 835, \ldots$ (sequence A001006 in OEIS).

Motzkin numbers and counting problems

Motzkin numbers are interesting primarily because they appear, often quite unexpectedly, in many counting problems. We will consider some of these examples.

Handshake problem

4.1.4. The *handshake problem* is a classic mathematical problem that involves finding the total number of handshakes between a finite number of people.

If there are n pairs of guests at a round table, and each guest shakes hands with one of the others so that no two pairs of joined arms intersect, the total number of handshake arrangements is equal to the nth Catalan number C_n.

But what happens if there are exactly n guests, and some of them (or even all of them) prefer to stay "lonely," without shaking anyone's hand? It turns out that in this case, the total number of handshake arrangements is equal to the nth Motzkin numbers M_n.

For the convenience of our consideration, we reformulate the problem by placing on a circle n points and trying to list the ways of construct disjoint chords connecting the specified n points of the circle, $n = 0, 1, 2, \ldots$.

Thus, if there are no selected points or if you specify exactly one point on a given circle, the number of possibilities is one: we do not have any chord.

If we specify two points, the number of possibilities is two: either we do not construct chords at all or we construct a single chord at the ends with the selected points. For $n = 3$, we have four possibilities: either there is no chord at all or one chord has been used, which can be done in $C_3^2 = 3$ ways.

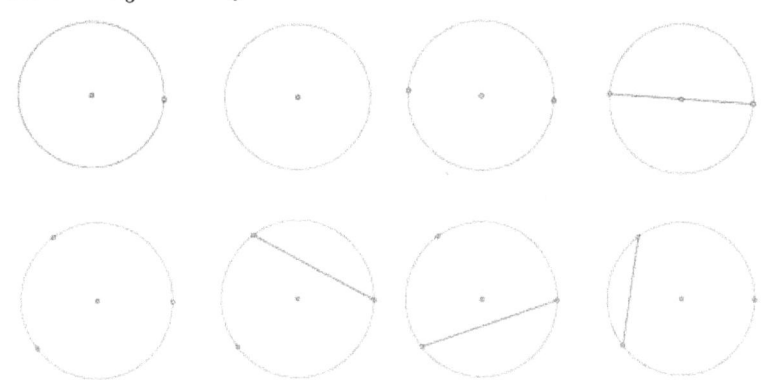

For $n = 4$, there are nine possibilities: one case without chords, $C_4^2 = 6$ cases with one chord, and two cases with two chords.

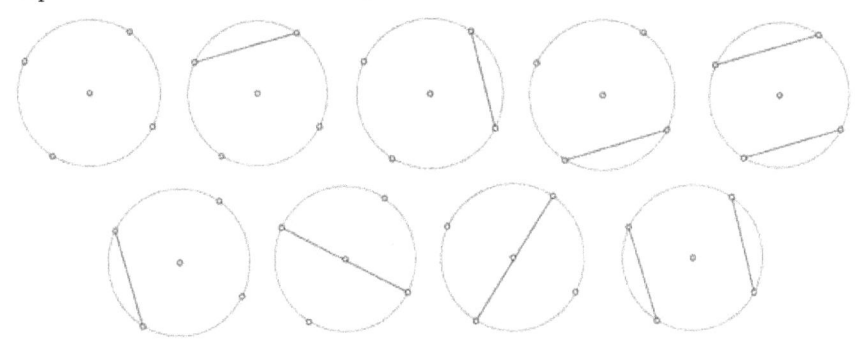

Now, it is easy to prove the following fact:

- *The number of ways for constructing disjoint chords connecting the specified n points on a circle is equal to the nth Motzkin number M_n, $n \geq 0$.*

☐ In fact, this is true for $n = 0$, $n = 1$, and $n = 2$: as was shown before, the corresponding number of ways for constructing disjoint chords connecting the specified n points on a circle is equal to 1 (i.e., M_0), 1 (i.e., M_1), and 2 (i.e., M_2).

Consider now a positive integer $n \geq 3$. Suppose that the number of ways for constructing disjoint chords connecting the specified k points on the circle, $0 \leq k \leq n-1$, is equal to M_k, and we try to prove that the number of ways for constructing disjoint chords connecting the specified n points on the circle is equal to the nth Motzkin number M_n. In order to do this, fix one of the n specified points. If we plan to use this point as a "lonely" point (i.e., there exists no chord with an end at this point), then there remain $n-1$ specified points that can be connected by disjoint chords in M_{n-1} ways. If we assume that there exists a chord with an end at the fixed point, then we choose one of the remaining $n-1$ specified points, and fix the corresponding chord. In this case, there remain $n-2$ specified points on the circle. The fixed chord divides the circle into two arcs. By the construction, we should construct chords on each arc independently of each other. If on one arc there are s specified points, $0 \leq s \leq n-2$, then on the other arc there are $n-s-2$ points, that is, it is possible to construct chords in $M_s M_{n-2-s}$ ways. Since we have exactly $n-1$ possibilities for fixing a chord, the in total we obtain $M_0 M_{n-2} + M_1 M_{n-3} + \cdots + M_{n-3} M_0$ possibilities. So, adding together the values obtained for the considered two cases ("without a fixed chord" and "with a fixed chord"), we have now exactly

$$M_{n-1} + M_0 M_{n-2} + M_1 M_{n-3} + \cdots + M_{n-3} M_0$$
$$= M_{n-1} + \sum_{i=0}^{n-2} M_i M_{n-2-i}$$

possibilities for constructing disjoint chords connecting the specified n points on the circle. As, by definition, $M_{n-1} + \sum_{i=0}^{n-2} M_i M_{n-2-i} = M_n$, we obtain that the number of ways of constructing disjoint chords connecting the specified n points on the circle is equal to the nth Motzkin number M_n. □

Brackets' problem

4.1.5. Recall that a *regular bracket* sequence is defined as a correct arrangement of n pairs of brackets: the number of opening brackets is equal to the number of closing brackets; in any initial segment of the arrangement, the number of closing brackets cannot be more than the number of opening brackets.

If there are n pairs of brackets, the number of regular bracket sequences is equal to the nth Catalan number C_n.

By adding several zeros to a regular bracket sequence, we obtain a *regular bracket sequence separated by zeros*.

Consider a regular bracket sequence separated by zeros consisting of n elements. How many such sequences of a given length n exist?

If the number of elements is zero, that is, $n = 0$, then there is exactly one such sequence: an "empty" sequence. For $n = 1$, we have exactly one such sequence: 0. For $n = 2$, there are two correct sequences: () and 00. For $n = 3$, there are exactly four such sequences: 0(), ()0, (0), and 000. For $n = 4$, we have nine possibilities: ()(), (()), ()00, 0()0, 00(), (0)0, 0(0), (00), and 0000. So, it turns out that:

- the number of regular bracket sequences separated by zeros of length n is equal to the nth Motzkin number M_n, $n \geq 0$.

□ The validity of the statement was already checked for $n = 0, 1, 2, 3, 4$: the number of correct sequences of length $n = 0$ is equal to $M_0 = 1$; the number of correct sequences of length $n = 1$ is equal to $M_1 = 1$; the number of correct sequences of length $n = 2$ is equal to $M_2 = 2$; the number of correct sequences of length $n = 3$ is equal to $M_3 = 4$; the number of correct sequences of length $n = 4$ is equal to $M_4 = 9$.

Consider now a positive integer, $n \geq 5$. Suppose that the number of regular bracket sequences separated by zeros and containing k elements is equal to the kth Motzkin number M_k for any $0 \leq k \leq n - 1$. Consider the set of all regular bracket sequences separated by zeros and consisting of n elements. Fix one of such sequences. If the first element of the sequence is 0, then in order to fill the remaining $n-1$ places, we can use any correct sequence of length $n-1$. By induction, there are exactly M_{n-1} such possibilities. If the first element of the fixed sequence is an opening bracket, then there exists a closing bracket that forms a pair with the first opening bracket of the sequence. It means that the initial subsequence, ending at this closing bracket, is correct; in particular, it has the same number of opening and closing brackets. Let us fix the initial opening bracket and its closing pair-bracket. In this case, the sequence is divided into two groups. The first one, having, say, s elements, is placed between ("in") two fixed brackets. The second one, having $n-s-2$ elements, is placed to the right ("out") of the two fixed brackets. Here, s is an integer, such that $0 \le s \le n-2$. By induction, the first group can be correctly arranged in M_s ways. The second group can be correctly arranged in M_{n-s-2} ways. So, we have, in total, exactly M_n possibilities:

$$M_{n-1} + (M_0 M_{n-2} + M_1 M_{n-3} + \cdots + M_s M_{n-s-2} + \cdots + M_{n-2} M_0)$$
$$= M_{n-1} + \sum_{i=0}^{n-2} M_i M_{n-i-2} = M_n. \qquad \square$$

For example, if we consider correct sequences of length 5, then the first group of sequences, starting with zero, can be obtained from nine correct sequences of length 4, as presented above: $()(), (()), ()00, 0()0, 00(), (0)0, 0(0), (00)$, and 0000. We can simply add an additional zero to the starting position: $0()(), 0(()), 0()00, 00()0, 000(), 0(0)0, 00(0), 0(00), 00000$. The second group consists of four subgroups. The subgroup, corresponding to $s = 0$, has the form $()abc$. There are M_3 possibilities (exactly, $0(), ()0, (0)$, and 000) to fill the three last positions. So, we obtain $M_3 = M_0 M_3 = 4$ sequences: $()0(), ()()0, ()(0)$, and $()000$. The subgroup, corresponding to $s = 1$, has the form $(a)ab$. There are M_1 possibilities (exactly, 0) to fill the one position within $()$ and M_2 possibilities (exactly, $()$ and 00) to fill the last two positions. So, we obtain $M_1 M_2 = 2$ sequences: $(0)00$ and $(0)()$. The subgroup, corresponding to $s = 2$, has the form $(ab)c$. There are M_2 possibilities (exactly, $()$ and 00) to fill the two positions within $()$ and M_1 possibilities (exactly, 0) to fill the last position. So, we obtain $M_2 M_1 = 2$ sequences: $(())0$ and $(00)0$. Finally, the subgroup, corresponding to $s = 3$, has the form (abc). There are M_3 possibilities (exactly, $0(), ()0, (0)$, and 000) to fill three positions within $()$. So, we obtain $M_3 = M_3 M_0 = 4$ sequences: $(0()), (()0), ((0))$, and (000).

However, the statement can be proven by constructing a one-to-one correspondence between the set of all arrangements of disjoint chords, connecting the specified n points of the circle, and the set of all correct sequences of length n. In fact, fixing a point on the circle, let us mark it by zero if it is "lonely," i.e., if there is no chord passing through this point, and mark it by an opening bracket otherwise. Moving clockwise along the circle, we mark any lonely point by zero, any point which is the beginning of a new chord by an opening bracket, and any point which is the end of an already considered chord by a closing bracket. In this way, we obtain a correct sequence of length n. Conversely, any correct sequence gives us an arrangement of chord on the circle with n points. □

4.1.6. There are several other formulations of the bracket problem.

Given positive integer n, define a *Dyck word with zeros of length n* as a sequence consisting of k symbols X, k symbols Y, and $n - 2k$ symbols 0, in which each initial segment contains at least as many symbols X as symbols Y.

For $n = 1$, we have exactly one Dyck word with zeros: 0; for $n = 2$, we have two such words: 00 and XY; for $n = 3$, four such words exist: $0000, 0XY, XY0$, and $X0Y$. For $n = 4$, we have nine possibilities: $00000, 00XY, 0XY0, 0X0Y, XY00, XYXY, X0Y0, X00Y$, and $XXYY$. It is natural to assume that there is exactly one Dyck word with zeros of length zero:

- *The number of Dyck words with zeros of length n is equal to the nth Motzkin number M_n, $n \geq 0$.*

□ In order to prove that the number of Dyck words with zeros of length n is M_n, it is enough to replace the symbol X with the symbol of an opening bracket, (, and the symbol Y with the symbol of a closing bracket,). Such a transform gives a bijection between the set of all Dyck words with zeros of length n and the set of all regular bracket sequences separated by zeros of length n. □

For a given positive integer n, consider the set of sequences of length n consisting of k "1", k "-1", and $n-2k$ zeros, all partial sums of which are non-negative. Here, k can take any possible value from zero up to $\lfloor \frac{n}{2} \rfloor$. Call any such sequence a *correct* $(1, -1)$-*sequence separated by zeros*.

For $n = 1$, there is only one such sequence: 0 (with the partial sum 0); for $n = 2$, there exist two such sequences: 0, 0 (with the partial sums 0, 0); 1, −1 (with the partial sums 1, $1 + (-1) = 0$); for $n = 3$, there exist four such sequences: 0, 0, 0 (with the partial sums 0, 0, 0); 0, 1, −1 (with the partial sums 0, 1, 0); 1, −1, 0 (with the partial sums 1, 0, 0); 1, 0, −1 (with the partial sums 1, 1, 0). It is easy to prove that:

- *the number of correct $(1, -1)$-sequences separated by zeros of length n is equal to the nth Motzkin number $M_n, n \geq 0$.*

□ In order to prove this statement, consider a Dyck word with zeros of length n. If instead of symbols X and Y, we use the symbols "1" and "−1", respectively, we get a sequence composed of k "1", k "−1", and $n-2k$ zeros, all partial sums of which are non-negative. However, we can use for the proof the correct bracket structures separated by zeros with k pairs of brackets: we should simply replace the symbol "(" with the symbol "1" and the symbol ")" with the symbol "−1".
■

An alternative interpretation is that Motzkin numbers are the ballot numbers (see Chapter 3) modified to allow abstentions.

For a given positive integer n, define a *mountain with k ascents, k descents, and $n - 2k$ plateau* as a broken line composed of k segments of the same length inclined at an angle of 45° to the positive direction of the OX-axis, of k segments of the same length inclined at an angle 135° to the positive direction of the OX-axis, and $n - 2k$ segments of the same length parallel to the OX-axis, such that the resulting "start" and "end" place themselves on the same level. Call any such construction a *mountain with plateau of length n*.

- *The number of mountains with plateau of length n is equal to the nth Motzkin number $M_n, n \geq 0$.*

□ The proof can be obtained using the correct bracket structures separated by zeros. In fact, it is enough to build a one-to-one correspondence between the set of all correct bracket structures separated by zeros with k pairs of brackets and the set of mountains with k ascents and k descents. The corresponding law is quite simple: each "ascending" segment corresponds to an opening bracket; each "descending" segment corresponds to a closing bracket; each plateau segment corresponds to zero. □

Similarly, the Motzkin number M_n gives the number of routes on the upper-right quadrant of the integer lattice $\mathbb{Z}^2$, from the point $(0,0)$ to the point $(n,0)$ in n steps, if one is allowed to move only to the right, right-up, and right-down at each step but is forbidden to move below the OX-axis. A one-to-one correspondence between the set of these routes of length n and the mountains with plateau of length n is obvious.

Equivalently, M_n is the number of monotonic lattice paths along the edges of a grid with $n \times n$ square cells which do not pass above the diagonal. A Motzkin's monotonic path is one which starts at the lowest-left corner, ends at the upper-right corner, and consists of the movements "two steps right," "two steps up," or "one step right-up." The counting of such paths is equivalent to the counting of Dyck words: X stands for moving "two steps right," Y stands for moving "two steps up," and zero stands for moving "one step right-up."

Two represented routes are equivalent up to a scaling. We consider this constructions in detail in Chapter 5.

Enumeration of quasi-trees

4.1.7. It is widely known that Catalan numbers appear quite often when counting the number of some classes of trees.

Recall that a *tree* is a connected graph that does not have cycles ([Ore80]). The degree of the vertex of a tree (in general, of any graph) is the number of edges included in it. The vertices of the tree, having the degree 1 are called *leaves*. The other vertices of a tree are called *internal vertices*.

In order to generalize the notion of tree, define a *loop* as an edge of a graph whose ends coincide. Formally, for a given graph $G = (V, E)$, the loop at a vertex $v \in V$ is the pair (v, v). Graphically, the loop at a vertex $v \in V$ is a line coming out and entering into the vertex v.

Define a *quasi-tree* as a graph, obtained from a tree by adding loops to some vertices. By definition, any tree is a quasi-tree with zero loops.

A *binary tree* is a tree whose internal vertices all have degree 3. As was shown in Chapter 3, it is convenient to present a binary tree on a plane without self-intersecting edges, i.e., as a *plane tree*. Moreover, any plane binary tree can be represented as a *planted binary tree*, i.e., with an edge connected to a hanging vertex chosen as a *trunk*; thus, the graph is drawn to simulate a tree growing up from the ground. The "ground" vertex can be called the *root*.

Define a *planted binary quasi-tree* as a planted binary tree with loops added to some vertices, excluding the lowest leaf (the root).

- *The number of planted binary quasi-trees with $n + 1$ edges is equal to the nth Motzkin number M_n, $n \geq 0$.*

□ In fact, the statement is true for $n = 0, 1, 2, 3$: there exists exactly one (i.e., M_0) planted binary quasi-tree (in fact, the complete graph on two vertices, K_2, that has one edge); there exists exactly one (i.e., M_1) planted binary quasi-tree with two edges (in fact, the complete graph on two vertices, K_2, equipped with one loop); there exist exactly two (i.e., M_2) planted binary quasi-trees with three edges: the graph K_2, equipped with two loops, and the only ordinary planted binary tree with one internal vertex; similarly, there exist exactly four planted binary quasi-trees with four edges.

We can prove the general statement using a one-to-one correspondence between the set of planted binary quasi-trees with $n + 1$ edges and the set of correct arrangements of brackets separated by zeros of length n. In fact, let as mark all leaves of a planted binary quasi-tree, except for the last right and the lowest, by closing brackets and all internal vertices by opening brackets; in addition, let us mark all existing loops by zeros. Now, let us imagine a worm crawling clockwise up the trunk and around the entire quasi-tree and collecting the available marks (zeros and brackets) on the way. If there is a "zero" mark at a vertex, the worm should first take zeros and then the bracket (if it exists). In this case, any plane binary quasi-three with $n+1$ edges corresponds to some correct bracket arrangement separated by zeros of length n.

Conversely, for each bracket arrangement separated by zeros of length n, we can restore the corresponding planted quasi-tree with $n+1$ edges. In fact, we should: count the number k of pairs of brackets; construct a corresponding planted binary tree with k internal vertices (see Chapter 3); finally, add loops to the corresponding vertices. As any plane (planted) binary tree with k internal vertices has exactly $k+2$ hanging vertices (one of which is grounded and called the root), we have a tree with $2k+2$ vertices and $2k+1$ edges. Adding $n-2k$ loop edges, we obtain a planted binary quasi-tree with $n+1$ edges.

Thus, we obtain a bijection between the set of all planted binary quasi-trees with $n+1$ edges and the set of all correct bracket sequences separated by zeros of length n. Since the number of such sequences is equal to the nth Motzkin number M_n, the number of planted binary quasi-trees with $n+1$ edges is also equal to the nth Motzkin number M_n.

However, the proof can be obtained by direct consideration. In fact, the statement was checked for $n=0,1,2,3$: there exist exactly M_n planted binary quasi-trees with $n+1$ edges if $n=0,1,2,3$. Consider now a positive integer $n \geq 4$. Suppose that the statement is true for any non-negative integer s, $0 \leq s \leq n-1$, and prove it for n. In fact, consider any planted binary tree with $n+1$ edges, $n \geq 4$. Consider the lowest internal vertex of the quasi-tree. If there is at least one loop at this vertex, fix it. Delete the fixed loop. In this way, we obtain a planted binary quasi-tree with n edges. By induction, there are M_{n-1} possibilities for such quasi-trees. If there is no loop at the lowest internal vertex, then we have exactly two edges, going up from this vertex. So, we get two planted binary quasi-trees which are subgraphs of the initial quasi-tree, with these two edges as their trunks. Consider the left such quasi-tree. It has $s+1$ edges, where s is a non-negative integer, such that $0 \leq s \leq n-2$, and $1 \leq s+1 \leq n-1$. Then, the right subtree has $n-s-1$ edges, $1 \leq n-s-1 \leq n-1$, and $0 \leq n-s-2 \leq n-2$. By induction, we can construct exactly M_s "left" quasi-trees and exactly M_{n-s-2} "right" quasi-trees, altogether $M_s M_{n-s-2}$ possibilities. So, for $s = 0$, we get $M_0 M_{n-2}$ possible planted binary quasi-trees with $n+1$ edges (in this case, the "left" subgraph is just the tree with two vertices); for $s = 1$, we get $M_1 M_{n-3}$ possible plane planted binary quasi-trees with n edges; ...; for $s = n-2$, we get $M_{n-2} M_0$ possible planted binary quasi-trees with n edges (in this case, the "right" subgraph is just the tree with two vertices). As a result, the total number of planted binary quasi-trees with $n+1$ edges is equal to the nth Motzkin number:

$$M_n + M_0 M_{n-2} + M_1 M_{n-3} + \cdots + M_{n-3} M_1 + M_{n-2} M_0 = M_n. \quad \Box$$

4.1.8. In terms of computer science, a *binary tree* is a tree data structure in which each node has at most two children, referred to as the left child and the right child. A *rooted binary tree* has a *root node*, and every node has at most two children. A *full* (or *proper, strict*) *binary tree* is a rooted binary tree in which every node has either zero or two children.

We call a *full binary quasi-tree* a graph, obtained by attached loops to some vertices of a full binary tree.

It is easy to see that any planted binary quasi-tree with $n+1$ edges corresponds to a full binary quasi-tree with n edges. In fact, it is enough to choose as the root the lowest internal vertex of the planted quasi-tree and delete the lowest ("planted") hanging vertex (i.e., the trunk). So, as a consequence, we obtain that:

- the number of all full binary quasi-trees with n edges is equal to the nth Motzkin number M_n, $n \geq 0$.

4.1.9. Consider now the set of all plane rooted trees. It was proven (see Chapter 3) that there are exactly C_n plane rooted trees with n edges.

Define a *plane rooted quasi-tree* as a plane rooted tree separated by loops, i.e., as a graph, obtained by attached loops to some vertices of a plane rooted tree. Note that any plane rooted quasi-tree can be presented as a *planted quasi-tree*, i.e., it is drawn to simulate a tree growing up from the root in the ground. In this case, the order of edges is important: for a given vertex of a tree with one or several "ordinary" edges, a loop can be attached "before," "between," or "after" existing edges.

For a given plane rooted quasi-tree, let us define its *length* as the number of steps of a worm, crawling clockwise around the quasi-tree. For each loop, we have exactly one "round" step; for each edge, we have exactly two steps, one up-step and one down-step.

In general, we can prove the following result:

- The number of plane rooted quasi-trees of length n is equal to the nth Motzkin number M_n, $n \geq 0$.

□ This statement for small n is already verified. The number of plane rooted quasi-trees of length n is equal to 1 (i.e., M_0) for $n = 0$, to 1 (i.e., M_1) for $n = 1$, to 2 (i.e., M_2) for $n = 2$, to 4 (i.e., M_3) for $n = 3$, and to 9 (i.e., M_4) for $n = 4$.

The simplest way to prove the general statement is to build a one-to-one correspondence between the set of all plane rooted quasi-trees of length n and the set of all planted binary quasi-trees with $n + 1$ edges.

Let us imagine that a given planted binary quasi-tree is depicted so that the edges, going from each internal vertex, are directed up, down, and to the right. Imagine that each horizontal edge shrinks to a point and disappears. If on the right end of the deleted horizontal edge were the vertex of degree 3, then it is carried to the left and merges with the point at the left. All the vertical edges remain different as well as all loops. This simple transformation translates any planted binary quasi-tree with $n + 1$ edges to a plane root quasi-tree of length n after discarding the "superfluous" lowest vertex (i.e., the trunk).

As we have exactly M_n planted binary trees with $n + 1$ edges, we have exactly M_n plane rooted quasi-trees of length n.

On the other hand, we can use for a proof the correct arrangements of brackets separated by zeros. In fact, there exists a simple one-to-one-correspondence between the set of all plane rooted quasi-trees of length n and the set of all correct bracket arrangements separated by zeros of length n. For a given (planted) plane rooted quasi-tree of length n, let as mark each of its loop by zero and each of its edge by one pair of brackets, putting the opening bracket to the left and the closing bracket to the right. Now, starting with the root and moving from the left to the right along the edges of the tree, we can collect all n attached symbols, obtaining a corresponding correct bracket arrangement. In fact, if there are k edges in a plane rooted quasi-tree, $0 \leq k \leq \lfloor \frac{n}{2} \rfloor$, then there are $n - 2k$ zeros, and we obtain a correct sequence containing $2k$ pairs of brackets and $n - 2k$ zeros.

A very beautiful correspondence between plane rooted quasi-trees and mountains with plateau exists. In fact, if we have a plane rooted quasi-tree with k ordinary edges and $n - 2k$ loops, then we can construct the corresponding mountain with k ascents, k descents, and $n - 2k$ plateau simply by moving clockwise around the quasi-tree and drawing instead of a loop-step a plateau, drawing instead of an up-step along an edge an ascent, and drawing instead of a down-step along an edge a descent.

However, the proof can be obtained by direct consideration. In fact, the statement was checked for $n = 0, 1, 2, 3, 4$: there exist exactly M_n plane rooted quasi-trees of length n if $n = 0, 1, 2, 3, 4$. Consider now a positive integer, $n \geq 5$. Suppose that the statement is true for any non-negative integer s, $0 \leq s \leq n - 1$, and prove it for n. In fact, consider a planted representation of any plane rooted quasi-tree of length n, $n \geq 5$. Find the left edge, going from the root, and delete it. If it is a loop, this operation gives us a plane rooted quasi-tree of length $n - 1$. By induction, we have M_{n-1} such possibilities. If it is an ordinary edge, this operation gives us two plane rooted quasi-trees, which are subgraphs of the initial quasi-tree. The "right" quasi-tree has the root of the initial quasi-tree as its root; the "left" quasi-tree has as its root the vertex adjacent to the root (by the delated edge) in the initial quasi-tree. Consider the "left" quasi-tree. It has a length of s, where s is a non-negative integer, such that $0 \leq s \leq n - 2$. Then, the "right" quasi-tree has a length of $n - s - 2$, $0 \leq n - s - 2 \leq n - 2$. By induction, we can construct exactly M_s "left" quasi-trees and exactly M_{n-s-2} "right" quasi-trees, altogether $M_s M_{n-s-2}$ possibilities. So, restoring the removed edge, for $s = 0$, we get $M_0 M_{n-2}$ possible plane rooted quasi-trees of length n (in this case, the "left" quasi-tree is just the graph with one vertex); for $s = 1$, we get $M_1 M_{n-3}$ possible plane binary quasi-trees of length n; ...; for $s = n - 2$, we get $M_{n-2} M_0$ possible plane binary quasi-trees of length n (in this case, the "right" subtree is just the graph with one vertex). As a result, the total number of plane rooted quasi-trees of length n is equal to the nth Motzkin number:

$$M_{n-1} + M_0 M_{n-2} + M_1 M_{n-3} + \cdots + M_{n-3} M_1 + M_{n-2} M_0 = M_n.$$
□

Generalized polygon division problem

4.1.10. *Euler's polygon division problem* is the problem of finding the number of ways a plane convex polygon of n sides can be divided into triangles by non-crossing diagonals.

It is well known that the n-th Catalan number C_n gives the number of Euler's triangulations of a convex $(n+2)$-gon.

Note that for any Euler's division of an $(n+2)$-gon, there are n triangles (internal faces), $2n + 1$ "edges" (sides and used for this division diagonals), and $n + 2$ vertices.

In order to generalize this problem, consider any Euler's triangulation of a convex $(n+2)$-gon as a planar graph. This graph has n internal faces of degree 3: all these faces are triangles of the considered Euler's division. Moreover, it has one face of degree $n + 2$ out of the interior of the $(n+2)$-gon. So, we can state that, for a given $(n+2)$-gon, there exist exactly C_n such planar graphs. As was noted before, this graph has n triangles (internal faces), $2n + 1$ edges, and $n + 2$ vertices.

Let us suppose now that, in addition to triangular faces, two-sided faces are allowed in the interior of a given polygon. It means that the corresponding graph can have parallel edges.

Define as a *polygon with parallel edges* a polygon obtained from such a graph, i.e., a convex polygon with its interior divided into triangles and two-angles by non-crossing *edges*. Here, an edge is an arc connecting two vertices of the polygon. It can be a side of the polygon, a diagonal of the polygon, or any additional arc "parallel"

232   Catalan Numbers

to a side or a diagonal of the polygon, i.e., passing from the same vertices and placed in the interior of the polygon.

It is easy to check that there exists exactly one polygon with one parallel edge; it is a segment represented as a complete graph K_2 on two vertices. There is exactly one polygon with two parallel edges: it is just a bigon with two parallel edges. There are exactly two polygons with three parallel edges: they are a triangle and a bigon with three parallel edges. There are exactly four polygons with four parallel edges: they are three triangles, such that one of the sides has a parallel edge, and a bigon with four parallel edges. There are nine polygons with five parallel edges: they are two classical Euler's divisions of a square; three triangles, in which a side is "normal," and two others have parallel edges; three triangles, in which two sides are "normal," and one has two parallel edges; finally, one bigon with five parallel edges.

- *The number of polygons with $n + 1$ parallel edges is equal to the nth Motzkin number M_n, $n \geq 0$.*

□ In fact, we have already checked that the statement is true for $n = 0, 1, 2, 3$: there exists exactly one (i.e., M_0) polygon with one parallel edge; there exists exactly one (i.e., M_1) polygon with two parallel edges; there exist exactly two (i.e., M_2) polygons with three parallel edges; there exist exactly four (i.e., M_3) polygons with four parallel edges; there exist exactly nine (i.e., M_4) polygons with five parallel edges.

Consider now a positive integer $n \geq 5$. Suppose that the statement is true for any non-negative integer s, $0 \leq s \leq n-1$, and prove it for n. In fact, consider any polygon with $n+1$ parallel edges, $n \geq 5$. Fix a side of the polygon. If this side has parallel edges, delete one of these edges. In this case, we obtain a polygon with $n-1$ parallel edges. By induction, we have exactly M_{n-1} such possibilities. If the chosen side is "normal," it belongs to a triangular face of the considered division. Delete this triangle, i.e., cut the considered polygon into two other polygons: "right" and "left." As we delete the initial side, in total there are n edges in the obtained two polygons. If the "left" polygon has $s + 1$ edges, $1 \leq s + 1 \leq n - 1$, or $0 \leq s \leq n - 2$, then the "right" polygon has $n-s-1$ edges, $1 \leq n-s-1 \leq n-1$, or $0 \leq n-s-2 \leq n-2$. By induction, there are M_s possibilities to obtain a "left" polygon as well as M_{n-s-2} possibilities to obtain a "right" polygon, altogether $M_s M_{n-s-2}$ possibilities for a fixed s. So, in total, we obtain M_n possibilities to construct a polygon with n parallel edges:

$$M_{n-1} + M_0 M_{n-2} + M_1 M_{n-3} + \cdots + M_{n-3} M_1 + M_{n-2} M_0 = M_n.$$

$\square$

However, we can obtain a proof by constructing a one-two-one correspondence between the set of all polygons with $n+1$ parallel edges and the set of all planted binary quasi-trees with $n+1$ edges. In fact, for a given polygon, fix one of its sides and draw the "trunk" of the corresponding quasi-tree to the triangle containing this side. Place the first internal vertex of the quasi-tree in this triangle. If the trunk goes through a two-angle face, add to the fixed vertex of the quasi-tree a loop. Draw the next two edges of the constructed quasi-three, interceded by two other sides of the considered triangle. These edges should "stop" in the next triangle or outside of the interior of the considered polygon. Place the corresponding vertex of the tree. If this new edge goes through one or several two-angle faces, add to the new vertex the required number of loops. If the obtained vertex in this step belong to the exterior of the polygon, it will be a hanging vertex of the quasi-tree. If it belongs to a triangular face, then continue the construction. It is easy to check that the considered algorithm can be easy inverted: if we have a planted binary quasi-tree with $n+1$ edges, we can obtain the corresponding polygon with $n+1$ parallel edges. First, construct a side of the polygon by intersecting the trunk of the quasi-tree. If there is a loop at the lowest internal vertex of the quasi-tree, add a parallel edge to the first constructed side of the polygon. Obtain a triangle by intersecting two other edges, going from the lowest internal vertex. Continue the algorithm.

Note that another bijection can be constructed. For a given polygon with $n+1$ parallel edges, fix one of its sides. Mark it by an opening bracket. If there are parallel edges at this side, mark all of them by zeros. Mark any of the "normal" diagonals of the polygon by an opening bracket and all other sides of the polygon, except the last "right" side, neighboring to the initial fixed side, by a closing bracket. If there are parallel edges at the sides and diagonals (including the side without any mark), mark all of them by zeros. Note that if we deal with a $(k+2)$-gon, $0 \le k \le \lfloor \frac{n}{2} \rfloor$, then there are $k+2$ sides and $k-1$ diagonals; so, we obtain exactly k pairs of brackets and $n-2k$ zeros, $0 \le k \le \lfloor \frac{n}{2} \rfloor$. Therefore, we can obtain a correct bracket sequence (with k pairs of brackets) separated by $(n-2k)$ zeros, moving clockwise along the considered polygon, starting from the fixed side, and gathering all marks. If there are parallel edges, first gather the zeros; if we are moving outside of the interior of the polygon, go back and continue.

Motzkin numbers also count the number of 2143-avoiding *involutions*, i.e., permutations σ, which are self-inverse: $\sigma(\sigma(x)) = x$ (see [Stan13] and [Stan15]).

Closed formula for Motzkin numbers

4.1.11. Considering recurrent approach for the construction of Motzkin numbers, we have proven that Motzkin numbers count, for example, the correct bracket sequences separated by zeros. In fact, the nth Motzkin number gives the number of all correct bracket sequences separated by zeros of length n. But it is easy to count such sequences using other algorithms. It gives us the following closed formula for Motzkin numbers:

- For the nth Motzkin number M_n, it holds that

$$M_n = \sum_{k=0}^{\lfloor \frac{n}{2} \rfloor} \binom{n}{2k} C_k,$$

where $\lfloor x \rfloor$ is the floor function, $\binom{n}{2k}$ are binomial coefficients, and C_k are Catalan numbers.

☐ Any correct bracket sequences separated by zeros has several pairs of brackets which, in turn, are correctly placed in this sequence. In other words, the subsequence, formed only from brackets, form a correct bracket arrangement. It is easy to see that the number of bracket pairs in a correct sequence separated by zeros of length n is at most $\lfloor \frac{n}{2} \rfloor$. If there are k pairs of brackets in a sequence of length n, $0 \leq k \leq \lfloor \frac{n}{2} \rfloor$, then there are exactly $n - 2k$ symbols 0. In order to obtain a correct arrangement of k pairs of brackets and $n - 2k$ zeros, we should choose from n possible places $2k$ places for brackets; it can be done in $\binom{n}{2k} = C_n^{2k}$ ways. Among the $2k$ chosen places, we can order k pairs of brackets in C_k ways, where C_k is the kth Catalan number. All other $n - 2k$ places will be occupied by zeros. So, for a fixed k, $0 \leq k \leq \lfloor \frac{n}{2} \rfloor$, we have $\binom{n}{2k} C_k$ possible correct sequences separated by zeros. Therefore, the total number of correct sequences separated by zeros of length n is $\sum_{k=0}^{\lfloor \frac{n}{2} \rfloor} \binom{n}{2k} C_k$. On the other hand, it was proven that the number of such sequences is equal to the nth Motzkin number. So, we have proven that $M_n = \sum_{k=0}^{\lfloor \frac{n}{2} \rfloor} \binom{n}{2k} C_k$. □

Thus,

$$M_0 = \binom{0}{0} C_0 = 1 \cdot 1 = 1; \quad M_1 = \binom{1}{0} C_0 = 1 \cdot 1 = 1;$$

$$M_2 = \binom{2}{0} C_0 + \binom{2}{2} C_1 = 1 \cdot 1 + 1 \cdot 1 = 2;$$

$$M_3 = \binom{3}{0} C_0 + \binom{3}{2} C_1 = 1 \cdot 1 + 3 \cdot 1 = 4;$$

$$M_4 = \binom{4}{0} C_0 + \binom{4}{2} C_1 + \binom{4}{4} C_2 = 1 \cdot 1 + 6 \cdot 1 + 1 \cdot 2 = 9.$$

Inversely, it can be proven that

$$C_{n+1} = \sum_{k=0}^{n} \binom{n}{k} M_k.$$

4.1.12. As the closed formula for the nth Catalan number is $C_n = \frac{1}{n+1} \binom{2n}{n}$, we can try to obtain other representations for the nth Motzkin number.

- For the nth Motzkin number, there exists the formula

$$M_n = \frac{1}{n+1} \binom{n+1}{1}_2,$$

where $\binom{n}{k}_2$, $-n \leq k \leq n$, is a *trinomial coefficient*, defined as the coefficient at x^{n+k} in the decomposition of $(1+x+x^2)^n$.

☐ In fact, the trinomial coefficients can be generated using the following recurrent formula (see Chapter 2):

$$\binom{0}{0}_2 = 1, \quad \text{and} \quad \binom{n+1}{k}_2 = \binom{n}{k-1}_2 + \binom{n}{k}_2 + \binom{n}{k+1}_2, \quad n \geq 0,$$

where $\binom{n}{k}_2 = 0$ for $k < -n$, and $k > n$. Thus, the trinomial triangle starts with the rows

$$\begin{array}{ccccccccc} & & & & 1 & & & & \\ & & & 1 & 1 & 1 & & & \\ & & 1 & 2 & 3 & 2 & 1 & & \\ & 1 & 3 & 6 & 7 & 6 & 3 & 1 & \\ 1 & 4 & 10 & 16 & 19 & 16 & 10 & 4 & 1 \end{array}$$

So, for $n = 0, 1, 2, 3, \ldots$, the elements $\binom{n+1}{1}_2$ are $1, 2, 6, 126, \ldots$, and we can check that $M_0 = 1 = 1 \cdot 1 = \frac{1}{1} \cdot \binom{1}{1}_2$, $M_1 = 1 = \frac{1}{2} \cdot 2 = \frac{1}{2} \cdot \binom{2}{1}_2$, $M_2 = 2 = \frac{1}{3} \cdot 6 = \frac{1}{3} \cdot \binom{3}{1}_2$, $M_3 = 4 = \frac{1}{4} \cdot 16 = \frac{1}{4} \cdot \binom{3}{1}_2$.

A general proof can be obtained using the closed formula for the trinomial coefficient $\binom{n}{k}_2$ (see Chapters 1 and 2):

$$\binom{n}{k}_2 = \sum_{\substack{n_1+n_2+n_3=n, \\ n_2+2n_3=n+k}} P_{(n_1,n_2,n_3)}.$$

In fact, as $C_n = \frac{1}{n+1}\binom{2n}{n}$,

$$M_n = \sum_{k=0}^{\lfloor \frac{n}{2} \rfloor} \binom{n}{2k} C_k = \sum_{k=0}^{\lfloor \frac{n}{2} \rfloor} \frac{n!}{(2k)!(n-2k)!} \cdot \frac{(2k)!}{(k+1)!k!}$$

$$= \frac{1}{n+1} \sum_{k=0}^{\lfloor \frac{n}{2} \rfloor} \frac{(n+1)!}{(k+1)!k!((n+1)-k-(k+1))!}$$

$$= \frac{1}{n+1} \sum_{k=0}^{\lfloor \frac{n}{2} \rfloor} P_{(n-2k,k,k+1)}$$

$$= \frac{1}{n+1} \sum_{\substack{n_1+n_2+n_3=n+1, \\ n_2+2n_3=n+2}} P_{(n_1,n_2,n_3)} = \frac{1}{n+1} \binom{n+1}{1}_2.$$

□

Generating function for the sequence of Motzkin numbers

4.1.13. In this section, we consider the generating function of the sequence of Motzkin numbers.

- *The generating function $f(x)$ of the sequence $1, 1, 2, 4, 9, \ldots$ of Motzkin numbers has the form $f(x) = \frac{1}{2x^2}(1 - x - \sqrt{1 - 2x - 3x^2})$, i.e., it holds that*

$$\frac{1}{2x^2}(1 - x - \sqrt{1 - 2x - 3x^2})$$
$$= M_0 + M_1 \cdot x + \cdots + M_n \cdot x^n + \cdots, \quad |x| < \frac{1}{3}.$$

□ Consider a formal decomposition:

$$f(x) = M_0 + M_1 x + M_2 x^2 + \cdots + + M_n x^n + \cdots .$$

Then, the following relationship can be obtained:

$$f(x) \cdot f(x) \cdot x^2 = (M_0 + M_1 x + M_2 x^2 + \cdots)$$
$$\cdot (M_0 + M_1 x + M_2 x^2 + \cdots) \cdot x^2$$
$$= (M_0 M_0 + (M_0 M_1 + M_1 M_0)x$$
$$+ (M_0 M_2 + M_1 M_1 + M_2 M_0)x^2 + \cdots)x^2$$
$$= ((M_2 - M_1) + (M_3 - M_2)x + +(M_4 - M_3)x^2 + \cdots)x^2$$
$$= (-M_1 + M_2(1 - x) + M_3(x - x^2) + \cdots)x^2$$
$$= (-M_1 x^2 + M_2(x^2 - x^3) + M_3(x^3 - x^4) + \cdots)$$
$$+ (M_1 x - M_1 x)$$
$$= (M_1(x - x^2) + M_2(x^3 - x^4)$$
$$+ M_3(x^3 - x^4) + + \cdots) - M_1 x$$

238 *Catalan Numbers*

$$\begin{aligned}
&= (M_1x(1-x) + M_2x^2(1-x) \\
&\quad + M_3x^3(1-x) + \cdots) - M_1x \\
&= (1-x)(M_1x + M_2x^2 + M_3x^3 + \cdots) - M_1x \\
&= (1-x)((M_0 + M_1x + M_2x^2 + M_3x^3 + \cdots) - M_0) \\
&\quad - M_1x \\
&= (1-x)(f(x) - 1) - x = (1-x)f(x) - 1.
\end{aligned}$$

It follows that $f(x) \cdot f(x) \cdot x^2 = (1-x)f(x) - 1$. Consider the quadratic equation

$$x^2 f^2(x) + (x-1)f(x) + 1 = 0,$$

and solve it with respect to $f(x)$.

Using the discriminant formula, we get that $D = (x-1)^2 - 4x^2 = 1 - 2x - 3x^2$. So, we obtain two possible solutions:

$$f(x)_{1,2} = \frac{-(x-1) \pm \sqrt{1 - 2x - 3x^2}}{2x^2} = \frac{1 - x \pm \sqrt{1 - 2x - 3x^2}}{2x^2}.$$

To choose the sign of the square root, note that the free term of the series $f(x) = M_0 + M_1x + M_2x^2 + \cdots + M_nx^n + \cdots$ is equal to unity. But obviously, it holds that

$$\lim_{x \to 0} \frac{1 - x + \sqrt{1 - 2x - 3x^2}}{2x^2} = \infty,$$

so we should choose the other possibility:

$$f(x) = \frac{1 - x - \sqrt{1 - 2x - 3x^2}}{2x^2}.$$

However, let us consider this situation more precisely.

At first, decompose the function $\sqrt{1 - 2x - 3x^2}$ into the Taylor series using the well-known result:

$$(1+z)^\alpha = 1 + \alpha z + \frac{\alpha(\alpha-1)}{2!}z^2 + \frac{\alpha(\alpha-1)(\alpha-2)}{3!}z^3$$

$$+ \cdots + \binom{\alpha}{n}z^n + \cdots, \quad |z| < 1, \quad \alpha \in \mathbb{R},$$

where $\binom{\alpha}{n} = \frac{\alpha(\alpha-1)\cdots(\alpha-n+1)}{n!}$ is a *generalized binomial coefficient*. In our case, $\alpha = \frac{1}{2}$ and $z = -2x - 3x^2$. So, we obtain the decomposition

$$\sqrt{1 - 2x - 3x^2} = (1 - 2x - 3x^2)^{\frac{1}{2}}$$

$$= 1 + \frac{1}{2}(-2x - 3x^2) + \frac{\frac{1}{2} \cdot (-\frac{1}{2})}{2!}(-2x - 3x^2)^2$$

$$+ \frac{\frac{1}{2} \cdot (-\frac{1}{2}) \cdot (-\frac{3}{2})}{3!}(-2x - 3x^2)^3$$

$$+ \cdots + \binom{\frac{1}{2}}{n}(-2x - 3x^2)^n + \cdots, \quad |-2x - 3x^2| < 1.$$

Since $|-2x - 3x^2| < 1$ for $1 < x < \frac{1}{3}$ and

$$1 + \frac{1}{2}(-2x - 3x^2) + \frac{\frac{1}{2} \cdot (-\frac{1}{2})}{2!}(-2x - 3x^2)^2$$

$$+ \frac{\frac{1}{2} \cdot (-\frac{1}{2}) \cdot (-\frac{3}{2})}{3!}(-2x - 3x^2)^3$$

$$+ \cdots = 1 - x - 2x^2 - 2x^3 - 4x^4 - \cdots,$$

we obtain that

$$\sqrt{1 - 2x - 3x^2} = 1 - x - 2x^2 - 2x^3 - 4x^4 - \cdots, \quad |x| < \frac{1}{3}.$$

Then, it holds that

$$\frac{1 - x - \sqrt{1 - 2x - 3x^2}}{2x^2}$$

$$= \frac{1}{2x^2}(1 - x - \sqrt{1 - 2x - 3x^2})$$

$$= \frac{1}{2x^2}(1 - x - (1 - x - 2x^2 - 2x^3 - 4x^4 - \cdots))$$

$$= \frac{1}{2x^2}(2x^2 + 2x^3 + 4x^4 + \cdots) = 1 + x + 2x^2 + \cdots, \quad |x| < \frac{1}{3}.$$

□

Properties of Motzkin numbers

Simplest properties of Motzkin numbers

4.1.14. Motzkin numbers have many interesting properties.

We have proven a closed formula for Motzkin numbers which connects three number sets, namely Motzkin numbers, Catalan numbers, and the elements of Pascal's triangle:

$$M_n = \sum_{k=0}^{\lfloor n/2 \rfloor} \binom{n}{2k} C_k.$$

So, we can state that:

- the nth Motzkin number can be obtaining by the multiplication of the elements of the nth row of Pascal's triangle with even indices by the first $\lfloor n/2 \rfloor$ elements of the right-hand side of Catalan's triangle.

But there exists an inversion. In fact, it holds that:

- the $(n+1)$th Catalan number can be obtained using the formula

$$C_{n+1} = \sum_{k=0}^{n} \binom{n}{k} M_k.$$

□ Let us check this formula for small values of n. In fact, for $n = 1$, we obtain that $C_1 = 1 = 1 \cdot 1 = \binom{0}{0} M_0$. For $n = 1$, we have that $C_2 = 2 = 1 \cdot 1 + 1 \cdot 1 = \binom{1}{0} M_0 + \binom{1}{1} M_1$. For $n = 2$, it holds that

$$C_3 = 5 = 1 \cdot 1 + 2 \cdot 1 + 1 \cdot 2 = \binom{2}{0} M_0 + \binom{2}{1} M_1 + \binom{2}{2} M_2.$$

A general proof can be obtained by induction, using recurrent relations for Catalan numbers, Motzkin numbers, and binomial coefficients. □

4.1.15. In turn, the last formula gives the following result:

- It holds the equality

$$\sum_{k=0}^{n} C_k = 1 + \sum_{k=1}^{n} \binom{n}{k} M_{k-1}.$$

☐ For a proof, it is enough to use the following combinatorial identity:
$$C_n^k = C_{n-1}^k + C_{n-2}^{k-1} + C_{n-3}^{k-2} + \cdots + C_{n-k-1}^0.$$

In order to prove it, let us use the classical recurrent relation for k-combinations of an n-set: $C_n^k = C_{n-1}^k + C_{n-1}^{k-1}$. In fact, we obtain that

$$\begin{aligned}
C_n^k &= C_{n-1}^k + C_{n-1}^{k-1} = C_{n-1}^k + C_{n-2}^{k-1} + C_{n-2}^{k-2} \\
&= C_{n-1}^k + C_{n-2}^{k-1} + C_{n-3}^{k-2} + C_{n-3}^{k-3} \\
&= \cdots = C_{n-1}^k + C_{n-2}^{k-1} + C_{n-3}^{k-2} + \cdots + C_{n-k+1}^2 + C_{n-k+1}^1 \\
&= C_{n-1}^k + C_{n-2}^{k-1} + C_{n-3}^{k-2} + \cdots + C_{n-k+1}^2 + C_{n-k}^1 + C_{n-k}^0 \\
&= C_{n-1}^k + C_{n-2}^{k-1} + C_{n-3}^{k-2} + \cdots + C_{n-k+1}^2 + C_{n-k}^1 + C_{n-k-1}^0.
\end{aligned}$$

Using the properties of the numbers C_n^k, we can rewrite this identity as

$$C_n^{n-k} = C_{n-1}^{n-k-1} + C_{n-2}^{n-k-1} + C_{n-3}^{n-k-1} + \cdots + C_{n-k-1}^{n-k-1}.$$

Now, we can consider the sum $\sum_{k=0}^n C_k$. In fact,

$$\sum_{k=0}^n C_k = C_0 + C_1 + C_2 + \cdots + C_n = C_0 + C_0^0 M_0$$
$$+ (C_1^0 M_0 + C_1^1 M_1) + (C_2^0 M_0 + C_2^1 M_1 + C_2^2 M_2)$$
$$+ \cdots + (C_{n-1}^0 M_0 + C_{n-1}^1 M_1 + \cdots + C_{n-1}^{n-1} M_{n-1})$$
$$= 1 + M_0(C_0^0 + C_1^0 + \cdots + C_{n-1}^0)$$
$$+ M_1(C_1^1 + C_2^1 + \cdots + C_{n-1}^1) + \cdots + M_{n-1} C_{n-1}^{n-1}$$
$$= 1 + M_0 C_n^1 + M_1 C_n^2 + \cdots + M_{n-1} C_n^n = 1 + \sum_{k=1}^n \binom{n}{k} M_{k-1}.$$

☐

4.1.16. Moreover, it is easy to see that the formulas $C_{n+1} = \sum_{k=0}^n \binom{n}{k} M_k$ and $M_n = \sum_{k=0}^{\lfloor n/2 \rfloor} \binom{n}{2k} C_k$ above allow us to obtain a new recurrent relation for Catalan numbers. In fact:

- *for the nth Catalan number, there exists the recurrent relation*

$$C_{n+1} = \sum_{k=0}^{\lfloor \frac{n}{2} \rfloor} \binom{n}{2k} C_k 2^{n-2k}.$$

□ We can simply combine the two previous relations, $C_{n+1} = \sum_{k=0}^{n} \binom{n}{k} M_k$, $M_n = \sum_{k=0}^{\lfloor n/2 \rfloor} \binom{n}{2k} C_k$, and the combinatorial identity

$$\sum_{k=q}^{n} \binom{n}{k}\binom{k}{q} = 2^{n-q}\binom{n}{q}, \quad n \geq q \geq 0.$$

For example,

$$C_5 = \binom{4}{0} M_0 + \binom{4}{1} M_1 + \binom{4}{2} M_2 + \binom{4}{3} M_3 + \binom{4}{4} M_4$$

$$= \binom{4}{0}\binom{0}{0} C_0 + \binom{4}{1}\binom{1}{0} C_0 + \binom{4}{2}\left(\binom{2}{0} C_0 + \binom{2}{2} C_1\right)$$

$$+ \binom{4}{3}\left(\binom{4}{0} C_0 + \binom{4}{2} C_2 + \binom{4}{4} C_3\right)$$

$$= C_0 \left(\binom{4}{0}\binom{0}{0} + \binom{4}{1}\binom{1}{0} + \binom{4}{2}\binom{2}{0} + \binom{4}{3}\binom{3}{0} + \binom{4}{3}\binom{4}{0}\right)$$

$$+ C_1 \left(\binom{4}{2}\binom{2}{2} + \binom{4}{3}\binom{3}{2} + \binom{4}{4}\binom{4}{2}\right) + C_2 \binom{4}{4}\binom{4}{4}$$

$$= C_0 \binom{4}{4} 2^4 + C_1 \binom{4}{2} 2^2 + C_2 \binom{4}{4} 2^0.$$

In order to prove the used combinatorial identity, note that its left-hand side is the number of ways to select some subset of the set $\{1, 2, \ldots, n\}$ which has $k \geq q$ elements and "mark" in this k-subset q elements. The right-hand side gives the same number of ways since there are exactly $\binom{n}{q}$ ways to choose a set with q "marked" elements and possibly add to them some additional elements that can be done in 2^{n-q} ways. □

4.1.17. As for Motzkin numbers, we can also obtain another recurrent relation. It based on the following result:

- *The three consecutive Motzkin numbers are connected by the following linear recurrent relation of the third order:*
$$(n+3)M_{n+1} = (2n+3)M_n + 3nM_{n-1}.$$

□ In fact, for $n = 1$, we obtain that $4M_2 = 4 \cdot 2 = 8 = 5 \cdot 1 + 3 \cdot 1 = 5M_1 + 3M_0$. For $n = 2$, it holds that $5M_3 = 5 \cdot 4 = 20 = 7 \cdot 2 + 6 \cdot 1 = 7M_1 + 6M_1$. For $n = 3$, we have $6M_4 = 6 \cdot 9 = 54 = 9 \cdot 4 + 9 \cdot 2 = 9M_3 + 9M_2$.

A general proof can be obtained using an approach that is similar to the combinatorial proof of the relation $(2n+1)C_n = (4n-2)C_{n-1}$ (see [Sula01] and [PWZ96]; see also Chapter 3). Alternatively, the *Motzkin triangle* can be used; we discuss this possibility in the following section. □

So, we can state that:

- *the nth Motzkin number can be calculated using the following recurrent relation:*
$$M_n = \frac{2n+1}{n+2}M_{n-1} + \frac{3n-3}{n+2}M_{n-2}, \quad M_0 = M_1 = 1.$$

Motzkin's triangle

4.1.18. An even simpler method for the construction of the sequence of Motzkin numbers involves using a special number triangle.

Motzkin's triangle is a number triangle in which there is a unity at the vertex and each of the remaining numbers are equal to the sum of the three numbers above it (left-up, up, and right-up) in the previous row. As always, if the element is missing, then it is considered to be equal to zero:

```
 1
 1   1
 2   2   1
 4   5   3   1
 9  12   9   4   1
21  30  25  14   5   1
51  76  69  44  20   6   1
 ⋮   ⋮   ⋮   ⋮   ⋮   ⋮   ⋮
```

Denoting the kth element of the nth row ($n \geq 0$, $0 \leq k \leq n$) of Motzkin's triangle with the symbol $m(n, k)$, we get the following recurrent relation for the elements of the triangle:

$$m(n+1, k) = m(n, k-1) + m(n, k) + m(n, k+1).$$

The initial conditions are $m(0,0) = 1$, $m(0, k) = m(0, -k) = 0$ for any positive integer k and $m(n, -k) = m(n, n+k) = 0$ for any positive integers n and k.

It is not difficult to see that the left-hand side of the constructed triangle is formed by Motskin numbers:

- *The left-hand side of the Motzkin's triangle consists of the Motzkin numbers* $1, 1, 2, 4, 9, \ldots$: $m(n, 0) = M_n$, $n \geq 0$.

☐ It is easy to see that the element $m(n, k)$ of the constructed triangle is equal to the number of paths along the triangle from the point corresponding to the element $m(0, 0)$ to the point corresponding to the element $m(n, k)$ if only the steps down, left-down, and right-down movements are allowed, and moving to the left of the left side of the triangle is forbidden. By definition, any path to the point corresponding to $m(n, k)$ can be obtained by a step from the point corresponding to $m(n-1, k-1)$ (if possible) by a step from the point corresponding to $m(n-1, k)$ and by a step from a point corresponding to $m(n-1, k+1)$ (if possible). In any case, the number of paths to the point corresponding to $m(n, k)$ is the sum of the numbers of paths to the points corresponding to $m(n-1, k-1)$, $m(n-1, k)$, and $m(n-1, k+1)$. In other words, $m(n, k) = m(n-1, k-1) + m(n-1, k) + m(n-1, k-1)$. It is obvious that the initial conditions also coincide.

It follows that all the elements on the right-hand side of the triangle are equal to 1: there exists only one path to the point $m(n, n)$; it consists only of the right-down steps.

As for the elements $m(n, 0)$, it is easy to see now that the number of paths from the point corresponding to $m(0, 0)$ to the point corresponding to $m(n, 0)$ is equal to the number of routes on the integer lattice $\mathbb{Z}^2$ from the point $(0, 0)$ to the point $(n, 0)$ if only the right-up, right, and right-down steps are allowed, and it is forbidden to go below the OX-axis. However, we know quite well that the number of such routes is equal to the nth Motzkin number M_n. So, we have proven that $m(n, 0) = M_n$ for any non-negative integer n.

Note that the numbers $m(n, k)$ count more general lattice paths. In terms of random walks on the number line, beginning at the origin $(0, 0)$, requiring that at each step we move from $(s, 0)$ to either $(s + 1, 0)$ or $(s - 1, 0)$, $s = 0, 1, 2, \ldots$, yields a Catalan family, but if we also allow moves in place, we get a Motzkin family. In fact, Catalan and Motzkin numbers count corresponding random walks in n steps, ending at the origin $(0, 0)$. It turns out that the elements $m(n, k)$ of the Motzkin triangle enumerate the number of random walks from $(0, 0)$ to $(k, 0)$ in n steps (see [DoSh77]). □.

4.1.19. Note that instead of the right Motzkin's triangle, we can construct the isosceles Motzkin's triangle using the same recurrent rule: $m(n, k) = m(n - 1, k - 1) + m(n - 1, k) + m(n - 1, k + 1)$. It means that in order to obtain any internal element, we should add two elements, placing them left-up and right-up, as well as placing the next right-up element of the previous row. The construction is presented as follows (sequence A064189 in the OEIS):

$$
\begin{array}{ccccccccccccc}
 & & & & & & 1 & & & & & & \\
 & & & & & 1 & & 1 & & & & & \\
 & & & & 2 & & 2 & & 1 & & & & \\
 & & & 4 & & 5 & & 3 & & 1 & & & \\
 & & 9 & & 12 & & 9 & & 4 & & 1 & & \\
 & 21 & & 30 & & 25 & & 14 & & 5 & & 1 & \\
51 & & 76 & & 69 & & 44 & & 20 & & 6 & & 1 \\
\vdots & & \vdots & & \vdots & & \vdots & & \vdots & & \vdots & & \vdots
\end{array}
$$

As before, Motzkin numbers form the left-hand side of the constructed isosceles triangle.

It is also easy to see that the numbers forming the first descending diagonal of the triangle are consecutive positive integers. In our notation, it means that $m(n, n - 1) = n$ for any positive integer n. This fact immediately follows from the recurrent definition of the triangle.

However, we can use for the construction of any internal element two elements right-up and left-up plus the previous element left-up. It leads to the recurrent relation $m(n,k) = m(n-1, k-2) + m(n-1, k-1) + m(n-1, k)$. The resulting triangle is presented as follows (sequence A026300 in the OEIS); it is simply a reflected version of the first isosceles triangle:

```
                          1
                      1       1
                  1       2       2
              1       3       5       4
          1       4       9      12       9
      1       5      14      25      30      21
  1       6      20      44      69      76      51
  ⋮       ⋮       ⋮       ⋮       ⋮       ⋮       ⋮
```

We find Motzkin numbers on the right-hand side of this construction.

4.1.20. Using the construction of Motzkin's triangle, it is possible to obtain some interesting connections between its elements. For example, we can prove (see [NeSz22]) that

- *for the nth row of Motzkin's triangle, it holds that*

$$k(n-k+3)m(n, n-k)$$
$$= (n-k+1)(n-k+3)m(n, n-k+1)$$
$$+ (n-k+1)(2n-k+4)m(n, n-k+2).$$

On the other hand, there exists a recurrent formula connecting the three consecutive elements $m(n, n-k)$, $m(n-1, n-k)$, and $m(n-2, n-k)$ of the kth vertical (see [NeSz22]):

$$k(2n-k+2)m(n, n-k) = n(2n+1)m(n-1, n-k)$$
$$+ 3n(n-1)m(n-2, n-k).$$

For $k = n$, we obtain that

$$n(n+2)m(n, 0) = n(2n+1)m(n-1, 0) + 3n(n-1)m(n-2, 0).$$

As $m(n,0) = M_n$, we get the recurrent relation for Motzkin numbers, as discussed above:

$$(n+2)M_n = (2n+1)M_{n-1} + (3n-3)M_{n-2}, \quad M_0 = M_1 = 1.$$

Number-theoretic properties of Motzkin numbers

4.1.21. Exactly four Motzkin primes are known: 2, 127, 15511, and 953467954114363 (sequence A092832 in the OEIS). They correspond to the indices $n = 2, 7, 12, 36$ (sequence A092831 in the OEIS). There are no other Motzkin prime with the indices $n \leq 2 \cdot 10^7$.

The sequence may be finite for the reason that with increasing n, the density of trivially composite Motzkin numbers approaches 1.

For $7 \cdot 10^6 < n < 20 \cdot 10^6$, all Motzkin numbers have a small factor, not exceeding 63809.

Rowland and Yassawi [RoYa13], and later Burns [Burn16], established the asymptotic densities of M_n modulo primes up to 29. In particular, the asymptotic densities of M_n, equal to zero modulo 3, 7, 17, or 19, are 1.

4.1.22. The number of decimal digits in the number $M(10^n)$, for $n = 0, 1, 2, 3, \ldots$, is equal to $1, 4, 45, 473, 4766, 47705, 477113, \ldots$ (sequence A114473 in the OEIS).

If we divide these values by 10^n, we obtain a sequence with a known limit at $n \to \infty$. In fact, $\lim_{n\to\infty} \frac{M_{10^n}}{10^n} = 3$, while $\lim_{n\to\infty} \frac{\lg M_{10^n}}{n} = \lg 3 = 0{,}4771212547\cdots$ (see sequence A114490 in the OEIS).

4.1.23. An integral representation of Motzkin numbers can be given by

$$M_n = \frac{2}{\pi} \int_0^\pi \sin(x)^2 (2\cos(x)+1)^n dx.$$

Moreover, Motzkin numbers show asymptotic behavior:

$$M_n \sim \frac{1}{2\sqrt{\pi}} \left(\frac{3}{n}\right)^{3/2} 3^n, \quad n \to \infty.$$

Motzkin numbers in the family of special numbers

4.1.24. As was shown before, Motzkin numbers are closely connected with Catalan numbers; in fact, these number sets have similar recurrent definitions and are connected with similar counting problems.

In 2014, Sun [Sun14] introduced the *generalized Motzkin numbers*:

$$M_n(b,c) = \sum_{k=0}^{\lfloor \frac{n}{2} \rfloor} \binom{n}{2k} C_k b^{n-2k} c^k, \quad b, c \in \mathbb{N}, \ n = 0, 1, 2, \ldots .$$

The generalized Motzkin numbers are common generalizations of Motzkin numbers and Catalan numbers: $M_n(1,1) = M_n$, $M_n(2,1) = C_{n+1}$. Moreover, the numbers $M_n(0,1)$ form the sequence $C_0, 0, C_1, 0, C_2, 0, \ldots$ of Catalan numbers separated by zeros.

For generalized Motzkin numbers, the following recurrent relation holds (see [WaZh15]):

$$M_{n+1}(b,c) = bM_n(b,c) + c(M_0(b,c)M_{n-1}(b,c)$$
$$+ M_1(b,c)M_{n-2}(b,c) + \cdots + M_{n-1}(b,c)M_0(b,c)),$$

with the initial condition $M_0(b,c) = 1$. However, there exists another recurrence for the sequence $M_n(b,c)$, $n = 0, 1, 2, \ldots$ (see [Sun14], [WaZh15], and [PWZ96]):

$$(n+3)M_{n+1}(b,c) = b(2n+3)M_n(b,c) - (b^2 - 4c)nM_{n-1}(b,c).$$

It allows us to obtain the generating function for the sequence $M_n(b,c)$, $n = 0, 1, 2, \ldots$ of generalized Motzkin numbers. It has the form $f(x) = \frac{1-bx-\sqrt{(1-bx)^2-4cx^2}}{2cx^2}$, i.e., it holds that

$$\frac{1-bx-\sqrt{(1-bx)^2-4cx^2}}{2cx^2} = M_0(b,c) + M_1(b,c)x + M_2(b,c)x^2$$
$$+ \cdots + M_n(b,c)x^n + \cdots .$$

It can also be proven (see [WaZh15]) that

$$\sum_{k=0}^{n} \binom{n}{k} M_k(b, c) = M_n(b+1, c).$$

In [Sun14], the reader can find many arithmetical properties of this number set.

Exercises

1. For $n = 2, 3, 4, 5, 6, 7$, calculate the nth Motzkin number M_n, using the recurrent relation $M_n = M_{n-1} + \sum_{k=0}^{n-2} M_k M_{n-2-k}$; calculate the nth Motzkin number M_n using the recurrent relation $M_n = \frac{3(n-1)M_{n-2} + (2n-1)M_{n-1}}{n+2}$.
2. For $n = 3, 4, 5, 6, 7, 8, 9, 10$, calculate the nth Motzkin number M_n using the closed formula $M_n = \sum_{k=0}^{\lfloor \frac{n}{2} \rfloor} \binom{n}{2k} C_k$.
3. Prove that the generating function $f(x)$ of the sequence of Motzkin numbers can be written as a continuous fraction,

$$f(x) = \cfrac{1}{1 - x - \cfrac{x^2}{1 - x - \cfrac{x^2}{1 - x - \cfrac{x^2}{\ldots}}}}.$$

4. Consider a number line with fixed integers. Prove that M_n gives the number of circular n-step routes along the number line from point 0 to the same point 0 if each step allows a movement by one unit to the right, by one unit to the left, or just the "empty movement" (a stop on the place), but it is forbidden to move to the left of point 0.
5. Prove that the nth Motzkin number M_n is the number of positive integer sequences of length $n - 1$, in which the beginning and ending elements are either 1 or 2, and the difference between any two consecutive elements is $-1, 0$, or 1.
6. Prove that the nth Motzkin number M_n is the number of positive integer sequences of length $n + 1$, in which the beginning and ending elements are 1, and the difference between any two consecutive elements is $-1, 0$, or 1.

7. Check that $M < 3M_{n-1}$ and $M_n^2 < M_{n-1}M_{n+1}$, $n = 1, 2, 3, 4, 5, 6, 7, 8, 9$. Prove these inequalities. Prove that $\lim_{n\to\infty} \frac{M_{n-1}}{M_n} = \frac{1}{3}$.

4.2. Schröder Numbers
A history of the question

4.2.1. In mathematics, the nth *Schröder number* S_n, also called a *large Schröder number*, or a *big Schröder number*, describes the number of lattice paths from the southwest corner $(0,0)$ of an $n \times n$ grid to the northeast corner (n,n) using only single steps north, $(0,1)$, northeast, $(1,1)$, or east, $(1,0)$, that do not rise above the southwest–northeast diagonal.

R. Stanley, professor at Massachusetts Polytechnic Institute, states that Hipparchus counted up to the 10th Schröder number 1,037,718. However, these set of numbers was named after a German mathematician Ernst Schröder (1841–1902).

E. Schröder learned mathematics at Heidelberg, Königsberg, and Zürich, under Otto Hesse, Gustav Kirchhoff, and Franz Neumann. After teaching school for a few years, he moved to the Technische Hochschule Darmstadt in 1874. Two years later, he took up the chair in mathematics at the Karlsruhe Polytechnische Schule, where he spent the remainder of his life.

E. Schröder is mainly known for his work on algebraic logic. He is best known for his monumental *Lectures on the Algebra of Logic* (*Vorlesungen über die Algebra der Logik*) in three volumes (1890–1905), which paved the way for the emergence of mathematical logic as a separate discipline in the 20th century by systematizing the various systems of formal logic of the day.

E. Schröder also made original contributions to algebra, set theory, lattice theory, ordered sets, and ordinal numbers. Along with Georg Cantor, he codiscovered the *Cantor–Bernstein–Schröder theorem*, although Schröder's proof (1898) is flawed.

Recurrent construction of Schröder numbers

4.2.2. Mainly, the nth *Schröder number* (or *large Schröder number*, *big Schröder number*) is defined as the number of lattice paths in the Cartesian plane that starts at $(0,0)$, ends at (n,n), contains no points above the line $y = x$, and is composed only of steps $(0,1)$, $(1,0)$, and $(1,1)$ ([Wiki24], [Sloa24], etc.).

However, according to the structure under consideration, we define the sequence of Schröder numbers recursively.

A *Schröder number* is a member of the sequence $S_0, S_1, S_2, \ldots, S_n, \ldots$, built according to the following recurrence law:

$$S_0 = 1, \quad \text{and} \quad S_{n+1} = S_n + S_0 S_n + S_1 S_{n-1}$$
$$+ S_2 S_{n-2} + \cdots + S_n S_0 \quad \text{for } n \geq 0.$$

So, we get

$$S_0 = 1, \quad S_1 = S_0 + S_0 S_0 = 1 + 1 \cdot 1 = 2,$$
$$S_2 = S_1 + S_0 S_1 + S_1 S_0 = 2 + 1 \cdot 2 + 2 \cdot 1 = 6,$$
$$S_3 = S_2 + S_0 S_2 + S_1 S_1 + S_2 S_0 = 6 + 1 \cdot 6 + 2 \cdot 2 + 5 \cdot 1 = 22,$$
$$S_4 = S_3 + S_0 S_3 + S_1 S_2 + S_2 S_1 + S_3 S_0$$
$$= 22 + 1 \cdot 22 + 2 \cdot 6 + 6 \cdot 2 + 1 \cdot 22 = 90,$$
$$S_5 = S_4 + S_0 S_4 + S_1 S_3 + S_2 S_2 + S_3 S_1 + S_4 S_0$$
$$= 90 + 1 \cdot 90 + 2 \cdot 22 + 6 \cdot 6 + 22 \cdot 2 + 90 \cdot 1 = 394.$$

This definition allows us to get a simple rule for the consecutive construction of the members of the sequence $S_0, S_1, S_2, \ldots, S_n, \ldots$. The corresponding algorithm is as follows: *in order to get the next member of the Schröder's sequence, write in the usual order the already built members of this sequence; under them, write the same numbers but in the reverse order; multiply each top number by the number below it, and add up all the obtained elements; add the result*

252   *Catalan Numbers*

with the last of the already built members of this sequence. For example, in order to find the number S_6, *it is enough to consider the numbers* $S_0 = 1, S_1 = 2, S_2 = 9, S_3 = 22, S_4 = 90, S_5 = 394$, *write one after another the first six Schröder numbers* $S_0, S_1, \ldots, S_5$, *and write the same numbers in the reverse order below:*

S_0	S_1	S_2	S_3	S_4	S_5
1	2	6	22	90	394
394	90	22	6	2	1

Multiplying each top number by the corresponding bottom number and adding all obtained products and the number S_5, we get $S_6 = 1806$:

$$\begin{array}{cccccc} 1 & 2 & 6 & 22 & 90 & 394 \\ \times & & & & & \\ 394 & 90 & 22 & 6 & 2 & 1 \end{array}$$

$1 \cdot 394 + 2 \cdot 90 + 6 \cdot 22 + 22 \cdot 6 + 90 \cdot 2 + 394 \cdot 1 + 394 = 1806$.

Using this simple algorithm, it is easy to check that the sequence S_n, $n = 0, 1, 2, 3, \ldots$, starts with the elements $1, 2, 6, 22, 90, 394, 1806, 8558, 41586, 206098, \ldots$ (sequence A006318 in OEIS).

As $S_0 = 1$, we can obtain an other variant of the above recurrent relation:

$$S_n = 3S_{n-1} + \sum_{k=1}^{n-2} S_k S_{n-k-1}.$$

Schröder numbers and counting problems

Lattice paths' problem

4.2.3. In this section, we show that our recurrent definition of Schröder numbers is equivalent to the classical definition based on the lattice path problem:

- *The nth Schröder number* S_n *is equal to the number of lattice paths from the southwest corner* $(0,0)$ *of an* $n \times n$ *grid to the northeast corner* (n,n), *using only single steps north,* $(0,1)$, *northeast,*

$(1, 1)$, or east, $(1, 0)$, that do not rise above the southwest–northeast diagonal.

☐ It is easy to check that the statement is true for small values of n.

Thus, for $n = 0$, there exists only one possibility: we just remain at the point $(0, 0)$. So, the number of paths is equal to S_0.

For $n = 1$, there are two possible paths: "right, up" and "right-up." If we denote a right (horizontal) step $(1, 0)$ by the letter h, an up (vertical) step $(0, 1)$ by the letter v, and a right-up (diagonal) step $(1, 1)$ by the letter d, then we obtain two paths: hv and d. So, the number of paths is equal to S_1.

For $n = 2$, there are six possible paths: dd, dhv, $hvhv$, hvd, hdv, and $hhvv$. So, the number of paths is equal to S_2.

Consider now a positive integer $n \geq 3$. Suppose that the number of considered lattice paths from $(0, 0)$ to (k, k), $0 \leq k \leq n - 1$, is equal to S_k, and we try to prove that the number of considered lattice paths from $(0, 0)$ to (n, n) is equal to the nth Schröder number S_n. In order to do this, fix the first step of a given path between $(0, 0)$ and (n, n). If it is a diagonal step, then we can move from the point $(1, 1)$ to the point (n, n) without additional restrictions; so, by induction, there are exactly S_{n-1} possible paths from $(0, 0)$ to (n, n), starting from a diagonal step. If the first step is horizontal, then find and fix its "pair": the first vertical step (from the point $(s + 1, s)$ to the point $(s + 1, s + 1)$ for some $0 \leq s \leq n - 1$), which brings us to the diagonal. Between these two fixed steps, there is a square $s \times s$ grid, $s = 0, 1, 2, \ldots, n - 1$ (with the southwest corner $(1, 0)$, and the northeast corner $(s + 1, s)$), which can be passed in S_s ways. After the second fixed step, there is a square $(n - s - 1) \times (n - s - 1)$ grid, $n - s - 1 = 0, 1, 2, \ldots, n - 1$ (with the southwest corner $(s+1, s+1)$ and the northeast corner (n, n)), which can be passed in S_{n-s-1} ways. By this construction, we go through the first and second obtained grids independently. Therefore, for a fixed s, $0 \leq s \leq n - 1$, we obtain $S_s S_{n-s-1}$ possible paths from $(0, 0)$ to (n, n), starting with a horizontal step. Thus, in total, there are $S_0 S_{n-1} + S_1 S_{n-2} + \cdots + S_{n-2} S_1 + S_{n-1} S_0$ paths from $(0, 0)$ to (n, n), starting with a horizontal step. So, adding together the values obtained for the considered two cases of the first step ("diagonal" and "horizontal"), we have now exactly

$$S_{n-1} + S_0 S_{n-1} + S_1 S_{n-2} + \cdots + S_{n-2} S_1 + S_{n-1} S_0$$
$$= S_{n-1} + \sum_{i=0}^{n-1} S_i S_{n-i-1} = S_n$$

possibilities for correct routes from the point $(0,0)$ to the point (n,n). □

4.2.4. There exists another representation of Schröder numbers as the numbers of special paths on the integer lattice. In fact, consider all routes from the point $(0,0)$ to the point $(2n,0)$ of the OX-axis if only right-up, right-down, and double right steps (i.e., $(1,1)$, $(1,-1)$, and $(2,0)$ steps) are allowed, and it is forbidden to go below the OX-axis.

It is easy to see that, for $n = 0$, there exists exactly one such (empty) path. For $n = 1$, we obtain two possibilities: either one double right step or one right-up and one right-down step. For $n = 2$, we have six possibilities.

- *The nth Schröder number gives the number of routes from the point $(0,0)$ to the point $(2n,0)$ of the OX-axis if only right-up, right-down, and double right steps are allowed, and it is forbidden to go below the OX-axis.*

□ As was shown above, the statement is true for small values of n.

One of the simplest ways to derive a general proof is to construct a one-to one correspondence between two considered sets of lattice paths. In fact, this correspondence is obvious. It is enough to use a rotation in which a horizontal step from the first model moves to a right-up step of the second model, a vertical step moves to a right-down step, and a diagonal step moves to a horizontal step. However, it should be done up to the natural scaling.

However, it is possible to give a direct consideration. In fact, we have proven the statement for $n = 0, 1, 2$. Consider now a positive integer, $n \geq 3$. Suppose that the number of the considered lattice paths from $(0,0)$ to $(2k,0)$, $0 \leq k \leq n-1$, is equal to S_k, and try to prove that the number of the considered lattice paths from $(0,0)$ to $(2n, 0)$ is equal to the nth Schröder number S_n. In order to do this, fix the first step of a given path between $(0,0)$ and $(2n, 0)$. If it is a double right step, then we can move from the point $(2, 0)$ to the point $(2n, 0)$ without additional restrictions; so, by induction, there are exactly S_{n-1} possible paths from $(0,0)$ to $(2n, 0)$, starting with a double right step. If the first step is a right-up step, then find and fix its "pair": the first right-down step, which brings us to the OX-axis. The corresponding point on the OX-axis has the form $(2(s + 1), 0)$ for some $0 \le s \le n-1$. For a fixed s, there are exactly $2s + 1$ points between the points $(0,0)$ and $(2(s+1), 0)$. On the other hand, there exist exactly $2(n - s - 1) + 1$ points from the point $(2(s + 2), 0)$ to the point $(2n, 0)$. In order to obtain a correct path from $(0,0)$ to $(2n, 0)$, containing two fixed steps, we can construct any correct path from the point $(1, 1)$ to the point $(2s + 1, 1)$ (which cannot go below the line $y = 1$) and any correct path from the point $(2(s + 2), 0)$ to the point $(2n, 0)$. As there exist S_s correct paths between the first pair of points and S_{n-1-s} correct paths between the second pair, we obtain in total $S_s S_{n-1-s}$ possibilities for a fixed s, $0 \le s \le n - 1$. Thus, in total there are $S_0 S_{n-1} + S_1 S_{n-2} + \cdots + S_{n-2} S_1 + S_{n-1} S_0$ paths from $(0, 0)$ to $(2n, 0)$, starting from a right-up step. So, adding together the values obtained for the considered two cases of the first steps ("double horizontal" and "right-up"), we have now exactly

$$S_{n-1} + S_0 S_{n-1} + S_1 S_{n-2} + \cdots + S_{n-2} S_1 + S_{n-1} S_0$$
$$= S_{n-1} + \sum_{i=0}^{n-1} S_i S_{n-i-1} = S_n$$

possibilities for the correct routes from the point $(0, 0)$ to the point $(2n, 0)$. □

Brackets' problem

4.2.5. As before, consider a *regular bracket* sequence, i.e., a correct arrangement of k pairs of brackets. These arrangements are numerated by Catalan numbers. By adding several zeros to a regular bracket sequence, we obtain a regular bracket sequence separated by zeros. These arrangements are numerated by Motzkin numbers.

Consider now a *regular bracket sequence separated by pairs of zeros*. It means that we can add to a correct bracket arrangement only 00-constructions. How many such sequences of length $2n$ exist?

In fact, there exists exactly one regular bracket sequence separated by pairs of zeros of length zero. Furthermore, there exist exactly two regular bracket sequences separated by pairs of zeros of length 2: () and 00. For $n = 2$, we obtain six sequences of length 4: 0000, 00(), ()00, ()(), (00), and (()). It is not difficult to obtain all 22 sequences of length 6. Each such sequence corresponds to some correct path from the point $(0,0)$ to the point $(6,0)$.

- *The number of regular bracket sequences separated by pairs of zeros of length $2n$ is equal to the nth Schröder number S_n, $n \geq 0$.*

☐ In order to prove this statement, it is enough to establish a one-to-one correspondence between the set of all Schröder's lattice paths from the point $(0,0)$ to the point $(2n,0)$ and the set af all regular bracket sequences separated by pairs of zeros of length $2n$. The correspondence is obvious: instead of a right-up step place an opening bracket; instead of a right-down step, place a closing bracket; instead of a double right step, place a pair of zeros.

However, the reader can easily obtain a direct consideration using the approach considered above: we should divide all possible sequences into two groups: any sequence of the first group starts with 00; any sequence of the second group starts with an opening bracket. ☐

4.2.6. There are many other formulations of the bracket problem.

For a given positive integer n, define a *Dyck word with pairs of zeros of length $2n$* as a sequence consisting of k symbols X, k symbols Y, and $n - k$ symbols 00, in which each initial segment contains at least as many symbols X as symbols Y.

For $n = 0$, we have exactly one Dyck word with pairs of zeros; for $n = 1$, we have two such words of length 2: XY and 00. For $n = 2$, six such words of length 4 exist: $0000, 00XY, XY00, XYXY, X00Y$, and $XXYY$. For $n = 3$, we have nine words of length 6, etc.

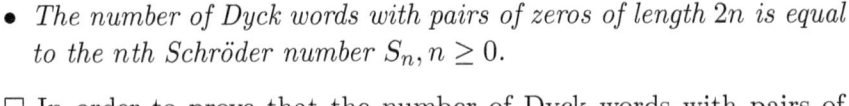

- The number of Dyck words with pairs of zeros of length $2n$ is equal to the nth Schröder number $S_n, n \geq 0$.

□ In order to prove that the number of Dyck words with pairs of zeros of length $2n$ is S_n, it is enough to replace the symbol X by an opening bracket and the symbol Y by a closing bracket. Such a transform gives a bijection between the set of all Dyck words with pairs of zeros of length $2n$ and the set of all regular bracket sequences separated by pairs of zeros of length $2n$. □

For a given positive integer n, consider the set of sequences of length $2n$, consisting of k "1", k "-1", and $n - k$ pairs of zeros, all partial sums of which are non-negative. Here, k can take any possible value from zero up to n. We call any such sequence a *correct $(1, -1)$-sequence separated by pairs of zeros*.

For $n = 0$, there is only one such sequence; for $n = 1$, we have two sequences of length 2: 0, 0 (with the partial sum 0) and 1, -1 (with the partial sums 1, 0); for $n = 2$, there exist six such sequences of length 4: 0, 0, 0, 0 (with the partial sums 0, 0, 0, 0), 0, 0, 1, -1, (with the partial sums 0, 0, 1, 0), 1, -1, 0, 0, (with the partial sums 1, 0, 0, 0), 1, -1, 1, -1 (with the partial sums 1, 0, 1, 0), 1, 0, 0, -1 (with the partial sums 1, 1,1, 0), and 1, 1, -1, -1 (with the partial sums 1, 2, 1, 0); for $n = 3$, there exist 22 such sequences of length 6, etc.

It is easy to prove that:

- the number of correct $(1, -1)$-sequences separated by pairs of zeros of length $2n$ is equal to the nth Schröder number $S_n, n \geq 0$.

□ In order to prove this statement, consider a Dyck word with pairs of zeros of length $2n$. If instead of the symbols X and Y, we use the symbols "1" and "-1", respectively, we get a sequence composed of k "1", k "-1", and $n - k$ pairs of zeros, all partial sums of which are non-negative. However, we can use for the proof correct bracket structures separated by pairs of zeros with k pairs of brackets: we can simply replace the symbol "(" with the symbol "1" and the symbol ")" with the symbol "-1". □

For a given positive integer n, define a *mountain with k ascents, k descents, and $n - k$ superplateau* as a broken line, composed of

k segments of the same length inclined at an angle of 45° to the positive direction of the OX-axis, by k segments of the same length inclined at an angle of 135° to the positive direction of the OX-axis, and $n - k$ segments of the double length, parallel to the OX-axis, such that the resulting "start" and "end" place themselves on the same level. Call any such construction a *mountain with superplateau of length* $2n$.

- The number of mountains with superplateaus of length $2n$ is equal to the nth Schröder number $S_n, n \geq 0$.

☐ The proof can be obtained using correct brackets structures separated by pairs of zeros. In fact, it is enough to build a one-to-one correspondence between the set of all correct bracket structures separated by pairs of zeros with k pairs of brackets and the set of mountains with superplateaus, having k ascents and k descents. The corresponding law is very simple: each "ascending" segment corresponds to an opening bracket; each "descending" segment corresponds to a closing bracket; each superplateau segment corresponds to a pair of zeros. ☐

However, an even more natural correspondence exists between the mountains with superplateaus of length $2n$ and Schröder's paths between the points $(0,0)$ and $(2n, 0)$. In fact, in both cases, we have the same structures.

Enumeration of quasi-trees

4.2.7. It is widely known that Catalan numbers appear quite often when counting the number of some classes of trees, while Motzkin numbers count certain classes of quasi-trees (i.e., trees with loops).

For a further generalization, define a *quasi-tree with pairs of loops* (or *quasi-tree with bows*) as a graph, obtained from a tree by adding pairs of loops to some vertices. By definition, any tree is a quasi-tree with zero bows.

We can use constructions, considered in the first section of this chapter, in order to obtain the following statement on the set of quasi-trees with pairs of loops with $2n + 1$ edges.

Define a *planted binary quasi-tree with bows* as a planted binary tree with pairs of loops added to some vertices, excluding the lowest leave (the root).

- *The number of planted binary quasi-trees with bows, having $2n+1$ edges, is equal to the nth Schröder number $S_n, n \geq 0$.*

□ In order to prove this statement, it is enough to obtain a bijection between the considered objects and the correct bracket sequences separated by pairs of zeros: for a given quasi-tree, mark any bow (a pair of loops) with a pair of zeros 00; mark any internal vertex with an opening bracket; mark any hanging vertex (except the root and the last right leaf of the quasi-tree) with a closing bracket; gather all marks, moving clockwise along the quasi-tree. □

Call a *full binary quasi-tree with bows* a graph, obtained by attached pairs of loops to some vertices of a full binary tree.

- *The number of full binary quasi-trees with bows, having $2n$ edges, is equal to the nthe Schröder number S_n, $n \geq 0$.*

□ It is easy to see that any planted binary quasi-tree with bows, having $2n + 1$ edges, corresponds to a full binary quasi-tree with bows, having $2n$ edges. In fact, it is enough to choose as the root the lowest internal vertex of the planted quasi-tree and delete the lowest hanging vertex (the root of the considered planted binary quasi-tree).
□

4.2.8. Define a *plane rooted quasi-tree with bows* as a plane rooted tree separated by pairs of loops, i.e., as a graph, obtained by attaching pairs of loops to some vertices of a plane rooted tree. Recall that in this case, the order of the obtained edges is important.

For a given plane rooted quasi-tree with bows, let us define its *length* as the number of steps of a worm crawling clockwise around the quasi-tree. For each loop, we have exactly one "round" step; for each edge, we have exactly two steps, one up-step and one down-step.

- *The number of plane rooted quasi-trees with bows of length $2n$ is equal to the nth Schröder number $S_n, n \geq 0$.*

☐ The simplest way to prove this statement is to build a one-to-one correspondence between the set of all plane rooted quasi-trees with bows of length $2n$ and the set of all correct bracket sequences separated by pairs of zeros of length $2n$. In fact, for a given plane rooted quasi-tree with bows of length $2n$, let us mark each of its bow by 00 and each of its edge by a pair of brackets, placing an opening bracket to the left and a closing bracket to the right. Now, starting from the root and moving from the left to the right along the edges of the graph, we can collect all $2n$ attached symbols, obtaining a corresponding correct arrangement. In fact, if there are k edges in a plane rooted quasi-tree, $0 \le k \le n$, then there are $n-k$ pairs of zeros, and we obtain a correct sequence, containing k pairs of brackets and $n-k$ pairs of zeros. ☐

However, it can be shown that:

- the nth Schröder number S_n gives the number of plane trees having n edges with leaves colored by one of two colors.

☐ For a proof, see, for example, [MaSu08]. ☐

Generalized polygon division problem

4.2.9. *Euler's polygon division problem* is the problem of finding the number of ways a plane convex polygon of n sides can be divided into triangles by non-crossing diagonals. It is well known that the nth Catalan number C_n gives the number of such triangulations for a convex $(n+2)$-gon.

On the other hand, we have shown that the number of polygons with $n+1$ parallel edges is equal to the nth Motzkin number M_n.

Here, a *polygon with parallel edges* is defined as a convex polygon with its interior divided into triangles and two angles by non-crossing *edges*. An edge is an arc connecting two vertices of the polygon. It can be a side of the polygon, a diagonal of the polygon, or any additional arc "parallel" to a side or a diagonal of the polygon, i.e., passing from the same vertices and placed in the interior of the polygon.

Now, choose from all *polygons with parallel edges* only those in which all two-angular faces are glued in pairs. Any such glued pair is a graph on two vertices with three parallel edges, i.e., it has the form of a *grain*. So, the only internal faces of a division of such polygon are triangles and grains. Define a *polygon with grains* as any polygon with parallel edges, having as internal faces only triangles and grains.

Let us assume that there exists exactly one polygon with grains, having one edge; it is a segment, represented as a complete graph on two vertices. Furthermore, there are exactly two such polygons with three edges; first, it is just a grain, i.e., a graph on two vertices with three parallel edges; second, it is a triangle. There are exactly six such polygons with five parallel edges; first, we have two glued grains, i.e., a graph on two vertices, having five parallel edges; second, we have three triangles with a grain, attached to one of the sides; at last, we have two Euler's divisions of a square. Similarly, there are 22 possible polygons with grains, having seven edges, etc.

- *The number of polygons with grains having $2n+1$ edges is equal to the nth Schröder number S_n, $n \geq 0$.*

□ In order to prove this fact, it is enough to establish a bijection between the set of these polygons and the set of all planted binary quasi-trees with bows. The algorithm is similar to the corresponding algorithm used in the previous section for the enumeration of polygons with parallel edges. □

4.2.10. On the other hand, there exists a classical problem connecting Schröder numbers with some special divisions of a rectangle.

In fact, the nth Schröder number S_n gives the number of ways to divide a rectangle into $n+1$ smaller rectangles using n cuts through n points given inside the rectangle in general positions, with each cut intersecting one of the points and dividing only a single rectangle into two (i.e., the number of structurally different guillotine partitions). This is similar to the process of triangulation, in which a shape is divided into non-overlapping triangles instead of rectangles.

- *The nth Schröder number S_n is equal to the number of ways to split a rectangle into $n+1$ smaller rectangles using n sections if* inside the rectangle is a set of n points in general position (no two of them lie on a straight line parallel to the sides of the rectangle), to the sides of the rectangle, and each cut must pass through one of these points and divide only one rectangle into two.

☐ Let $s(n)$ be the number of rectangulations of a rectangle with n fixed internal points. Clearly, $\frac{s(n)}{2}$ rectangulations contain a vertical segment cutting the bounding rectangle into two rectangles, while the remaining $\frac{s(n)}{2}$ rectangulations contain a horizontal segment cutting the bounding rectangle into two rectangles. Considering only the first set and denoting by k the first point left-to-right, through which passes a vertical segment cutting the bounding rectangle into two, we derive the following recurrent formula for $s(n)$:

$$\frac{s(n)}{2} = s(n-1) + \sum_{k=2}^{n} \frac{s(k-1)}{2} s(n-k), \quad s(0) = 1.$$

The formula holds since, for $k = 1$, there are $s(n-1)$ rectangulations, while for $2 \le k \le n$, the segment through the kth point splits the bounding rectangle into two rectangles: the right one has $s(n-k)$ rectangulations, while the left one has $\frac{s(k-1)}{2}$ rectangulations, as it must be cut into two rectangles by a horizontal segment. So, it holds that

$$s(n) = 2s(n-1) + \sum_{k=2}^{n} s(k-1)s(n-k)$$
$$= s(n-1) + s(0)s(n-1) + s(1)s(n-2)$$
$$+ \cdots + s(n-1)s(0), \quad s(0) = 1.$$

Therefore, the sequence $s(n)$, $n = 0, 1, 2, 3, \ldots$, coincides with the sequence S_n, $n = 0, 1, 2, 3, \ldots$, and the statement is proven (see [ABP06]). ☐

The nth Schröder number S_n also counts the *separable permutations* of length $n-1$. Such permutations may be characterized by the forbidden permutation patterns 2413 and 3142. For a proof, see, for example, [West95].

Closed formula for Schröder numbers

4.2.11. Considering a recurrent approach to the construction of Schröder numbers, we have proven that Schröder numbers count, for example, the correct bracket sequences separated by pairs of zeros. In fact, the nth Schröder number gives the number of all correct bracket sequences separated by pairs of zeros of length $2n$. But it is easy to count such sequences by another algorithm. It gives us the following closed formula for Schröder numbers:

- For the nth Schröder number S_n, it holds that

$$S_n = \sum_{k=0}^{n} \binom{n+k}{2k} C_k,$$

where $\binom{n+k}{k}$ are binomial coefficients and C_k are Catalan numbers.

☐ Any correct bracket sequence separated by pairs of zeros has several pairs of brackets which, in turn, are correctly placed in this sequence. In other words, the subsequence consisting of only brackets forms a correct bracket arrangement. It is easy to see that the number k of bracket pairs in a correct sequence of length $2n$ satisfies the conditions $0 \leq k \leq n$. If there are k pairs of brackets in a sequence of length $2n$, $0 \leq k \leq n$, then there are exactly $2n - 2k$ zeros, i.e., more precisely, exactly $n-k$ symbols 00. In order to obtain a correct arrangement of k pairs of brackets and $n-k$ pairs of zeros, we should choose some places for 00. In order to do this, let us first choose $n-k$ places for the "second zeros" of each zero pair. Note that in this case, we cannot choose the first place; moreover, the chosen places cannot be neighbors. So, let us block the first place. Then, there are $2n - 1$ positions, and $n - k$ of them are occupied by the "second zeros." So, there are $n - k - 1$ other symbols. The "second zeros" can be placed before, between, or after other symbols. So, there exist $n + k$ possible places for the "second zeros." We can choose $n - k$ of them in $\binom{n+k}{n-k} = \binom{n+k}{2k}$ ways. Thus, we arrange $n-k$ "second zeros"; hence, we have arranged all $n-k$ pairs of zeros using $2n - 2k$ places from the initial $2n$ places. The remaining $2k$ places will be used for a correct arrangement of k pairs of brackets.

Among the $2k$ places, we can order k pairs of brackets in C_k ways, where C_k is the kth Catalan number. So, for a fixed $0 \le k \le n$, we have $\binom{n+k}{2k} C_k$ possible correct sequences separated by pairs of zeros. Therefore, the total number of correct sequences of length $2n$ is $\sum_{k=0}^{n} \binom{n+k}{2k} C_k$. On the other hand, it was proven that the number of such sequences is equal to the nth Schröder number. So, we have proven that $S_n = \sum_{k=0}^{n} \binom{n+k}{2k} C_k$. □

In fact, we have that

$$S_0 = \binom{0}{0} C_0 = 1 \cdot 1 = 1; \quad S_1 = \binom{1}{0} C_0 + \binom{2}{2} C_1 = 1 \cdot 1 + 1 \cdot 1 = 1;$$

$$S_2 = \binom{2}{0} C_0 + \binom{3}{2} C_1 + \binom{4}{4} C_2 = 1 \cdot 1 + 3 \cdot 1 + 1 \cdot 2 = 6;$$

$$S_3 = \binom{3}{0} C_0 + \binom{4}{2} C_1 + \binom{5}{4} C_2 + \binom{6}{6} C_3$$
$$= 1 \cdot 1 + 6 \cdot 1 + 5 \cdot 2 + 1 \cdot 4 = 22.$$

Using the well-known properties of binomial coefficients and the closed formula $C_k = \frac{1}{k+1} \binom{2k}{k}$ for the nth Catalan number, we can obtain the following variants of the closed formula for Schröder numbers:

- *For the nth Schröder number S_n, it holds that*

$$S_n = \sum_{k=0}^{n} \binom{2n-k}{k} C_{n-k} = \sum_{k=0}^{n} \frac{\binom{n}{k}\binom{n+k}{k}}{k+1}.$$

□ The first variant can be obtained from the proved equality by a change of the summation's index: $k \to n-k$. On the other hand, we can simply repeat the proof under the condition that there are exactly k pairs of zeros. The second variant follows from simple transformations:

$$S_n = \sum_{k=0}^{n} \binom{n+k}{2k} C_k = \sum_{k=0}^{n} \frac{1}{k+1} \cdot \frac{(2k)!}{k!k!} \cdot \frac{(n+k)!}{(2k)!(n-k)!}$$

$$= \sum_{k=0}^{n} \frac{1}{k+1} \cdot \frac{n!}{k!(n-k)!} \cdot \frac{(n+k)!}{n!k!} = \sum_{k=0}^{n} \frac{\binom{n}{k}\binom{n+k}{k}}{k+1}. \quad □$$

Generating function for the sequence of Schröder numbers

4.2.12. In this section, we consider the generating function of the sequence of Schröder numbers.

- *The generating function $f(x)$ of the sequence $1, 2, 6, 22, 90, \ldots$ of Schröder numbers has the form $f(x) = \frac{1-x-\sqrt{1-6x+x^2}}{2x}$, i.e., it holds that*

$$\frac{1-x-\sqrt{1-6x+x^2}}{2x}$$
$$= S_0 + S_1 \cdot x + \cdots + S_n \cdot x^n + \cdots, \quad |x| \leq 3 - 2\sqrt{2}.$$

☐ Consider a formal decomposition:

$$f(x) = S_0 + S_1 x + S_2 x^2 + \cdots + S_n x^n + \cdots.$$

Then, the following relationships can be obtained:

$$(f(x))^2 = f(x)f(x) = (S_0 + S_1 x + S_2 x^2 + \cdots + S_n x^n + \cdots)$$
$$\times (S_0 + S_1 x + S_2 x^2 + \cdots + S_n x^n + \cdots)$$
$$= S_0 \cdot S_0 + (S_0 \cdot S_1 + S_1 \cdot S_0) x$$
$$+ (S_0 \cdot S_2 + S_1 \cdot S_1 + S_2 \cdot S_0) x^2 + \cdots$$
$$+ (S_0 \cdot S_n + S_1 \cdot S_{n-1} + \cdots + S_n \cdot S_0) x^n + \cdots;$$
$$x(f(x))^2 = S_0 \cdot S_0 x + (S_0 \cdot S_1 + S_1 \cdot S_0) x^2 + \cdots$$
$$+ (S_0 \cdot S_n + S_1 \cdot S_{n-1} + \cdots + S_n \cdot S_0) x^{n+1} + \cdots;$$
$$xf(x) = S_0 x + S_1 x^2 + \cdots + S_n x^n + \cdots;$$
$$x(f(x))^2 + xf(x) = (S_0 + S_0 \cdot S_0) x + (S_1 + S_0 \cdot S_1 + S_1 \cdot S_0) x^2$$
$$+ \cdots + (S_n + S_0 \cdot S_n + S_1 \cdot S_{n-1} + \cdots$$
$$+ S_n \cdot S_0) x^{n+1} + \cdots$$
$$= S_1 x + S_2 x^2 + \cdots + S_{n+1} x^{n+1} + \cdots$$
$$= (S_0 + S_1 x + S_2 x^2 + \cdots + S_{n+1} x^{n+1} + \cdots)$$
$$- S_0 = f(x) - 1.$$

So, we have the following quadratic equation with respect to $f(x)$:
$$x(f(x))^2 + (x-1)f(x) + 1 = 0.$$
As $D = (x-1)^2 - 4x = 1 - 6x + x^2$, then
$$f_1(x) = \frac{1 - x - \sqrt{1 - 6x + x^2}}{2x}, \quad f_2(x) = \frac{1 - x + \sqrt{1 - 6x + x^2}}{2x}.$$
In order to choose the sign of the square root, let us note that
$$xf(x) = S_0 x + S_1 x^2 + \cdots + S_n x^{n+1} + \cdots,$$
and at $x = 0$, we should obtain zero. But $xf_2(x) = 0.5(1 - x + \sqrt{1 - 6x + x^2})$, which gives unity at $x = 0$. On the other hand, $xf_1(x) = 0.5(1 - x - \sqrt{1 - 6x + x^2})$ gives zero at $x = 0$. Therefore, we obtain that the generating function of the sequence S_n, $n = 0, 1, 2, 3, \cdots$, has the form
$$f_1(x) = \frac{1 - x - \sqrt{1 - 6x + x^2}}{2x}.$$

In order to find the radius of convergence, note that the condition $|x^2 - 6x| < 1$ holds for $|x| < 3 - 2\sqrt{2}$. Therefore, we obtain that
$$\frac{1 - x - \sqrt{1 - 6x + x^2}}{2x}$$
$$= S_0 + S_1 x + S_2 x^2 + \cdots + S_n x^n + \cdots, \quad |x| \leq 3 - 2\sqrt{2}. \quad \square$$

Properties of Schröder numbers

Simplest properties of Schröder numbers

4.2.13. Schröder numbers have many interesting properties.

We have proven a closed formula for Schröder numbers which connects three number sets, namely Schröder numbers, Catalan numbers, and the elements of the Pascal's triangle:
$$S_n = \sum_{k=0}^{n} \binom{n+k}{2k} C_k = \sum_{k=0}^{n} \binom{2n-k}{k} C_{n-k} = \sum_{k=0}^{n} \frac{\binom{n}{k}\binom{n+k}{k}}{k+1}.$$

We can also obtain another recurrent relation for Schröder numbers. It is based on the following result:

- *The three consecutive Schröder numbers are connected by the following linear recurrent relation of the third order:*

$$(n+2)S_{n+1} = (6n+3)S_n - (n-1)S_{n-1}, \quad n \geq 1.$$

□ In fact, for $n = 1$, we obtain that $3S_2 = 3 \cdot 6 = 18 = 9 \cdot 2 - 0 \cdot 1 = 18 = 9 \cdot S_1 - 0 \cdot S_0$. For $n = 2$, it holds that $4S_3 = 4 \cdot 22 = 88 = 15 \cdot 6 - 1 \cdot 2 = 15S_2 - 1 \cdot S_1$. For $n = 3$, we have $5S_4 = 5 \cdot 90 = 450 = 21 \cdot 22 - 2 \cdot 6 = 21S_3 - 2S_2$.

A general proof can be obtained using an approach that is similar to the combinatorial proof of the relation $(2n+1)C_n = (4n-2)C_{n-1}$ (see [Sula01] and [PWZ96]; see also Chapter 3). □

So, we can state that:

- *the nth Schröder number S_n can be calculated using the following recurrent relation:*

$$S_n = \frac{6n-3}{n+1}S_{n-1} - \frac{n-2}{n+1}S_{n-2}, \quad n \geq 2.$$

Number triangle associated with Schröder numbers

4.2.14. There is a triangular array associated with Schröder numbers. The first few of its terms are

```
1
1   2
1   4    6
1   6   16   22
1   8   30   68    90
1  10   48  146   304   394
1  12   70  264   714  1412  1806
⋮   ⋮    ⋮    ⋮     ⋮     ⋮     ⋮
```

(sequence A033877 in the OEIS).

The recurrence relation for the kth element of the nth row, $n = 0, 1, 2, \ldots, k = 0, 1, 2, \ldots, n$, is

$$T^S(n,k) = T^S(n, k-1) + T^S(n-1, k-1) + T^S(n-1, k),$$

with $T(0, k) = 1$, and $T(n, k) = 0$ for $k > n$.

- Schröder numbers form the right-hand side of the constructed triangular array: $S_n = T^S(n, n)$.

□ It is easy to see that the element $T^S(n, k)$ of the constructed triangle is equal to the number of paths along the triangle from the point corresponding to the element $T^S(0, 0)$ to the point corresponding to the element $T^S(n, k)$ if only the down, right-down, and right steps are allowed, and moving to the right from the right-hand side of the triangle is forbidden. By definition, any path to the point corresponding to $T^S(n, k)$ can be obtained: by a step from the point corresponding to $T^S(n, k-1)$ (if possible), by a step from the point corresponding to $T^S(n-1, k)$, and by a step from a point corresponding to $T^S(n-1, k+1)$ (if possible). In any case, the number of paths to the point corresponding to $T^S(n, k)$ is the sum of the numbers of paths to the points corresponding to $T^S(n, k-1), T^S(n-1, k)$, and $T^S(n-1, k+1)$. In other words, $T^S(n, k) = T^S(n, k-1) + T^S(n-1, k) + T^S(n-1, k-1)$. It is obvious that the initial conditions also coincide.

It follows that all the elements on the left-hand side of the triangle are equal to 1: there exist only one path to the point $T^S(n, 0)$; it consists only of the down steps.

As for the elements $T^S(n, n)$, it is easy to see now that the number of paths from the point corresponding to $T^S(0, 0)$ to the point corresponding to $T^S(n, n)$ is equal to the number of routes on the integer lattice from the point $(0, 0)$ to the point (n, n) if only the right-up, right, and up steps are allowed, and it is forbidden to cross the diagonal $y = x$. But we know quite well that the number of such routes is equal to the nth Schröder number S_n. So, we have proven that $T^S(n, n) = S_n$ for any non-negative integer n. □

Another interesting observation is that the sum of the elements of the nth row is the $(n+1)$-st *little Schröder number*: $\sum_{k=0}^{n} T^S(n,k) = \frac{S_{n+1}}{2} = SH_{n+1}$.

Number-theoretic properties of Schröder numbers

4.2.15. Despite its combinatorial basis, Schröder numbers, surprisingly, have some interesting number-theoretic properties.

At first, it is easy to see that, starting with $n = 1$, all Schröder numbers are even.

- *The nth Schröder number S_n, $n \geq 1$, is an even positive integer.*

☐ This fact follows from the recurrent definition of Schröder numbers; as $S_2 = 2$ is even, $S_{n+1} = 3S_{n-1} + S_1 S_{n-2} + \cdots + S_{n-2} S_1$ is even as a sum of even numbers. ☐

As a trivial consequence, we obtain that:

- *the only prime Schröder number is $S_1 = 2$.*

There are some congruences modulo $p, p \in P$, related to Schröder numbers. For example:

- *if $p > 3$ is a prime, then*

$$\sum_{k=1}^{p-1} \frac{S_k}{6^k} \equiv 0 \pmod{p}.$$

☐ In fact, if $p = 5$, then $6 \equiv 1 \pmod 5$, and

$$\sum_{k=1}^{p-1} \frac{S_k}{6^k} = S_1 + S_2 + S_3 + S_4 = 2 + 6 + 22 + 90 \equiv 0 \pmod 5.$$

If $p = 7$, then $6 \equiv -1 \pmod 7$, and

$$\sum_{k=1}^{p-1} \frac{S_k}{6^k} = -S_1 + S_2 - S_3 + S_4 - S_5 + S_6$$
$$= \equiv -2 + 6 - 22 + 90 - 394 + 1806 \equiv 0 \pmod 7.$$

The reader can find the general proof in Chapter 8. ☐

4.2.16. Consider the behavior of the sequence of Schröder numbers on infinity. It can be proven that:

- *for large positive integers n, it holds that*

$$S_n \sim \frac{1}{2\sqrt{\pi n^3}}(3 - 2\sqrt{2})^{-n-\frac{1}{2}}.$$

☐ In order to determine the principal term of the asymptotic of the nth Schröder number S_n, we can use the closed formula $S_n = \sum_{k=0}^{n} \frac{\binom{n}{k}\binom{n+k}{k}}{k+1}$ and the Stirling formula (see Chapter 1): $k! \sim \sqrt{2\pi k}(\frac{k}{e})^k$. ☐

Schröder numbers in the family of special numbers

4.2.17. As was shown earlier, Schröder numbers are closely connected with Catalan and Motzkin numbers; in fact, all three number sets have similar recurrent definitions and are connected with similar counting problems.

However, there exists one other class of special numbers, which is closely related to Schröder numbers and Catalan numbers. They are the so-called *Schröder–Hipparchus numbers*.

In combinatorics, *Schröder–Hipparchus numbers* (or *super-Catalan numbers, little Schröder numbers, Hipparchus numbers*) SH_n, $n = 0, 1, 2, 3, \ldots$, form an integer sequence that can be used to count the number of ways of dissecting a convex polygon into smaller polygons by inserting diagonals. They are named after Eugène Charles Catalan and his Catalan numbers, Ernst Schröder and the closely related Schröder numbers, and the ancient Greek mathematician Hipparchus (circa 190 – circa 120 BC), who appears from evidence in Plutarch to have known these numbers.

The sequence SH_n, $n = 0, 1, 2, 3, \ldots$, of Schröder–Hipparchus numbers begins with the numbers 1, 1, 3, 11, 45, 197, 903, 4279, 20793, 103049, ... (sequence A001003 in the OEIS), as there exists exactly one way of dissecting a bigon into smaller polygons, exactly one way of dissecting a triangle into smaller polygons (just the triangle itself), exactly three ways of dissecting a square into smaller polygons (two Euler's triangulations and the square itself), exactly eleven ways of dissecting a convex pentagon into smaller polygons (five Euler's triangulations, five decompositions into a triangle and a quadrangle, and the pentagon itself), etc.

The Schröder–Hipparchus numbers may be used to count several closely related combinatorial objects. In fact, the nth Schröder–Hipparchus number SH_n gives:

- the number of different plane trees with n leaves, with all internal vertices having two or more children;
- the number of different ways of inserting parentheses into a sequence of $n+1$ symbols, with each pair of parentheses surrounding two or more symbols or parenthesized groups and without any parentheses surrounding the entire sequence;
- the number of lattice paths from $(0,0)$ to (n,n) with steps $(1,0), (0,1)$, and $(1,1)$, that never rise above the diagonal $y = x$ and, moreover, have no $(1,1)$ steps on the diagonal itself;
- the number of lattice paths from $(0,0)$ to $(2n,n)$ with steps $(1,1), (1,-1)$, and $(2,0)$, that never fall below the OX-axis and, moreover, have no $(2,0)$ steps on the OX-axis itself.

It is not difficult to see that, starting with the number SH_1, the Schröder–Hipparchus numbers are equal to the half of corresponding Schröder numbers. In this context, a second name for these objects is *little Schröder numbers*, while ordinary Schröder numbers are sometimes called *large Schröder numbers*.

- It holds that $S_n = 2SH_n$ for any $n \geq 1$.

☐ In fact, if we consider the numbers S_n and SH_n as the numbers of special paths between the points $(0,0)$ and $(2n,0)$, we can say that there are (large) Schröder's paths and little Schröder's paths. A little Schröder's path is a Schröder's path that has no horizontal steps on the OX-axis. To prove the equality $S_n = 2SH_n$, we build a bijection between Schröder's paths in which there is a step lying on the OX-axis and paths of the same length in which there is no such step. If there is at least one horizontal step in a Schröder's path lying on the same level as the beginning of the path, consider the leftmost such step and, without changing the path's previous part, put the path's next part on the "legs." □

- Schröder–Hipparchus numbers satisfy the recurrent relation

$$SH_{n+1} = 2(SH_0 SH_n + SH_1 SH_{n-1} + \cdots + SH_{n-1} SH_1 + SH_n SH_0) \\ - SH_n.$$

☐ This fact can be obtained using the recurrent definition of Schröder numbers and the relationship $S_n = 2SH_n$, $n \geq 1$. In fact,

$$\begin{aligned} 2SH_n = S_{n+1} &= S_n + (S_0 S_n + S_1 S_{n-1} + \cdots + S_{n-1} S_1 + S_n S_0) \\ &= 3S_n + (S_1 S_{n-1} + \cdots + S_{n-1} S_1) \\ &= 6SH_n + 4(SH_1 SH_{n-1} + \cdots + SH_{n-1} SH_1), \end{aligned}$$

and we obtain that

$$\begin{aligned} SH_n &= 3SH_n + 2(SH_1 SH_{n-1} + \cdots + SH_{n-1} SH_1) \\ &= 2(SH_0 SH_n + SH_1 SH_{n-1} + \cdots + SH_{n-1} SH_1 + SH_n SH_0) \\ &- SH_n. \end{aligned}$$
□

It can also be shown that:

- the Schröder–Hipparchus numbers satisfy the recurrent relation

$$SH_n = \frac{1}{n+1}((6n-3)SH_{n-1} - (n-2)SH_{n-2}), \quad SH_0 = SH_1 = 1.$$

☐ In fact, $SH_0 = 1$, $SH_1 = 1$, and $SH_2 = 3 = \frac{1}{3}(9 \cdot 1 - 0 \cdot 1) = \frac{1}{3}(9SH_1 - 0 \cdot SH_0)$; $SH_3 = 11 = \frac{1}{4}(15 \cdot 3 - 1 \cdot 1) = \frac{1}{4}(15SH_2 - 1 \cdot SH_1)$, etc.

The general statement follows from the property $S_n = 2SH_n$, $n \geq 1$, as Schröder numbers have the same recurrent relation. ☐

- the Schröder–Hyparchus numbers can also be calculated using the formula

$$SH_{n+1} = \sum_{i=0}^{n} SH_i S_{n-i}.$$

☐ In fact, $SH_1 = 1 = 1 \cdot 1 = SH_0 S_0$; $SH_2 = 3 = 1 \cdot 2 + 1 \cdot 1 = SH_0 S_1 + SH_1 S_0$; $SH_3 = 11 = 1 \cdot 6 + 1 \cdot 2 + 3 \cdot 1 = SH_0 S_1 + SH_1 S_0 + SH_2 S_0$. The general result can then be obtained using direct consideration:

$$SH_{n+1} = 2 \sum_{i=0}^{n} SH_i SH_{n-i} - SH_n$$

$$= 2 SH_n SH_0 + \sum_{i=1}^{n} SH_i (2 SH_{n-i}) - SH_n$$

$$= SH_n S_0 + \sum_{i=1}^{n} SH_i S_{n-i} = \sum_{i=0}^{n} SH_i S_{n-i}. \qquad \Box$$

It is easy to obtain several closed formulas for the Schröder–Hyparchus numbers. In fact:

- for $n \geq 1$, the nth Schröder–Hyparchus number SH_n can be calculated using the formulas

$$SH_n = \frac{1}{2} \sum_{k=0}^{n} \binom{n+k}{2k} C_k = \frac{1}{2} \sum_{k=0}^{n} \binom{2n-k}{k} C_{n-k} = \sum_{k=0}^{n} \frac{\binom{n}{k}\binom{n+k}{k}}{2k+2}.$$

☐ We obtain these equalities from the existing closed formulas for S_n. ☐

Moreover, it holds that:

- for $n \geq 1$, the nth Schröder–Hyparchus number SH_n can be calculated using the formula

$$SH_n = \sum_{k=1}^{n} \frac{1}{n} \binom{n}{k} \binom{n}{k-1} 2^{k-1}.$$

☐ In fact, it holds that

$$SH_1 = 1 = \frac{1}{1} \binom{1}{1} \binom{1}{0} 2^0;$$

$$SH_2 = 3 = \frac{1}{2} \left(\binom{2}{1} \binom{2}{0} 2^0 + \binom{2}{2} \binom{2}{1} 2^1 \right);$$

$$SH_3 = 11 = \frac{1}{3} \left(\binom{3}{1} \binom{3}{0} 2^0 + \binom{3}{2} \binom{3}{1} 2^1 + \binom{3}{3} \binom{3}{2} 2^2 \right).$$

274   Catalan Numbers

This is a special case of a more general identity:

$$x_n = \sum_{k=1}^{n} N(n,k) m^{k-1} = \sum_{k=1}^{n} \frac{1}{n} \binom{n}{k} \binom{n}{k-1} m^{k-1}$$

for sequences $x_0, x_1, x_2, \ldots$, which are calculated as the sums of Narayana numbers multiplied by the powers of m. Substituting $m = 1$ into this formula gives the Catalan numbers, and substituting $m = 2$ into this formula gives the Schröder–Hipparchus numbers (see [Coke04]).

4.2.18. As the generating function of the sequence S_n, $n = 0, 1, 2, 3, \ldots$, of Schröder numbers has the form $\frac{1-x-\sqrt{1-6x+x^2}}{2x}$, we can obtain the following fact:

- *The generating function of the sequence SH_n, $n = 0, 1, 2, 3, \ldots$, of the Schröder–Hyparchus numbers has the form $g(x) = \frac{1+x-\sqrt{1-6x+x^2}}{4x}$, i.e., it holds that*

$$\frac{1+x-\sqrt{1-6x+x^2}}{4x} = SH_0 + SH_1 x + SH_2 x^2$$

$$+ \cdots + SH_n x^n + \cdots, \quad |x| \leq 3 - 2\sqrt{2}.$$

☐ This follows from the relationship $S_n = 2SH_n$, $n \geq 1$. In fact, if $f(x) = \sum_{n=0}^{\infty} S_n x^n$, then $f(x) = 2 \sum_{n=0}^{\infty} SH_n x^n - 1$, i.e., $\sum_{n=0}^{\infty} SH_n x^n = \frac{f(x)+1}{2}$. As $f(x) = \frac{1-x-\sqrt{1-6x+x^2}}{2x}$,

$$\sum_{n=0}^{\infty} SH_n x^n = SH_0 + SH_1 x + SH_2 x^2 + \cdots + SH_n x^n + \cdots$$

$$= \frac{1+x-\sqrt{1-6x+x^2}}{4x}. \qquad \square$$

Note that the obtained function $g(x) = \frac{1+x-\sqrt{1-6x+x^2}}{4x}$ is a solution to the quadratic equation

$$2xg^2(x) - (1+x)g(x) + 1 = 0,$$

considered with respect to the variable $g(x)$. As

$$\left(\sum_{n=0}^{\infty} SH_n x^n \right)^2 = \sum_{n=0}^{\infty} \left(\sum_{k=0}^{n} SH_k SH_{n-k} \right) x^n,$$

$$2x\left(\sum_{n=0}^{\infty} SH_n x^n\right)^2 = \sum_{n=0}^{\infty}\left(2\sum_{k=0}^{n} SH_k SH_{n-k}\right) x^{n+2},$$

$$(1+x)\sum_{n=0}^{\infty} SH_n x^n = 1 + \sum_{n=0}^{\infty}(SH_n + SH_{n+1})x^{n+1},$$

we obtain the recurrent relation

$$SH_{n+1} + SH_n = 2\sum_{k=0}^{n} SH_k SH_{n-k};$$

it gives another proof of the formula

$$SH_{n+1} = 2(SH_0 SH_n + SH_1 SH_{n-1} + \cdots + SH_{n-1} SH_1 + SH_n SH_0)$$
$$- SH_n.$$

Exercises

1. Calculate the first five Schröder numbers: using their recurrent definition; using Schröder's paths from $(0,0)$ to (n,n); using Schröder's paths from $(0,0)$ to $(2n,0)$.
2. For $n = 3, 4, 5$, find the Schröder–Hyparchus number SH_n using the definition. Construct the tables of the corresponding decompositions of a convex $(n+2)$-gon.
3. For $n = 2, 3, 4, 5, 6, 7, 8$, find the Schröder–Hyparchus number SH_n using the recurrent relation. Prove this recurrent relation using a combinatorial approach.
4. Prove that, for $n = 1, 2, 3, 4$, there is a relationship $S_n = 2SH_n$, considering the correspondences between Schröder's paths and the Schröder–Hyparcus paths from $(0,0)$ to $(2n,0)$.
5. For $n = 2, 3, 4, 5$, find all plane trees with n leaves in which any internal vertex has at least two children. How is the number of such trees associated with the Schröder-Hyparchus numbers? Prove that SH_n is equal to the number of such trees with $n+1$ leaves.
6. Using the result of the previous problem, prove that the nth Schröder–Hyparchus number gives the number of possibilities of classical placement of brackets in a sequence of $n+1$ characters

with an additional condition: each pair of brackets must surround two or more characters or a group of characters enclosed in brackets (as usual, the brackets surrounding the entire sequence are not used).

4.3. Delannoy Numbers

A history of the question

4.3.1. Delannoy numbers $D(m,n)$ describe the number of lattice paths from the southwest corner $(0,0)$ of a rectangular grid to the northeast corner (m,n) using only single steps north, northeast, or east.

They are named after the French officer and amateur mathematician Henri-Auguste Delannoy (1833–1915).

H. Delannoy grew up in Guéret, France. After taking the baccalaureate in 1849, he studied mathematics in Bourges, and after continuing his studies in Paris, he entered the École Polytechnique in 1853.

In 1879, H. Delannoy began a correspondence with Édouard Lucas on the subject of recreational mathematics and probability theory; he eventually published 11 mathematical articles. Along with his mathematical interests, H. Delannoy wrote about local history and painted, and from 1896 to 1915, he served as the president of the Société des Sciences Naturelles et archéologiques de la Creuse.

Recurrent construction of Delannoy numbers

4.3.2. Define the *Delannoy number* $D(m,n)$ as the number of lattice paths from the lowest-left corner $(0,0)$ of the rectangular grid of size $m \times n$ to its upper-right corner (m,n) using only up, right, or right-up movements.

A direct calculation shows that $D(1,1) = 3$, $D(1,2) = D(2,1) = 5$, $D(1,3) = D(3,1) = 7$, $D(2,2) = 25$, etc. It is natural to assume also that $D(0,0) = 1$ and $D(0,n) = D(m,0) = 1$ for any positive integers n and m.

So, we obtain a two-dimension array, starting with the elements $1, 1, 1, 1, 3, 1, 1, 5, 5, 1, \ldots$, listed by antidiagonals (sequence A008288 in the OEIS).

4.3.3. Already, these simple calculations allow us to find and prove the recurrent relation for Delannoy numbers.

In fact, by considering a route from the bottom-left to the top-right corner of the grid, we can take last step up to a point (m, n) with positive integer coordinates from one of the following three possible positions: $(m-1, n), (n-1, n)$, or $(m-1, n-1)$. Thus, the number $D(m, n)$ of allowable routes is equal to the sum of the number $D(m-1, n)$ of allowable routes leading to the point $(m-1, n)$, the number $D(m, n-1)$ of allowable routes leading to the point $(m, n-1)$, and the number $D(m-1, n-1)$ of allowable routes leading to the point $(m-1, n-1)$. The initial conditions of this recurrent structure were discussed earlier: $D(0, 0) = D(0, n) = D(m, 0) = 1$. So:

- for the Delannoy number $D(m, n)$, the following recurrent relation holds:

$$D(m, n) = D(m-1, n) + D(m-1, n-1) + D(m, n-1),$$
$$D(0, 0) = D(0, n) = D(m, 0) = 1.$$

4.3.4. Using this recurrent relation, we can build the rectangle of Delannoy numbers, guided by simple rules: the zeroth row and the zeroth column consist of unities, and each internal element of the rectangle is calculated as the sum of its three neighbors located on the left in the same row, on the left in the previous row, and directly above in the previous row. In other words, we find the sum of the three elements that form the angle that borders the calculated element on the left:

m/n	0	1	2	3	4	5	6	7	8
0	1	1	1	1	1	1	1	1	1
1	1	3	5	7	9	11	13	15	17
2	1	5	13	25	41	61	85	113	145
3	1	7	25	63	129	231	377	575	833
4	1	9	41	129	321	681	1289	2241	3649
5	1	11	61	231	681	1683	3653	7183	13073
6	1	13	85	377	1289	3653	8989	19825	40081
7	1	15	113	575	2241	7183	19825	48639	108545
8	1	17	145	833	3649	13073	40081	108545	265729
9	1	19	181	1159	5641	22363	75517	224143	598417

By studying this table, we can see that it is symmetrical.

- *The nth column of the Delannoy rectangle coincides with the nth row:*

$$D(m, n) = D(n, m).$$

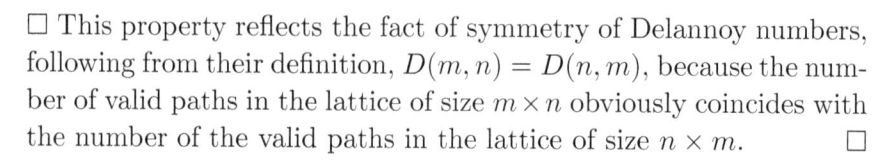

□ This property reflects the fact of symmetry of Delannoy numbers, following from their definition, $D(m, n) = D(n, m)$, because the number of valid paths in the lattice of size $m \times n$ obviously coincides with the number of the valid paths in the lattice of size $n \times m$. ∎

- *The zeroth row, as well as the zeroth column, of the Delannoy rectangle consists of unities:* $D(m, 0) = D(0, n) = 1$, $m, n = 0, 1, 2, 3, \ldots$.

□ Obviously, going from the point $(0, 0)$ to the point $(m, 0)$, we can use only right steps; there exist only one such path. Similarly, going to the point $(0, n)$, we can use only the up steps; there exists only one such path. ∎

- *All Delannoy numbers are odd.*

□ This simple property follows from the recurrent construction of the considered number set. In fact, since the sum of any three odd numbers is odd, then, starting with odd initial entries $1, 1, 1, \ldots$, we can obtain only odd elements in the constructed table. ∎

- *The first row of the rectangle consists of consecutive odd numbers:*

$$D(1, n) = 2n + 1, n = 0, 1, 2, 3, \ldots.$$

□ This can be proved by induction: for $n = 0, 1$, it holds that $D(1, 0) = 2 \cdot 0 + 1 = 1$, $D(1, 1) = 2 \cdot 1 + 1 = 3$. Furthermore, if $D(1, n) = 2n+1$, then $D(1, n+1) = D(0, n+1) + D(1, n) + D(0, n) = 2n + 1 + 2 = 2(n + 1) + 1$. ∎

The second row is more difficult to deal with, but readers who are interested in the theory of figurate numbers can recognize in the sequence 1, 5, 13, 25, 41,... the *centered square numbers*, resulting from the construction of the desired number of square frames for a given center point, and counting the resulting elements. The sequence of centered square numbers starts with the elements 1, 5, 13, 25, 41, 61, 85, 113, 145, 181, ... (sequence A001844 in the OEIS). The nth element is calculated using the formula $\frac{(2n-1)^2+1}{2}$, $n = 1, 2, 3, \ldots$. In other words:

- the second row of the Delannoy array consists of the centered square numbers:

$$D(2, n) = \frac{(2n+1)^2 + 1}{2}.$$

☐ This statement can be proved by induction, too. For $n = 0, 1$, it holds that $D(2, 0) = 1 = \frac{(2 \cdot 0+1)^2+1}{2} = 1$, and $D(2, 1) = 5 = \frac{(2 \cdot 1+1)^2+1}{2}$. Furthermore, if $D(2, n) = \frac{(2n+1)^2+1}{2}$, then

$$D(2, n+1) = D(1, n+1) + D(2, n) + D(1, n)$$

$$= (2(n+1)+1) + \frac{(2n+1)^2 + 1}{2} + (2n+1)$$

$$= \frac{((2n+1)^2 + 4(2n+1) + 4) + 1}{2} = \frac{(2n+3)^2 + 1}{2}$$

$$= \frac{(2(n+1)+1)^2 + 1}{2}.$$
☐

The third row of the rectangle is formed by the *centered octahedral numbers* 1, 7, 25, 63, 129, 231, 377, 575, 833, 1159, ... (sequence A001845 in the OEIS), which are similarly defined and can be calculated using the formula $\frac{(2n-1)(2n^2-2n+3)}{3}$, $n = 1, 2, 3, \ldots$. In other words:

- the elements of the third row of the Delannoy array are centered octahedral numbers:

$$D(3, n) = \frac{(2n+1)(2n^2 + 2n + 3)}{3}.$$

☐ It can also be proved by induction. Thus, $D(3,0) = 1 = \frac{13}{3}$; $D(3,1) = 7 = \frac{37}{3}$. Furthermore, if $D(3,n) = \frac{(2n+1)(2n^2+2n+3)}{3}$, then
$$D(3, n+1) = D(3,n) + D(2, n+1) + D(2,n)$$
$$= \frac{(2n+1)(2n^2+2n+3)}{3} + \frac{(2n+3)^2+1}{2}$$
$$+ \frac{(2n+1)^2+1}{2} = \frac{(2n+3)(2(n+1)^2+2n+5)}{3}. \quad \square$$

4.3.5. Alternatively, the same numbers can be arranged in a triangular array, resembling Pascal's triangle:

$$D(0,0)$$
$$D(1,0) \qquad D(0,1)$$
$$D(2,0) \qquad D(1,1) \qquad D(0,2)$$
$$\vdots \qquad \vdots \qquad \vdots$$

The law of its construction is as follows: the vertex and the sides of the triangle are formed by unities, while each internal element is equal to the sum of the three elements placed immediately above it and forming a triangle. Call this number triangle *Delannoy's triangle*:

$$1$$
$$1 \quad 1$$
$$1 \quad 3 \quad 1$$
$$1 \quad 5 \quad 5 \quad 1$$
$$1 \quad 7 \quad 13 \quad 7 \quad 1$$
$$1 \quad 9 \quad 25 \quad 25 \quad 9 \quad 1$$
$$1 \quad 11 \quad 41 \quad 63 \quad 41 \quad 11 \quad 1$$
$$\vdots \quad \vdots \quad \vdots \quad \vdots \quad \vdots \quad \vdots \quad \vdots$$

Looking closely at the constructed triangle, we see that we have already comes across it: this is the *tribonacci triangle*, having the following property: the sums of the elements of descending (ascending) diagonals form the sequence of the *tribonacci numbers* 1, 1, 2, 4, 7, 13, 24, 44, 81, 149, ... (sequence A000073 in the OEIS), which are defined by the recurrent relation
$$t_{n+3} = t_{n+2} + t_{n+1} + t_n, \quad t_0 = t_1 = 1, t_2 = 2.$$

- The sum of the elements of the nth ascending diagonal of Delannoy's triangle is equal to the $(n+1)$th tribonacci number t_{n+1}:

$$D(n,0) + D(n-2,1) + D(n-4,2) + \cdots + D(n-2k,k) + \cdots = t_n.$$

□ It can be easy checked for small values of n: $D(0,0) = 1$, $D(1,0) = 1$, $D(2,0) + D(0,1) = 1 + 1 = 2$, $D(3,0) + D(1,1) = 1 + 3 = 4$, $D(4,0) + D(2,1) + D(0,2) = 1 + 5 + 1 = 7$. Denoting $D(n,0) + D(n-2,1) + D(n-4,2) + \cdots + D(n-2k,k) + \cdots$ by $X(n)$, it is easy to check that $X(n+3) = X(n+2) + X(n+1) + X(n)$. In fact,

$$X(n+2) + X(n+1) + X(n)$$
$$= (D(n+2,0) + D(n,1) + D(n-2,2)$$
$$+ \cdots + D(n+2-2k,k) + \cdots) + (D(n+1,0) + D(n-1,1)$$
$$+ D(n-3,2) + \cdots + D(n+1-2k,k) + \cdots)$$
$$+ (D(n,0) + D(n-2,1) + D(n-4,2)$$
$$+ \cdots + D(n-2k,k) + \cdots) = D(n+2,0)$$
$$+ (D(n,1) + D(n+1,0) + D(n,0)) + (D(n-2,2)$$
$$+ D(n-1,1) + D(n-2,1)) + (D(n-4,3) + D(n-3,2)$$
$$+ D(n-4,2)) + \cdots = D(n+3,0) + D(n+1,1) + D(n-1,2)$$
$$+ D(n-3,3) + \cdots = X(n+3).$$

So, we obtain the same recurrent relation with the same initial conditions. Thus, $X(n) = t_n$, $n = 0, 1, 2, 3, \ldots$. □

4.3.6. Our constructions and calculations lead to the idea that these algorithms are very similar to those used in the construction of Pascal's triangle — it is no coincidence. The recurrent relation for Delannoy numbers containing three terms is a natural generalization of the recurrent relation for elements of Pascal's triangle containing two of the three such terms. This situation becomes completely obvious if we interpret the value C_{n+m}^n as the number $C(m,n)$ of paths between the vertices $(0,0)$ and (m,n) in the rectangular lattice of size $m \times n$ if only the up and right moves are allowed. In fact, $C_{n+m}^n = C(m,n)$ since the recurrent relation $C(m,n) = C(m-1,n) + C(m,n-1)$ for the numbers $C(m,n)$ coincides with the recurrent relation $C_{n+m-1}^n + C_{n+m-1}^{n-1}$ for the numbers C_{n+m}^n, and the initial conditions $C(0,0) = C(m,0) = C(0,n) = 1$ match the initial conditions $C_0^0 = C_m^0 = C_n^n = 1$.

Delannoy numbers and counting problems

4.3.7. Combinatorial applications of Delannoy numbers are numerous. For example, the Delannoy number $D(m,n)$ counts:

- the number of alignments of two sequences of lengths m and n;
- the number of points in an m-dimensional integer lattice located no further than n steps from the origin;
- in cellular automata, the number of cells in an m-dimensional von Neumann neighborhood of radius (rank) n.

Thus, an *alignment* is a way of pairing up elements of two strings, optionally skipping some elements but preserving the order.

For example, the following presents three (from 63) alignments of the two 3-sequences abc and xyz:

$$\begin{array}{ccc} a & b & c \\ x & y & z \end{array} \qquad \begin{array}{ccc} a & b & c \\ x & y & z \end{array} \qquad \begin{array}{ccc} a & b & c \\ x & y & z \end{array}$$

It is easy to check that for the two 1-sequences a and x, we obtain exactly three (i.e., ($D(1,1)$)) alignments:

$$\begin{array}{c} a \\ x \end{array} \qquad \begin{array}{c} a \\ x \end{array} \qquad \begin{array}{c} a \\ x \end{array}$$

For the 2-sequence ab and the 1 one-sequence x, we have five (i.e., $D(2,1)$) alignments:

$$\begin{array}{cc} a & b \\ x & \end{array} \quad \begin{array}{cc} a & b \\ & x \end{array} \quad \begin{array}{cc} a & b \\ x & \end{array} \quad \begin{array}{cc} a & b \\ & x \end{array} \quad \begin{array}{cc} a & b \\ & x \end{array}$$

Now, it is easy to prove that the numbers $A(m,n)$ of alignments of two-sequences of lengths m and n satisfy the same recurrent relation with the same initial conditions as Delannoy numbers. So, the Delannoy number $D(m,n)$ counts the alignments of an m-sequence and an n-sequence.

Closed formula for Delannoy numbers

4.3.8. For Delannoy numbers, one can also obtain a closed formula:

- The Delannoy number $D(m,n)$ can be calculated using the formula

$$D(m,n) = \sum_{k=0}^{m} P_{(k,m-k,n-k)} = \sum_{k=0}^{m} \frac{(m+n-k)!}{k!(m-k)!(n-k)!}.$$

□ In order to prove this statement, it is enough to note that, for a fixed number k of diagonal steps, any path between the vertices $(0,0)$ and (m,n) will contain exactly $m-k$ horizontal steps and $n-k$ vertical steps.

A sequence, containing exactly k diagonal, $m-k$ horizontal, and $n-k$ vertical steps, will uniquely define a route. The number of such sequences is the number $P_{(k,m-k,n-k)}$ of permutations with repetitions, in which the first element (diagonal step's symbol) is repeated k times, the second element (horizontal step's symbol) is repeated $m-k$ times, and the third element (vertical step's symbol) is repeated $n-k$ times. As $P_{(k,m-k,n-k)} = \frac{(m+n-k)!}{k!(m-k)!(n-k)!}$, the Delannoy number $D(m,n)$ can be calculated using the formula

$$D(m,n) = \sum_{k=0}^{m} P_{(k,m-k,n-k)} = \sum_{k=0}^{m} \frac{m+n-k)!}{k!(m-k)!(n-k)!}.$$ ■

This result can be presented in another form:

- The Delannoy number $D(m,n)$, $m \leq n$, can be calculated by the formula

$$D(m,n) = \sum_{k=0}^{m} \binom{m}{k} \cdot \binom{m+n-k}{m}.$$

☐ In fact, it holds that

$$P_{(k,m-k,n-k)} = \frac{(m+n-k)!}{k!(m-k)!(n-k)!}$$

$$= \frac{m!}{k!(m-k)!} \cdot \frac{m+n-k)!}{m!(m+n-k-m)!}$$

$$= C_m^k \cdot C_{m+n-k}^m = \binom{m}{k} \cdot \binom{m+n-k}{m}. \qquad \square$$

It is not difficult to get another similar equality:

- The Delannoy number $D(m,n)$ can be calculated using the formula

$$D(m,n) = \sum_{k=0}^{n} 2^k \cdot \binom{m}{k} \cdot \binom{n}{k}, \quad m \leq n.$$

☐ The proof can be easily obtained using a combinatorial approach. Note that $\binom{m}{k} = \binom{m}{m-k}$ and $\binom{n}{k} = \binom{n}{n-k}$. $\qquad \square$

- The Delannoy number $D(m,n)$ can be represented by the infinite series

$$D(m,n) = \sum_{k=0}^{\infty} \frac{1}{2^{k+1}} \binom{k}{n} \binom{k}{m}.$$

☐ The proof can be obtained by direst transformation of the previous formula. $\qquad \square$

Generating function of the sequence of Delannoy numbers

4.3.9. Using the recurrent relation for Delannoy numbers, we can find the two-dimensional generating function of the Delannoy array.

- The generating function of the array of Delannoy numbers has the form $f(x,y) = \frac{1}{1-x-y-xy}$, i.e., it holds that

$$\frac{1}{1-x-y-xy} = \sum_{m,n=0}^{\infty} D(m,n) x^m y^n.$$

□ In fact,

$$\sum_{m,n=0}^{\infty} D(m,n)x^m y^n = 1 + \sum_{m,n=1}^{\infty} D(m,n)x^m y^n$$

$$= 1 + \sum_{m,n=1}^{\infty} D(m-1,n)x^m y^n$$

$$+ \sum_{m,n=1}^{\infty} D(m,n-1)x^m y^n$$

$$+ \sum_{m,n=1}^{\infty} D(m-1,n-1)x^m y^n$$

$$= 1 + x \sum_{k,n=0}^{\infty} D(k,n)x^k y^n$$

$$+ y \sum_{m,k=0}^{\infty} D(m,k)x^m y^k + xy \sum_{k,l=0}^{\infty} D(k,l)x^k y^l.$$

So, we have

$$1 = \sum_{m,n=0}^{\infty} D(m,n)x^m y^n - x \sum_{m,n=0}^{\infty} D(m,n)x^m y^n$$

$$- y \sum_{m,n=0}^{\infty} D(m,n)x^m y^n - xy \sum_{m,n=0}^{\infty} D(m,n)x^m y^n.$$

By converting, we get

$$(1 - x - y - xy) \sum_{m,n=0}^{\infty} D(m,n)x^m y^n = 1.$$

That is,

$$\sum_{m,n=0}^{\infty} D(m,n)x^m y^n = \frac{1}{1 - x - y - xy}.$$

□

Central Delannoy numbers

4.3.10. *Central Delannoy Numbers* $D(n) = D(n,n)$ calculate the number of paths from the lowest-left to the upper-right angle of a square lattice of size $n \times n$ if only the steps $(1,0)$, $(1,1)$, and $(1,-1)$ are allowed.

The sequence $D(n), n = 0, 1, 2, 3, \ldots$, of central Delannoy numbers begins with the elements 1, 3, 13, 63, 321, 1683, 8989, 48639, 265729, 1462563, ... (sequence A001850 in OEIS).

It can be shown that for the central Delannoy numbers, the following recurrent relation holds:

$$nD(n) = 3(2n-1)D(n-1) - (n-1)D(n-2).$$

In other words:

- the nth central Delannoy number can be calculated using the linear recurrent relation of the second order:

$$D(n) = \frac{3(2n-1)}{n}D(n-1) - \frac{n-1}{n}D(n-2), \quad D(0)=1, D(1)=3.$$

☐ In fact, $D(2) = 13 = \frac{9}{2} \cdot 3 - \frac{1}{2} \cdot 1 = \frac{9}{2} \cdot D(1) - \frac{1}{2} \cdot D(0)$, $D(3) = 63 = \frac{15}{3} \cdot 13 - \frac{2}{3} \cdot 3 = \frac{15}{3} \cdot D(2) - \frac{2}{3} \cdot D(1)$, etc. A general result can be obtained using a combinatorial approach. ☐

4.3.11. Using closed formulas for the Delannoy number $D(m,n)$, we can obtain several closed formulas for $D(n)$.

- For the nth central Delannoy number $D(n)$, the following closed formulas exist:

$$D(n) = \sum_{k=0}^{n} \binom{n}{k}\binom{2n-k}{n} = \sum_{k=0}^{n} \binom{n}{k}\binom{n+k}{k} = \sum_{k=0}^{n} \binom{n+k}{2k}\binom{2k}{k}.$$

☐ The first result is simply the closed formula for $D(m,n)$, taken for $m = n$. As $\binom{n}{k} \cdot \binom{2n-k}{n} = \binom{n}{n-k} \cdot \binom{2n-k}{n-k}$, then substituting into the closed formula for the Delannoy number $D(m,n)$ the value $m = n$ and moving from the summation's variable k to the summation's variable $n - k$, we get the second result. The third result can be

obtained by direct checking:
$$\binom{n}{k}\binom{n+k}{k} = \frac{(n+k)!}{k!k!(n-k)!} = \binom{n+k}{2k}\binom{2k}{k}. \qquad \square$$

So, we should note that the nth central Delannoy number and the nth Schröder number S_n can be given, respectively, by the formulas

$$D_n = \sum_{k=0}^{n} C_n^k C_{n+k}^k, \quad S_n = \sum_{k=0}^{n} C_n^k C_{n+k}^k \frac{1}{k+1}.$$

Also, the central Delannoy number $D(n)$ and the nth Schröder number S_n can be given, respectively, by the formulas

$$D_n = \sum_{k=0}^{n} C_{n+k}^{2k} C_{2k}^k, \quad S_n = \sum_{k=0}^{n} C_{n+k}^{2k} C_k = \sum_{k=0}^{n} C_{n+k}^{2k} C_{2k}^k \frac{1}{k+1}.$$

The second closed formula for $D(m,n)$ gives that:

- for the nth central Delannoy number $D(n)$, the following closed formula exists:

$$D(n) = \sum_{k=0}^{n} 2^k \cdot \binom{n}{k}^2.$$

Finally, we can state that:

- the nth central Delannoy number $D(n)$ can be represented by the infinite series

$$D(n) = \sum_{k=0}^{\infty} \frac{1}{2^{k+1}} \binom{k}{n}^2.$$

There are also more exotic ways to calculate the central Delannoy numbers. For example,

$$D(n) = \frac{1}{\pi} \int_{3-2\sqrt{2}}^{3+2\sqrt{2}} \frac{1}{\sqrt{(t-3+2\sqrt{2})(3+2\sqrt{2}-t)}} \frac{1}{t^{n+1}} dt.$$

4.3.12. Consider now the generating function of the sequence $D(n)$, $n = 0, 1, 2, \ldots$.

- The generating function of the sequence $D(0), D(1), D(2), \ldots$ of the central Delannoy numbers has the form $f(x) = \frac{1}{\sqrt{1-6x+x^2}}$, i.e., it holds that

$$\frac{1}{\sqrt{1-6x+x^2}} = D(0) + D(1)x + D(2)x^2$$
$$+ \cdots + D(n)x^n + \cdots, \quad |x| < 3 - 2\sqrt{2}.$$

☐ If we take $\alpha = -\frac{1}{2}$ and $z = x^2 - 6x$ in the well-known decomposition

$$(1+z)^\alpha = 1 + \alpha z + \frac{\alpha(\alpha-1)}{2!}z^2$$
$$+ \frac{\alpha(\alpha-1)(\alpha-2)}{3!}z^3 + \cdots, \quad |z| < 1, \alpha \in \mathbb{R},$$

then we obtain that

$$\frac{1}{\sqrt{1-6x+x^2}} = (1 + x^2 - 6x)^{-\frac{1}{2}}$$
$$= 1 - \frac{1}{2}(x^2 - 6x) + \frac{3}{8}(x^2 - 6x)^2 + \cdots$$
$$= 1 + 3x - \frac{1}{2}x^2 + \frac{3}{8} \cdot 36x^2 + \cdots$$
$$= 1 + 3x + 13x^2 + \cdots .$$

As the condition $|x^2 - 6x| < 1$ implies the condition $|x| < 3 - 2\sqrt{2}$, we obtain that

$$\frac{1}{\sqrt{1-6x+x^2}} = 1 + 3x + 13x^2 + \cdots = D(0)$$
$$+ D(1)x + D(2)x^2 + \cdots, \quad |x| < 3 - 2\sqrt{2}.$$

Formal transformations can be based on the recurrent approach. See the corresponding results obtained above for other number sets. ☐

4.3.13. The first prime central Delannoy numbers are the numbers 3, 13, and 265729 (sequence A092830 in the OEIS), corresponding to the indices 1, 2, and 8; no other primes are found up to $n = 10^{30}$.

The asymptotic behavior of the central Delannoy numbers $D(n)$ for large n has the form
$$D(n) = \frac{c\alpha^n}{\sqrt{n}}(1 + O(n^{-1})),$$
where $\alpha = 3 + 2\sqrt{2} \approx 5.828$, and $c = \frac{1}{\sqrt{4\pi(3\sqrt{2}-4)}} \approx 0.5727$.

The number $N(n)$ of decimal digits in the number $D(10^n)$ for $n = 0, 1, 2, 3, \ldots$ is equal to $1, 7, 76, 764, 7654, 76553, 765549, 7655510, \ldots$ (sequence A114470 in the OEIS).

We can prove that at $n \to \infty$, the value $\frac{N(n)}{10^n}$ tends to the limit $\lg(3 + 2\sqrt{2}) = 0,765551\cdots$ (sequence A114491 in the OEIS).

Delannoy polynomials

4.3.14. The nth *Delannoy polynomial* $d_n(x)$ is defined as
$$d_n(x) = \sum_{k=0}^{n} D(n-k, k)x^k.$$
It is easy to get for $d_n(x)$ a simple recurrent relation:

- There exists the following recurrent relation for Delannoy polynomials:

$$d_n(x) = (x+1)d_{n-1}(x) + xd_{n-2}(x), \quad d_0(x) \equiv 1, d_1(x) = x + 1.$$

□ As we have already shown, Delannoy numbers can be given by the recurrent formula
$$D(n,k) = D(n-1,k) + D(n-1,k-1) + D(n,k-1).$$
So, it holds that $d_0(x) = D(0,0)x^0 \equiv 1$, $d_1(x) = D(1,0)x^0 + D(0,1)x^1 = 1 + x$, and

$$d_n(x) = \sum_{k=0}^{n} D(n-k, k)x^k$$
$$= \sum_{k=0}^{n} D(n-k, k-1)x^k + \sum_{k=0}^{n} D(n-k-1, k-1)x^k$$
$$+ \sum_{k=0}^{n} D(n-k-1, k-1)x^k = (x+1)d_{n-1}(x) + xd_{n-2}(x).$$
∎

4.3.15. From the obtained recurrent relation, it follows that:

- the generating function for the sequence $d_0(x), d_1(x), d_2(x), \ldots$ of Delannoy polynomials has the form $f(x,y) = \dfrac{1}{1-(x+1)y-xy^2}$.
  i.e., it holds that

$$\frac{1}{1-(x+1)y-xy^2} = \sum_{n \geq 0} d_n(x) y^n.$$

□ The proof can be obtained using the classical scheme; see Chapter 1. □

The following result can also be proven:

- The zeros of the Delannoy polynomial $d_n(x)$ are real, distinct, and belong to the open interval $(-3-2\sqrt{2}, -3+2\sqrt{2})$.

□ Using the recurrent relation for Delannoy polynomials and their initial values, we can obtain for the polynomials a Binet-like formula:

$$d_n(x) = \frac{\lambda_1^{n+1} - \lambda_2^{n+1}}{\lambda_1 - \lambda_2},$$

where $\lambda_{1,2} = \dfrac{1+x \pm \sqrt{1+6x+x^2}}{2}$ are the roots of the characteristic equation $\lambda^2 - (x+1)\lambda - x = 0$.

Using this formula, we can obtain a *trigonometrical representation* of $d_n(x)$:

$$d_n(x) = \prod_{k=1}^{n} \left(x + \left(\sqrt{1+\cos^2 \frac{k\pi}{n+1}} + \cos \frac{k\pi}{n+1} \right)^2 \right).$$

Let $r_{n,k} = -\left(\sqrt{1+\cos^2 \frac{k\pi}{n+1}} + \cos \frac{k\pi}{n+1}\right)^2$. Then, the obtained trigonometric formula for Delannoy polynomials allows us to draw the following conclusions:

- the polynomial $d_n(x)$ has different real zeros $r_{n,1} < r_{n,2} < \cdots < r_{n,n}$;
- the sequence $(r_{n,1})_{n \geq 1}$ strictly decreases, and $\lim\limits_{n \to +\infty} r_{n,1} = -(\sqrt{2}+1)^2 = -3-2\sqrt{2}$;

- the sequence $(r_{n,n})_{n\geq 1}$ strictly increases, and $\lim\limits_{n\to +\infty} r_{n,n} =$
  $-(\sqrt{2}-1)^2 = -3+2\sqrt{2}$.

Thus, our statement is proven. □

4.3.16. There are also some polynomials connected with the central Delannoy numbers.

The *Lendre polynomial* $P_n^{(a,b)}(x)$ is defined as

$$P_n^{(a,b)}(x) = \sum_k \binom{n+a+b+k}{k}\binom{n-k}{n-a}\cdot\left(\frac{x-1}{2}\right)^k.$$

So, we can state that:

- *for the nth central Delannoy number $D(n)$, it holds that*

$$D(n) = P_n^{(0,0)}(3).$$

☐ As $P_n^{(a,b)}(x) = \sum_k \binom{n+a+b+k}{k}\binom{n-k}{n-a}\cdot(\frac{x-1}{2})^k$, and $D(n) = \sum_{k=0}^n \binom{n}{k}\cdot\binom{n+k}{k}$, we get that $D(n) = P_n^{(0,0)}(3)$. □

Delannoy numbers in the family of special numbers

Delannoy numbers and hyperoctahedron numbers

4.3.17. Consider another set of positive integers which satisfy the same recurrent rule as Delannoy numbers.

In fact, we define *hyperoctahedron numbers* $HO(m,n)$, $n,m \geq 1$, by the following recurrent relation:

$$HO(m,n) = HO(m-1,n) + HO(m,n-1)$$
$$+ HO(m-1,n-1), \text{ and}$$
$$HO(m,1) = 1, HO(1,n) = n.$$

So, starting with the first row $1,2,3,\ldots,n,\ldots$, and with the first column $1,1,1,\ldots$, we obtain the rectangular array of this number set, represented as follows (sequence A142978 in the OEIS):

m/n	1	2	3	4	5	6	7	8	9
1	1	2	3	4	5	6	7	8	9
2	1	4	9	16	25	36	49	64	81
3	1	6	19	44	83	146	231	344	145
4	1	8	33	96	225	456	833	575	833
5	1	10	51	180	501	1182	1289	2241	3649
6	1	12	73	304	985	1683	3653	7183	13073
7	1	14	99	476	1289	3653	8989	19825	40081
8	1	16	129	575	2241	7183	19825	48639	108545
8	1	18	145	833	3649	13073	40081	108545	265729
9	1	19	181	1159	5641	22363	75517	224143	598417

4.3.18. The mth row's entries of this array are the *regular polytope numbers* for the m-dimensional *hyperoctahedron* (or *cross polytope*), as defined in [DeDe12].

In fact, for $m = 3$, we obtain the ordinary *octahedral numbers*: space figurate numbers that represent an octahedron, or two pyramids placed together, one upside-down underneath the other. The nth octahedral number $O(n) = HO(3, n)$ is the sum of two consecutive square pyramidal numbers:

$$O(n) = S_4^3(n-1) + S_4^3(n).$$

As $S_4^3(n) = \frac{n(n+1)(2n+1)}{6}$, it implies the following closed formula for the octahedral numbers:

$$O(n) = HO(3, n) = \frac{n(2n^2 + 1)}{3}.$$

The first few octahedral numbers are 1, 6, 19, 44, 85, 146, 231, 344, 489, 670, ... (Sloane's A005900).

For $m = 4$, we obtain *hyperoctahedral numbers* (or *hexadecahoron numbers*, *4-cross-polytope numbers*, *4-orthoplex numbers*, *16-cell numbers*, β^4-numbers, $\{3, 3, 4\}$ numbers) $HO(n)$, corresponding to a hyperoctahedron in $\mathbb{R}^4$, i.e., four-dimensional hyperoctahedron numbers $HO(4, n)$.

It can be proven ([DeDe12]) that the closed formula for hyperoctahedral numbers has the form

$$HO(n) = HO(4, n) = \frac{n^2(n^2 + 2)}{3}.$$

The sequence of the hyperoctahedral numbers starts with the elements $1, 8, 33, 96, 225, 456, 833, 1408, 2241, 3400, \ldots$ (Sloane's A014820).

It is easy to see that the *square numbers* 1, 4, 9, 16, 25, 36, 49, 64, 81, 100,... (sequence A000290 in the OEIS) are the two-dimensional analog of the three-dimensional octahedral and four-dimensonal hyperoctahedral numbers; they can be represented as the sum of two consecutive triangular numbers. So, $HO(2, n) = n^2$.

Similarly, the positive integers 1, 2, 3, 4, 5, 6, 7, 8, 9, 10,... (sequence A000027 in the OEIS) are one-dimensional representatives of hyperoctahedron numbers.

4.3.19. Hyperoctahedron numbers are closely connected with Delannoy numbers. In fact:

- *any hyperoctahedron number can be represented as a sum of Delannoy numbers:*

$$HO(m+1, n) = D(m, 0) + D(m, 1) + \cdots + D(m, n-1).$$

☐ In order to prove this statement, it is sufficient to compare two similar recurrent constructions. In fact, the statement is true for the first row and the first column of the rectangle of hyperoctahedron numbers:

$$HO(m+1, 1) = 1 = D(m, 0);,$$
$$HO(1, n) = n = 1 + 1 + \cdots + 1$$
$$= D(0, 0) + D(0, 1) + \cdots + D(0, n-1).$$

Suppose that the statement is true for a positive integer m, and we prove it for $m + 1$. First,

$$HO(m+1, 1) = HD(m, 0),$$
$$HO(m+1, 2) = HO(m+1, 1) + HO(m, 2) + HO(m, 1)$$
$$= 1 + (D(m-1, 0) + D(m-1, 1)) + D(m-1, 0)$$
$$= D(m, 0) + (D(m, 0) + D(m-1, 1) + D(m-1, 0))$$
$$= D(m, 0) + D(m, 1).$$

If $HO(m+1, n-1) = D(m, 0) + D(m, 1) + \cdots + D(m, n-2)$, then

$$\begin{aligned}
HO(m+1, n) &= HO(m+1, n-1) + HO(m, n) + HO(m, n-1) \\
&= (D(m, 0) + D(m, 1) + \cdots + D(m, n-2)) \\
&\quad + (D(m-1, 0) + D(m-1, 1) \\
&\quad + \cdots + D(m-1, n-1)) + (D(m-1, 0) \\
&\quad + D(m-1, 1) + \cdots + D(nm-1, n-2)) \\
&= D(m-1, 0) + (D(m, 0) + D(m-1, 1) \\
&\quad + D(m-1, 0)) + (D(m, 1) + D(m-1, 2) \\
&\quad + D(m-1, 1)) + \cdots + (D(m, n-2) \\
&\quad + D(m-1, n-1) + D(m-1, n-2)) \\
&= D(m, 0) + D(m, 1) + \cdots + D(m, n-1). \quad \square
\end{aligned}$$

Conversely, summing the neighboring hyperoctahedron numbers, we can obtain a Delannoy number:

- *The sum of two consecutive members of the mth row of the rectangle of hyperoctahedron numbers is a Delannoy number:*

$$HO(m, n) + HO(m, n+1) = D(m, n).$$

□ This statement can also be proven by induction: if $HO(m, n-1) + HO(m, n) = D(m, n-1)$, $HO(m-1, n) + HO(m-1, n+1) = D(m, n)$, and $HO(m-1, n-1) + HO(m-1, n) = D(m-1, n-1)$, then

$$\begin{aligned}
HO(m, n) + HO(m, n+1) &= (HO(m, n-1) + HO(m, n)) \\
&\quad + (HO(m-1, n) + HO(m-1, n+1)) \\
&\quad + (HO(m-1, n-1) + HO(m-1, n)) \\
&= D(m, n-1) + D(m-1, n) \\
&\quad + D(m-1, n-1) = D(m, n). \quad \square
\end{aligned}$$

4.3.20. Now, it is easy to obtain a closed formula for the considered number set.

- For the number $HO(m,n)$, it holds that

$$HO(m+1,n) = \sum_{k=0}^{m} \binom{n+k}{m+1} \cdot \binom{m}{k} = \sum_{k=0}^{m} \binom{m+n-k}{m+1} \cdot \binom{m}{k}.$$

☐ We have proven that

$$D(m,n) = \sum_{k=0}^{m} C_m^k \cdot C_{m+n-k}^m = \sum_{k=0}^{\min\{m,n\}} P_{(k,m-k,n-k)}.$$

Then, using the well-known combinatorial identity $\binom{n+k+1}{i} = \binom{n+k}{i} + \binom{n+k-1}{i-1} + \cdots + \binom{n+k-i}{0}$, we can obtain that

$$HO(m+1,n) = \sum_{i=0}^{n-1} D(m,i) = \sum_{k=0}^{\min\{m,n-1\}} \sum_{i=k}^{n-1} P_{(k,m-k,i-k)}$$

$$= \sum_{k=0}^{\min\{m,n-1\}} \sum_{i=0}^{n-1} \frac{(m+i-k)!}{k!(m-k)!(i-k)!}$$

$$= \sum_{k=0}^{\min\{m,n-1\}} \frac{m!}{k!(m-k)!} \sum_{i=0}^{n-1} \frac{(m+i-k)!}{m!(i-k)!}$$

$$= \sum_{k=0}^{\min\{m,n-1\}} \binom{m}{k} \sum_{i=0}^{n-1} \binom{m+i-k}{i-k}$$

$$= \sum_{k=0}^{\min\{m,n-1\}} \binom{m}{k} \binom{m+n-k}{n-1-k}$$

$$= \sum_{k=0}^{\min\{m,n-1\}} \binom{m}{k} \binom{m+n-k}{m+1}$$

$$= \sum_{k=0}^{\min\{m,n-1\}} \binom{m}{j} \binom{n+j}{m+1}. \quad \square$$

Edwards and Griffiths ([EdGr16]) found two similar formulas for hyperoctahedron numbers.

- For the number $HO(m,n)$, the following equalities hold:

$$HO(m,n) = \sum_{k=0}^{\lfloor \frac{n}{2} \rfloor} \binom{n}{2k} \cdot \binom{m+n-2k-1}{n-1} \quad \text{for even } n;$$

$$HO(m,n) = \sum_{k=0}^{\lfloor \frac{n-1}{2} \rfloor} \binom{n}{2k+1} \cdot \binom{m+n-2j-2}{n-1} \quad \text{for odd } n.$$

□ For example, $HO(3,2) = 6 = 4+2 = \binom{2}{0}\binom{4}{1} + \binom{2}{2}\binom{2}{1}$, while $HO(3,3) = 19 = 3 \cdot 6 + 1 = \binom{3}{1}\binom{4}{2} + \binom{3}{3}\binom{2}{2}$. For a general proof and other interesting combinatorial identities, see [EdGr16]. □

Generalized Delannoy numbers

4.3.21. In this section, we introduce two types of *generalized Delannoy numbers*. They count Delannoy paths but with some constraints on the diagonal steps.

More precisely, for a given non-negative integer k, define the number $D^k(m,n)$, $m,n \geq 0$, as the number of lattice paths from the point $(0,0)$ to the point $(m, n+k)$, such that diagonal steps are forbidden after height n, i.e., after the first n vertical steps.

By definition,

$$D(m,n) = D^0(m,n).$$

It is easy to obtain a closed formula for this new number set.

- For the number $D^k(m,n)$, it holds that

$$D^k(m,n) = \sum_{j=0}^{n} C_n^j \cdot C_{m+n-j+k}^{n+k}.$$

□ Recall that for the Delannoy number $D(m,n)$, it was proven that $D(m,n) = D(n,m) = \sum_{j=0}^{n} \binom{n}{j} \cdot \binom{m+n-j}{n}$. We obtained this formula using simple combinatorial reasoning. In fact, any path can be represented as a string containing symbols r ("right"), u ("up"), and d ("diagonally"). If there are j diagonal steps, $0 \leq j \leq \min\{m,n\}$, then there should be $m-j$ horizontal steps and $n-j$ vertical steps, altogether $m+n-j$ positions. Let us choose from $m+n-j$ places n places for the symbol u; it can be done in $\binom{m+n-j}{n}$ ways. The remaining $m-j$ places are used for the symbol r. Then, let us choose from those fixed at the previous step n symbols u exactly j symbols and replace them with the symbols d. It can be done in C_n^j ways. So, we obtain a string containing j symbols d, $n-j$ symbols u, and $m-j$ symbols r.

A similar consideration can be realized in the general case.

In fact, consider the paths counted by the number $D^k(m,n)$. Any such path has j diagonal steps, where $0 \le k \le \min\{m,n\}$. If there are exactly j diagonal steps, then there should be $m-j$ horizontal steps and $n+k-j$ vertical steps, altogether $m+n+k-j$ positions. Let us choose from $m+n+k-j$ places $n+k$ places for the symbol u; it can be done in $\binom{m+n+k-j}{n+k}$ ways. The remaining $m-j$ places are used for the symbol r. Then, let us choose from the first n symbols fixed at the previous step $n+k$ symbols u exactly j symbols and replace them with the symbols d. In can be done in C_n^j ways. So, we obtain a string containing j symbols d, $n+k-j$ symbols u, and $m-j$ symbols r; by this construction, the symbol d cannot be used after n symbols u. $\square$

Similarly, it is possible to define the numbers $D_k(m,n)$; they can be given by the formula

$$D_k(m,n) = \sum_{j=0}^{n} C_n^j \cdot C_{m+n-j-k}^{n-k}.$$

Again, it holds that $D(m,n) = D_0(m,n)$.

We consider these and other relative constructions in detail in Chapter 5.

Exercises

1. Calculate the Delannoy numbers $D(m,n)$, $n,m \in \{1, 2, \ldots, 10\}$ using the recurrent relation. Build several next elements in the rectangle of Delanoy numbers and in Delannoy's triangle.
2. Calculate the Delannoy numbers $D(m,n)$, $n,m \in \{1, 2, \ldots, 10\}$ using the closed formula. Compare the complexity of the calculation by the closed formula and by the recurrent relation.

3. Build a table of Delannoy numbers modulo 2. What have you obtained? Prove that all Delannoy numbers are odd.
4. Build a table of Delannoy numbers modulo 3. Compare the obtained picture with the *Sierpiński carpet*, obtained by dividing a given square into nine equal squares, removing the central square and repeating the procedure for each of the remaining squares an infinite number of times.
5. Build a table of Delannoy numbers modulo 4. What do you have? Prove that the Delannoy numbers $D(2k, n)$ are congruent to 1 modulo 4 and the numbers $D(2k-1, n)$ are congruent to 1 and 3 modulo 4, alternated. What can be said about the behavior of the Delannoy numbers modulo 8?
6. Construct the triangle of Delannoy numbers modulo k, $k \in \{2, 3, 4\}$. Formulate the known properties of Delannoy's rectangle, considered modulo k, in terms of Delannoy's triangle.

4.4. Narayana Numbers

A history of the question

4.4.1. *Narayana numbers*, which appear in many counting problems, are named after Tadepalli Venkata Narayana (1930–1987).

T. V. Narayana was an Indo-Canadian statistician and mathematician, known for his contributions to combinatorics, lattice theory, and mathematical statistics.

He was born in Madras (now Chennai). He studied at the Madras and Bombay universities in India before joining the North Carolina University to pursue his PhD studies. He was awarded the PhD degree in mathematical statistics in 1954. After securing his PhD, T. V. Narayana did postdoctoral work at the Indian Council of Agricultural Research, New Delhi, and the Henri Poincare Institute, Paris. He was appointed assistant professor at McGill University, Canada, in 1955. In 1958, he joined the University of Alberta, Canada, as an associate professor, and was promoted to professor in 1966.

He made seminal contributions to the theory of tournaments, compositions, sampling plans, and lattice path combinatorics.

Much of his work was collected together in a monograph, *Lattice Path Combinatorics with Statistical Applications*, published by the University of Toronto Press in 1979.

A certain sequence of numbers, called *Narayana numbers*, and a certain class of polynomials, called *Narayana polynomials*, which were both named after him for his work in bringing out their importance in lattice path theory, have found extensive and varied applications in combinatorics and lattice theory.

Construction of Narayana numbers

4.4.2. Like many other combinatorial numbers, Narayana numbers occur at a study of a special type of partitions, namely, the so-called *non-crossing partitions*.

Recall, that a *partition* of the set $V_n = \{1, 2, \ldots, n\}$ is a representation V_n as a combination of q disjoint non-empty sets $S_1, \ldots, S_q$:

$$V_n = S_1 \cup \cdots \cup S_q, \quad \text{where} \quad S_i \neq \emptyset, \quad \text{and} \quad S_i \cap S_j = \emptyset,$$

$$i \neq j \in \{1, 2, \ldots, q\}.$$

To define non-crossing partitions, let us consider the set $\{1, 2, 3, \ldots, n\}$, and place it in a cyclic order, for example, placing the numbers $1, 2, 3, \ldots, n$ in the specified order on a circle.

In this case, a *non-crossing partition* is such a partition of the set $\{1, 2, 3, \ldots, n\}$, in which no two partition blocks "cross" each other, i.e., if a and b belong to one partition block, and x and y to another, they are not arranged in the order $axby$. For example, the partition $\{1, 2\} \cup \{3, 4, 5\}$ of the set $\{1, 2, 3, 4, 5\}$ is non-crossing, but the partition $\{1, 3\} \cup \{2, 4, 5\}$ is not.

If a set is represented by the points of the circle, then its partition will be non-crossing, if the polygons, whose vertices are elements of the blocks of the partition, do not intersect (a point and a line are viewed in such a model as an one-gon and an bigon, respectively).

The *Narayana Number* $N(n, k)$, $n = 1, 2, 3, \ldots, k = 1, 2, \ldots, n$, is defined as the number of non-crossing partitions of an n-set into k parts. So, for example, $N(4, 2) = 2$, since we have three partitions of an 4-set into two 2-sets, and exactly two of them are non-crossing: $\{1,2\}\cup\{3,4\}$, $\{1,4\}\cup\{2,3\}$, and $\{2,3\}\cup\{1,4\}$; on the other hand, all four partitions of the 4-set into an 3-set and an 1-set are non-crossing.

Direct calculations allow us to construct the *Narayana's triangle*, consisting of the numbers $N(n,k)$ (sequence A001263 in the OEIS).

n\k	1	2	3	4	5	6	7	8
1	1							
2	1	1						
3	1	3	1					
4	1	6	6	1				
5	1	10	20	10	1			
6	1	15	50	50	15	1		
7	1	21	105	175	105	21	1	
8	1	28	196	490	490	196	28	1

Obviously, $N(n,1) = 1$, since any partition, consisting of one subset, will be non-crossing. Similarly, $N(n,n) = 1$. So,

- *the sides of the Narayana's triangle consist from unities.*

Since, as was previously shown, the n-th *Catalan number* C_n gives the number of all disjoint partitions of an n-set, $n = 0, 1, 2, 3, \ldots$, then

- *the sum of elements of the n-th row of the Narayana's triangle is equal to the n-the Catalan number C_n:*

$$N(n,1) + N(n,2) + N(n,3) + \cdots + N(n,n) = C_n.$$

So, it holds

$$N(1,1) = 1 = C_1; \quad N(2,1) + N(2,2) = 1 + 1 = C_2;$$

$$N(3,1) + N(3,2) + N(3,3) = 1 + 3 + 1 = 5 = C_3;$$

$$N(4,1) + N(4,2) + N(4,3) + N(4,4) = 1 + 6 + 6 + 1 = 14 = C_4.$$

By this reason, this triangle is sometimes called the *Catalan's triangle*.

Considering this triangle as an isosceles triangle, we can see, that

- *the Narayana's triangle is symmetrical with respect to the principal diagonal:*

$$N(n,k) = N(n, n-k+1).$$

☐ In fact, it holds for small values of n.

$$
\begin{array}{ccccccccc}
 & & & & 1 & & & & \\
 & & & 1 & & 1 & & & \\
 & & 1 & & 3 & & 1 & & \\
 & 1 & & 6 & & 6 & & 1 & \\
1 & & 10 & & 20 & & 10 & & 1 \\
\vdots & \vdots & \vdots & \vdots & \vdots & \vdots & \vdots & \vdots & \vdots
\end{array}
$$

We will give the general proof of this fact later. ☐

Counting problems related to Narayana numbers

Rhyme schemes' problem

4.4.3. In the previous section, we have defined the Narayana number $N(n,k)$ as the number of non-crossing partitions of an n-set into k blocks. Recall that the number of all partitions of an n-set is the nth *Bell number* $B(n)$, the number of all partitions of an n-set into k blocks is the Stirling number of the second kind $S(n,k)$, and the number of all non-crossing partitions of an n-set is the nth Catalan number C_n.

The set of partitions of an n-set can be represented in different ways. So, for a poet, the number of all partitions of an n-set (i.e., the nth Bell number $B(n)$) is the number of all kinds of rhyme schemes of an n-line poem.

Such a transition from partitions to rhymes is easily possible if, instead of each element k of the set $V_n = \{1, 2, \ldots, n\}$, we write a symbol of the class S_i into which the element k falls in a given partition $V_n = S_1 \cup \cdots \cup S_q$: symbol a for S_1, symbol b for S_2, etc. Thus, the partition $V_4 = \{1,2\} \cup \{3,4\}$ corresponds to the rhyme *aabb*, and the partition $V_4 = \{1,2\} \cup \{3\} \cup \{4\}$ corresponds to the rhyme *aabc*.

Connecting the rhyme's elements by the same arcs, we see that in some rhymes, the arcs intersect, and in some rhymes, they do not. Let us call the rhymes in which arcs do not intersect *plane rhymes*.

It turns out that the number of plane rhymes of an n-line poem is equal to the nth Catalan number C_n (see Chapter 3).

In such a representation, the partition having exactly k blocks are represented by a rhyme schemes with exactly k symbols. So, we obtain that:

- The Narayana number $N(n,k)$ counts the number of plane rhyme schemes of an n-line poem, which has exactly k symbols.

Let us illustrate the behavior of $N(n,k)$ in the considered model for small values of n. For $n = 1$, $N(1,1) = 1$: the only (non-crossing) rhyme scheme for a one-line poem is a; it contains exactly one symbol. For $n = 2$, there are exactly two (non-crossing) rhyme schemes for a two-line poem: aa and ab. One of them has one symbol; the other has two symbols. So, $N(2,1) = 1$ and $N(2,2) = 1$. For a 3-line poem, there are exactly five rhyme schemes: aaa, aab, aba, abb, and abc. All these rhyme schemes are non-crossing. Exactly three of them are represented by two symbols: aab, aba, abb, so $N(3,2) = 2$. Moreover, $N(3,1) = 1$: we count only aaa; $N(3,3) = 1$: we count only abc. Similarly, $N(4,2) = 6$: for a four-line poem, there are exactly 15 rhyme schemes: $abbb$, $abbc$, $abcb$, $abcc$, $abcd$, $aabb$, $aabc$, $aaab$, $abab, abac$, $aaaa$, $aaba$, $abaa$, $abba$, and $abca$; 14 of them are plane (excluding $abab$); 6 plane schemes consist of two symbols: $aaab$, $aaba$, $aabb$, $abaa$, $abba$, $abbb$. Moreover, $N(4,1) = 1$: we count only $aaaa$; $N(4,3) = 6$: we count the rhymes $abbc$, $abcb$, $abcc$, $aabc$, $abac$, $abca$; finally, $N(4,4) = 1$: we count only $abcd$.

Brackets' problem

4.4.4. It is well known that the number of correct arrangements of n pairs of brackets is equal to the nth Catalan number C_n. Define a *nesting* as the placement () of one pairs of brackets "glued together" without other pairs of brackets between them. Now, it is easy to show that:

- the Narayana number $N(n,k)$ counts the number of correct bracket arrangements of n pairs of brackets which contain k distinct nestings.

□ For example, $N(2,2) = 1$, and there is one arrangement with two pairs of brackets and two nestings: ()(); $N(3,2) = 3$, and there are three arrangements with three pairs of brackets and two nestings: ()()(), ()(()), (())(); $N(4,2) = 6$, and there are six arrangements with three pairs of brackets and two nestings: (()(())), ((()())), ((())()), ()((())), (())(()), ((()))().

In order to prove this fact, it is enough to build a one-to-one correspondence between the set of all plane rhymes for an n-line poem, represented by k different symbols, and the set of all correct arrangements of n pairs of brackets with k nestings. For a given plane rhyme, the correspondence is constructed according to the following law:

- put opening bracket before each new symbol of the rhyme;
- close the bracket (the required number of times) after the last symbol of the rhyme, i.e., in the case, when the set of symbols of this type is completely exhausted.

So, for $n = 1$, only rhyme a corresponds to the bracket arrangement (). For $n = 2$, the rhyme aa corresponds to the arrangement (()), and the rhyme ab corresponds to the arrangement ()(). For $n = 3$, the rhyme aaa corresponds to the arrangement ((())), the rhyme aab corresponds to the arrangement (())(), the rhyme aba corresponds to the arrangement (()()), the rhyme abb corresponds to the arrangement ()(()), and the rhyme abc corresponds to the arrangement ()()(). For $n = 4$, the rhyme $aaaa$ corresponds to the arrangement ((())), the rhyme $abba$ corresponds to the arrangement ((())()), etc. This construction shows that any symbol a_i used in the rhyme corresponds to exactly one nesting: we get it when we reach the last item of a_i.
□

Lattice path problem

4.4.5. For a given positive integer n, define a *mountain with n ascents and n descents* as a broken line, composed of n segments of the same length, inclined at an angle of 45° to the positive direction of the OX-axis, and of n segments of the same length, inclined at an angle 135° to the positive direction of the OX-axis, such that the resulting "start" and "end" place themselves on the same level. Any such mountain has one or several *peaks* ⟋⟍.

- *The number of mountains with n ascents, n descents, and k peaks are equal to the Narayana number $N(n,k)$.*

☐ The proof can be obtained using correct bracket structures. In fact, it is enough to build a one-to-one correspondence between the set of all correct arrangements of n pairs of brackets (with k nestings) and the set of mountains with n ascents and n descents (and k peaks). The corresponding law is very simple: each "ascending" segment corresponds to an opening bracket; each "descending" segment corresponds to a closing bracket. Obviously, a peak ⟋⟍ corresponds to a nesting (). ☐

4.4.6. Similarly, Narayana numbers give the numbers of certain routes on the upper-right quadrant of the integer lattice. Recall that the nth Catalan number C_n gives the number of routes on the upper-right quadrant of the integer lattice $\mathbb{Z}^2$, from the point $(0,0)$ to the point $(2n,0)$, if we can only move right-up and right-down, but it is forbidden to move below the OX-axis. It is natural to call such paths *Catalan's paths*.

- *The number of Catalan's paths from the point $(0,0)$ to the point $(2n,0)$ with k peaks is equal to the Narayana number $N(n,k)$.*

☐ Obviously, the counting of such paths is equivalent to the counting of mountains with n ascents, n descents, and k peaks. ☐

Equivalently, Narayana numbers give the number of certain monotonic lattice paths in a grid with $n \times n$ square cells. In fact, the nth Catalan number C_n is the number of monotonic lattice paths along the edges of a grid with $n \times n$ square cells, which do not pass above the diagonal. A monotonic path is one which starts at the lowest left corner, ends at the upper right corner, and consists entirely of edges pointing rightward or upward.

- The Narayana number $N(n,k)$ is the number of monotonic lattice paths along the edges of a grid with $n \times n$ square cells which do not pass above the diagonal and have exactly k "lower-right angles."

□ The counting of such paths is equivalent to the counting of correct bracket sequences: an open bracket stands for "move right," a closing bracket stands for "move up"; so, a nesting () stands for a "lower-right angle," formed by two consecutive steps: "move right" and "move up." □

4.4.7. There are many other equivalent formulations of bracket and lattice path problems.

For a given positive integer n, consider a *Dyck word of length* $2n$: a sequence consisting of n symbols X and n symbols Y, in which each initial segment contains at least as many symbols X as symbols Y.

- The number of Dyck words of length $2n$ with k entries XY is equal to the Narayana number $N(n,k)$.

□ In order to prove this statement, it is enough to replace the symbol X with an opening bracket and the symbol Y with a closing bracket. Such a transform gives a bijection between the set of all Dyck words of length $2n$ and the set of all correct arrangements of n pairs of brackets. Obviously, we have an entry XY if and only if we have a nesting (). □

For a given positive integer n, consider the set of *correct* $(1,-1)$-*sequences* of length $2n$, i.e., sequences consisting of n "1" and n "−1", all partial sums of which are non-negative.

It is easy to see that:

- the number of $(1,-1)$-sequences of length $2n$, having exactly k entries $1-1$, is equal to the Narayana number $N(n,k)$.

□ In order to prove this statement, consider a Dyck word of length $2n$. If instead of the symbols X and Y we use the symbols "1" and "−1", respectively, we get a sequence composed of n "1" and n "−1", all partial sums of which are non-negative. Moreover, there is a segment XY if and only if there is a segment $1-1$. However, we can use for the proof the correct bracket structures with n pairs of brackets and k nestings: we should replace the symbol "(" with the symbol "1" and the symbol ")" with the symbol "−1". □

Narayana numbers are related also to *polyomino* enumerations, which are planar geometric shapes formed by connecting several unicellular squares by their sides.

- The number $N(n,k)$ expresses the number of polyominoes with perimeter $2k+2$, having exactly k columns.

Enumeration of quasi-trees

4.4.8. It is widely known that Catalan numbers appear quite often when counting the number of certain classes of trees.

One of the simplest connection with Narayana numbers can be made for the set of all *plane rooted trees*. It is easy to check that there is exactly one plane rooted tree with one edge, exactly two plane rooted trees with two edges, exactly five plane rooted trees with three edges, and exactly 14 plane rooted trees with four edges. In general, the number of plane rooted trees with n edges is equal to the nth Catalan number C_n.

The classical proof is based on the one-to-one-correspondence between the set od all plane rooted trees with n edges and the set of all correct bracket arrangements with n pairs of brackets. For a given (planted) plane rooted tree with n edges, let us mark each edge by one pair of brackets, putting an opening bracket to the left and a closing bracket to the right. Now, starting from the root and moving from the left to the right along the edges of the tree, we can collect all n pairs of brackets, obtaining a corresponding correct arrangement.

Noting that any nesting () corresponds to a leaf of the tree, we can prove the following result:

- The number of plane rooted trees with n edges and k leaves is equal to the Narayana number $N(n,k)$.

□ We have already proven this result using correct bracket sequences with n pairs of brackets and k nestings (). However, the reader can obtain a similar bijection between the set of all plane rooted trees with n edges and k leaves and the set of all mountains of length $2n$ with k peaks (see Chapter 8). □

4.4.9. When it comes to plane binary trees, the situation is not so obvious. We know that the number of plane binary trees with n internal vertices is equal to the nth Catalan number C_n.

We can prove the statement using a one-to-one correspondence between the set of plane binary trees with n internal vertices and the set of correct arrangements of n pairs of brackets. In fact, we mark all leaves of a (planted) plane binary tree, except for the last right and the lowest (the root), by closing brackets and all internal vertices by opening brackets; imagine a worm crawling up the trunk and around the entire tree and collecting the available brackets on the way. In this case, any plane binary three with n internal vertices corresponds to some correct arrangement of n pairs of brackets. Moreover, we obtain the nestings () exactly for the "left" leaves of the tree. More precisely, in the planted representation of a plane binary tree, any internal vertex has two children, placing in left-to-right order on the plane. If one of these children is a leaf, call it *left* or *right* depending on this position. It is easy to check that the considered Łukasiewicz's bijection gives a nesting if and only if the worm is passing through a "left" leaf of the tree. So:

- the number of planted binary trees with n internal vertices and k left leaves is equal to the Narayana number $N(n, k)$.

Enumeration of permutations

4.4.10. Narayana numbers can be used for the enumeration of some classes of *pattern-avoiding permutations*.

A *pattern* is a smaller permutation embedded in a larger permutation. More precisely, a permutation σ of an n-set contains a permutation α of a k-set, $k \leq n$, as a pattern if we can find k digits of σ that appear in the same relative order as the digits of α. For example, the permutation $\sigma = 18274653$ contains the patterns 123, 132, 231, 312, and 321 but does not contain the pattern 213. In this case, we say that the permutation σ *avoids* the pattern 213.

Pattern-avoiding permutations have provided many counting problems. In particular, the number of permutations of an n-set that avoid the pattern 132 is equal to the nth Catalan number C_n.

Similarly, it can be proven that:

- the Narayana number $N(n, k)$ gives the number of permutations of an n-set which avoid the pattern 132 and have $k - 1$ descents.

□ Recall (see Chapter 1) that, for a given permutation σ of the set $\{1, 2, 3, \ldots, n\}$, a *descent* is a position $i < n$ with $\sigma_i > \sigma_{i+1}$; if $\sigma = \sigma_1 \sigma_2 \cdots \sigma_n$, then i is a descent if $\sigma_i > \sigma_{i+1}$.

For example, the permutation $\sigma = 3452167$ has descents (at the positions) 3, 4 (and has ascents at the positions 1, 2, 5, and 6). Every i with $1 \leq i < n$ either is an ascent or is a descent of σ.

For example, if $n = 2$, $N(2, 1) = 1$ and $N(2, 2) = 1$. On the other hand, there are two permutations of the set $\{1, 2\}$: 12 and 21. Both permutations are 132-avoiding: the first one has zero descents, while the second one has one descent. If $n = 3$, $N(3, 1) = 1$, $N(3, 2) = 3$, and $N(3, 3) = 1$. On the other hand, there are six permutations of the set $\{1, 2, 3\}$: 123, 132, 213, 213, 312, 321. Five of them are 132-avoiding (except 132); one of them has no descents: 123; three of them has one descent: 132, 213, 231; finally, one of them has two descents: 321.

The general proof can be obtained using combinatorial considerations similar to the proof of the corresponding statement about Catalan numbers. □

Recurrences related to Narayana numbers

4.4.11. A recurrent relation for Narayana numbers should be similar to the recurrent relation for Catalan numbers. However, the additional restriction, connected with the parameter k, makes the consideration more difficult. In fact, supposing that $N(0,0) = 1$ and $N(n,0) = N(0,k)$ for the positive integers n and k, we can prove that:

- for Narayana numbers, the following recurrent relation holds:

$$N(n+1,k) = N(n,k-1) + \sum_{t=1}^{n} \sum_{\substack{l+m=k,\\ l\geq 0, m\geq 0}} N(t,l)N(n-t,m).$$

☐ For example, $N(2,2) = N(1,1) + \sum_{t=1}^{1} \sum_{\substack{l+m=2,\\ l\geq 1, m\geq 0}} N(t,l)N(1-t,m)$. In this case, the double sum is simply an empty sum. On the other hand, for $N(4,3)$, the double sum has three summands. For $t=1$, we have exactly one construction $N(t,l)N(1-t,m)$; it is $N(1,1)N(2,2) = 1 \cdot 1 = 1$. For $t=2$, we have one summand $N(2,2)N(1,1) = 1 \cdot 1 = 1$; For $t=3$, the only summand has the form $N(3,3)N(0,0) = 1 \cdot 1 = 1$. As $N(3,2) = 3$, we obtain that

$$N(4,3) = 6 = 3 + (1+1+1) = N(3,2)$$

$$+ \sum_{t=1}^{3} \sum_{\substack{l+m=3,\\ l\geq 0, m\geq 0}} N(t,l)N(3-t,m).$$

Let us give a general proof considering the Narayana number $N(n,k)$ as the number of Catalan's paths from the point $(0,0)$ to the point $(2n,0)$, having exactly k peaks.

It is easy to check that the statement is true for small values of n. Assuming, that the statement is true for all positive integers $k \leq n$, try to prove it for a given positive integer $n+1$:

I. If a path from the point $(0,0)$ to the point $(2n+2,0)$ starts from the first right-up and the second right-down steps, i.e., with the construction ↗↘, then we already have one peak, and the remaining part of the considered path should be a Catalan's path from $(2,0)$ to $(2n+2,0)$, containing exactly $k-1$ peaks. There are exactly $N(n, k-1)$ such possibilities.

II. If the path, always starting from an up-step ↗, has the down step associated with the first up-step, ending at the point $(2s,0)$,

$2 \leq s \leq n+1$, let us fix both these steps. Then, we obtain between the points $(0,0)$ and $(2s,0)$ exactly $2s - 1$ points, while "out" we have $2(n - s + 1) + 1$ points, starting from the point $(2s, 0)$.

If we want to count the number of possible paths of this configuration, we should consider the number of possible "between" and "out" constructions. If a "between" construction has l peaks, $l \geq 1$, then the corresponding "out" construction should have $m = k - l$ peaks, $m \geq 0$ (zero is possible if $s = n+1$). For a fixed $1 \leq l \leq k$, we obtain $N(s-1, l)$ "between" possibilities and $N(n-(s-1), k-l)$ "out" possibilities. Denoting $t = s - 1$ and $m = k - l$, we obtain the result. □

The obtained recurrent formula can be used for the construction of Narayana's triangle. For example,

$$N(5,2) = N(4,1) + N(1,1)N(3,1) + N(2,1)N(2,1)$$
$$+ N(2,2)N(2,0) + N(3,1)N(1,1) + N(3,2)N(1,0)$$
$$+ N(4,1)(N(0,1) + N(4,2)N(0,0)$$
$$= 1 + 1 \cdot 1 + 1 \cdot 0 + 1 \cdot 1 + 3 \cdot 0 + 1 \cdot 0 + 6 \cdot 1 = 10.$$

The above algorithm is rather complicated; for small values of the parameters n, k, it is more convenient to use the well-known closed formula: $N(n, k) = \frac{1}{n}\binom{n}{k}\binom{n}{k-1}$. We prove this formula in the following section.

However, using the closed formula $N(n, k) = \frac{1}{n}\binom{n}{k}\binom{n}{k-1}$, it is easy to obtain that:

- the Narayana number $N(n, k)$ can be calculated using the number $N(n-k, k-1)$ using the formula

$$N(n, k) = \frac{n(n-1)}{k(k-1)} N(n-1, k-1).$$

☐ In fact,

$$N(n, k) = \frac{1}{n}\binom{n}{k}\binom{n}{k-1} = \frac{n!(n-1)!}{k!(k-1)!(n-k)!(n-k-1)!}$$

$$= \frac{n(n-1)}{k(k-1)} \cdot \frac{(n-1)!}{(n-k)!(k-1)!} \cdot \frac{(n-2)!}{(k-2)!(n-k+1)!}$$

$$= \frac{n(n-1)}{k(k-1)} N(n-1, k-1).$$ □

This formula allows us to construct Narayana's triangle using *weights*: for the construction of the kth element of the nth row, we should use the element placed left-up in the previous row, with the multiplicative weight $\frac{n(n-1)}{k(k-1)}$: as $N(3,2) = 3$, $N(4,3) = \frac{4 \cdot 3}{3 \cdot 2} \cdot N(3,2) = 2N(3,2) = 6$.

So, for the construction of the only internal element of the third row, we should use the second row $1, 1$ (in fact, its first element 1) and just one weight $\frac{3 \cdot 2}{2 \cdot 1}$: $3 = \frac{3 \cdot 2}{2 \cdot 1} \cdot 1$. For the construction of the two internal elements of the fourth row, we should use the first two elements of the third row $1, 3, 1$ and the weights $\frac{4 \cdot 3}{2 \cdot 1}, \frac{4 \cdot 3}{3 \cdot 2}$: $6 = \frac{4 \cdot 3}{2 \cdot 1} \cdot 1$, $6 = \frac{4 \cdot 3}{3 \cdot 2} \cdot 3 = 6$, etc.

This rule is relatively compact; however, we should use multiplications and divisions, which is not convenient for big values of n, k.

4.4.12. In the special numbers literature, we can find other recurrent formulas related to Narayana numbers. For example:

- *for a recurrent calculation of the Narayana number $N(n, k)$, i.e., the kth element of the nth row of Narayana's triangle, we can use only two consecutive elements, $N(n-1, k-1)$ and $N(n-1, k)$, of the previous $(n-1)$th row, and three consecutive elements $N(n-2, k-2), N(n-2, k-1), N(n-2, k)$ of the $(n-2)$th row:*

$$N(n, k) = \frac{1}{n+1}((2n-1)N(n-1, k-1) + (2n-1)N(n-1, k)$$
$$- (n-2)N(n-2, k-2) + 2(n-2)N(n-2, k-1)$$
$$- (n-2)N(n-2, k)).$$

□ This formula has six summands, but we should realize only one division. In order to obtain the kth element of the nth row of Narayana's triangle, we should consider the following small

triangle:
$$N(n-2,k-2) \quad N(n-2,k-1) \quad N(n-2,k)$$
$$N(n-1,k-1) \quad N(n-1,k)$$
$$N(n,k)$$

In order to find $N(n,k)$, we should take five elements in the upper trapezium with corresponding multiplicative weights, add the obtained five numbers, and divide the sum by $n+1$. For example, the small triangle for the number $N(5,3)$ has the following form:
$$N(3,1) = 1 \quad N(3,2) = 3 \quad N(3,3) = 1$$
$$N(4,2) = 6 \quad N(4,3) = 6$$
$$N(5,3)$$

So, we obtain $N(5,3)$ using the calculations
$$N(5,3) = \frac{1}{6}(9 \cdot 6 + 9 \cdot 6 - 3 \cdot 1 + 6 \cdot 3 - 3 \cdot 1) = 20.$$

A general proof can be based on direct calculations, using the formula $N(n,k) = \frac{1}{n}\binom{n}{k}\binom{n}{k-1}$. □

4.4.13. One more recurrence involving Narayana numbers, has the following form:

- *The number $N(n+1, k+1)$, i.e., the $(k+1)$th element of the $(n+1)$th row of Narayana's triangle, can be calculated using three consecutive elements, $N(n,k)$, $N(n,k+1)$, and $N(n,k+2)$, of the previous row:*

$$\binom{n+2}{2}N(n+1,k+1) = \binom{2n+2-k}{2}N(n,k)$$
$$+ (k+1)(2n+1-k)N(n,k+1)$$
$$+ \binom{k+2}{2}N(n,k+2).$$

☐ For example, for $n = 5, k = 2$, we have
$$300 = 15 \cdot 20 = \binom{6}{2}N(5,3) + \binom{8}{2}N(4,2) + 3 \cdot 7 N(4,3)$$
$$+ \binom{4}{2}N(4,4) = 28 \cdot 6 + 21 \cdot 6 + 6 \cdot 1.$$

In fact, this recurrence is a generalization of the previous formula and can also be proved by direct checking. □

Closed formula for Narayana numbers

4.4.14. As was already mentioned above, there exists a simple closed formula for Narayana numbers.

- *For the Narayana number $N(n,k)$ the following formula holds:*

$$N(n,k) = \frac{1}{n}\binom{n}{k}\binom{n}{k-1} = \frac{1}{k}\binom{n}{k-1}\binom{n-1}{k-1}$$
$$= \frac{1}{n+1}\binom{n+1}{k+1}\binom{n-1}{k}.$$

□ Let us prove this statement by representing the number $N(n,k)$ as the number of Catalan's paths between the points $(0,0)$ and $(2n,0)$ (or, equivalently, as the number of correct $(1,-1)$-sequences). In order to construct such a path, we should use n right-up steps ↗, n right-down steps ↘, and k peaks ↗↘.

If there are exactly k peaks, then there are exactly k ascents and exactly k descents in the sequence of steps. In order to obtain such constructions, we add to n right-up steps ↗ an additional step and divide the set containing $n+1$ symbols ↗ in k ordered non-empty groups. The number of such partitions is equal to the number of positive integer solutions of the equation $x_1 + x_2 + \cdots + x_k = n+1$, i.e., $P_{(n+1-k,k-1)} = \binom{n}{k-1}$. In fact, in order to obtain a positive integer solution of this equation, we should simply place $k-1$ symbols ⊕ between $n+1$ symbols 1, i.e., choose $k-1$ places from n possible places between symbols 1. Similarly, divide the set containing n symbols ↘ in k ordered non-empty groups; it can be done in $P_{(n-k,k-1)} = \binom{n-1}{k-1}$ ways.

The first obtained group of ↗ gives the first ascent, the first obtained group of ↘ gives the first descent, etc. Formally, if $((a_1, a_2, \ldots, a_k), (b_1, b_2, \ldots, b_k))$ is a positive solution to the system

$$\begin{cases} x_1 + x_2 + \cdots + x_k = n+1, \\ y_1 + y_2 + \cdots + y_k = n, \end{cases}$$

then the sequence $(a_1, b_1, a_2, b_2, \ldots, a_k, b_k)$ completely defines the structure of the corresponding sequence with $n+1$ symbols ↗ and n symbols ↘, which has exactly k peaks. By this construction, there are exactly $\binom{n}{k-1}\binom{n-1}{k-1}$ such sequences.

However, some of such sequences can have parts placed below the OX-axis. In terms of $(1,-1)$-sequences, some sequences can have negative partial sums.

In order to avoid this problem, let us consider, instead of the constructed set of sequences of length $2n+1$, the relative set of infinite sequences. In fact, for a given sequence containing $n+1$ symbols ↗, n symbols ↘, and having exactly k peaks, construct an infinite sequence by periodical repeating the elements of the considered sequence. On this way, we obtain $\binom{n}{k-1}\binom{n-1}{k-1}$ infinite sequences, some of which coincide.

In fact, for a fixed solution $((a_1, a_2, \ldots, a_k), (b_1, b_2, \ldots, b_k))$ of the system
$$\begin{cases} x_1 + x_2 + \cdots + x_k = n+1, \\ y_1 + y_2 + \cdots + y_k = n, \end{cases}$$

exactly k solutions $((a_i, a_{i+1}, \ldots, a_k, a_1, \ldots, a_{i-1}), (b_i, b_{i+1}, \ldots, b_k, b_1, \ldots, b_{i-1})$, $i = 1, 2, \ldots, k$, lead to the same infinite sequence.

So, we obtain $\frac{1}{k}\binom{n}{k-1}\binom{n-1}{k-1} = \frac{1}{n}\binom{n}{k}\binom{n}{k-1}$ classes of infinite sequences. Any class has exactly one subsequence of length $2n$, which can be considered a Catalan's sequence; in terms of $(1,-1)$-sequences, there exists only one subsequence (of length $2n$), starting with 1, which has exactly n symbols 1, exactly n symbols -1, and all non-negative partial sums.

For example, if $n = 3$ and $k = 2$, we obtain six infinite sequences, divided into two groups. Each group produces only one correct subsequence of length six.

Precisely, the solution $((1, 3), (1, 2))$ of the system $\begin{cases} x_1 + x_2 = 4, \\ y_1 + y_2 = 3 \end{cases}$ produces the infinite sequence

↗↘↗↗↗↘↘↗↘↗↗↘↘ ⋯ .

Relatives of Catalan Numbers 315

The solution $((1,3),(1,2))$ produces the infinite sequence

↗↗↗↘↘↗↘↗↗↗↘↘↗↘ ··· .

These two sequences, considered as two-sided infinite sequences, coincide. Their only Catalan's subsequence (of length 6 with two peaks) is ↗↗↘↘↗↘ .

The solution $((2,2),(1,2))$ produces the infinite sequence

↗↗↘↗↗↘↘↗↗↘↗↗↘↘ ··· .

The solution $((2,2),(2,1))$ produces the infinite sequence

↗↗↘↘↗↗↘↗↗↘↘↗↗↘ ··· .

These two sequences, considered as two-sided infinite sequences, coincide. Their only Catalan's subsequence (of length 6 with two peaks) is ↗↘↗↗↘↘ .

The solution $((3,1),(1,2))$ produces the infinite sequence

↗↗↗↗↘↘↘↗↗↗↘ ··· .

The solution $((2,2),(2,1))$ produces the infinite sequence

↗↗↗↘↗↘↘↗↗↘↗↘↘ ··· .

These two sequences, considered as two-sided infinite sequences, coincide. Their only Catalan's subsequence (of length 6 with two peaks) is ↗↗↘↗↘↘ .

So, we obtain that the number of Catalan's paths of length n with k peaks is equal to the number of the obtained classes of sequences, i.e., $\frac{1}{k}\binom{n}{k-1}\binom{n-1}{k-1} = \frac{1}{n}\binom{n}{k}\binom{n}{k-1}$. The statement is proven. □

Generating functions for the sequences of Narayana numbers

4.4.15. If we consider Narayana numbers as a two-dimensional array, we can state that:

- *the two-dimensional generating function for the sequence of Narayana numbers has the form* $f(x,y) = \frac{1}{2x}\Big(1 - x(1+y) - $

$\sqrt{(1-x(1+y))^2 - 4yx^2}$, i.e., it holds that

$$\frac{1}{2x}\left(1-x(1+y) - \sqrt{(1-x(1+y))^2 - 4yx^2}\right) = \sum_{n>0, k>0} N(n,k) x^n y^k.$$

☐ In fact, let $F = F(x,t) = \sum_{n>0, k>0} N(n,k) x^n y^k$. For fixed n and k, consider the set $X_{n,k}$ of all Catalan's paths of length $2n$ with k peaks; so, $|X_{nk}| = N(n,k)$. Let $X_k = X_{1k} \cup X_{2k} \cup \cdots \cup X_{nk} \cup \cdots$. In other words, X_k is the set of all possible Catalan's paths with k peaks. Any path from X_k has one of four possible forms:

- ↗ u ↘, where $u \in X_j$, and $v \in X_{k-j}$;
- ↗↘ u, where $u \in X_{k-1}$;
- ↗ u ↘, where $u \in X_k$;
- ↗↘ (if $k = 1$).

So, we obtain that

$$|X_k| = \sum_{j=1}^{k-1} |X_j| \cdot |X_{k-j}| + |X_{k-1}| + |X_k| + \delta_{1k}.$$

In other words, we obtain a recurrent formula for the sequence $|X_k| = \sum_{n \geq 1} N(n,k) = \sum_{n \geq k} N(n,k)$, $k = 1, 2, 3, \ldots$.

Consider now the generating function $y_k = y_k(x) = \sum_{n \geq 1} N(n,k) x^n$ of the sequence $N(1,k), N(2,k), \ldots, N(n,k), \ldots$. Taking $y_0 = y_0(x) \equiv 0$, we obtain that

$$y_k = x \sum_{j=0}^{k} y_j y_{k-j} + x y_{k-1} + x y_k + \delta_{1k} x.$$

As $F = F(x,t) = \sum_{k \geq 1} y_k t^k = \sum_{k \geq 0} y_k t^k$, we have that

$$xF = \sum_{k \geq 1} x y_k t^k, \quad F^2 = \sum_{k \geq 0} \sum_{j=0}^{k} y_j y_{k-j} t^k,$$

$$xF^2 = \sum_{k \geq 0}\left(x \sum_{j=0}^{k} y_j y_{k-j} t^k\right), \quad tF = \sum_{k \geq 1} y_{k-1} t^k, \quad xtF = \sum_{k \geq 1} x y_{k-1} t^k.$$

Since $y_k = x\sum_{j=0}^{k} y_j y_{k-j} + xy_{k-1} + xy_k + \delta_{1k}x$, we get that

$$F = \sum_{k\geq 1} y_k t^k = \sum_{k\geq 1} \left(x\sum_{j=0}^{k} y_j y_{k-1} \right) t^k$$

$$+ \sum_{k\geq 1} xy_{k-1} t^k + \sum_{k\geq 1} xy_k t^k + \sum_{k\geq 1} \delta_{1k} xt^k.$$

Noting that $\sum_{k\geq 1} \delta_{1k} xt^k = xt$, we obtain for F the equation $F = xF^2 + xtF + xF + tx$, or, which is the same, the equation

$$xF^2 + (xt + x - 1)F + xt = 0.$$

Solving this quadratic equation with respect to F, we obtain that

$$F_{1,2} = \frac{1}{2x}(1 - x(1+y) \pm \sqrt{(1 - x(1+y))^2 - 4yx^2}).$$

The standard procedure of choosing the sign gives us the result. $\square$

4.4.16. For the finite sequence $N(n, k)$, $k = 1, 2, \ldots, n$, the situation is more simple.

In fact, for a fixed positive integer n, the generating function of the sequence $N(n, k)$, $k = 1, 2, \ldots, n$, has the form $\sum_{k=1}^{n} N(n, k)x^k$. Thus, we obtain an integer polynomial of degree n. Such polynomials possess many interesting properties and are called *Narayana polynomials*.

Formally, the *n*th *Narayana polynomial* is defined as

$$N_n(x) = \sum_{k=1}^{n} N(n, k)x^k.$$

Using Narayana's triangle, we make sure that

$$N_1(x) = x, \quad N_2(x) = x + x^2, \quad N_3(x) = x + 3x^3 + x.$$

In general:

- *for the sequence of Narayana polynomials, the following recurrent relation holds:*

$$(n+1)N_n(x) = (2n-1)(1+x)N_{n-1}(x) - (n-2)(x-1)^2 N_{n-2}(x).$$

☐ The result can be obtained using one of the above recurrent relations for Narayana numbers. In fact,

$$(n+1)N(n,k) = (2n-1)N(n-1,k-1) + (2n-1)N(n-1,k)$$
$$- (n-2)N(n-2,k-2) + 2(n-2)N(n-2,k-1)$$
$$- (n-2)N(n-2,k).$$

By definition of $N_{n-1}(x)$, it holds that

$$(1+x)N_{n-1}(x) = \sum_k N(n-1,k)x^k + \sum_k N(n-1,k-1)x^k.$$

Similarly, by definition of $N_{n-2}(x)$, it holds that

$$(x-1)^2 N_{n-2}(x) = kN(n-2,k-2)x^k + \sum_k 2N(n-2,k-1)x^k$$
$$+ \sum_k N(n-2,k)x^k.$$

Therefore, we can state that

$$(2n-1)(1+x)N_{n-1}(x) - (n-2)(x-1)^2 N_{n-2}(x)$$
$$= \sum_k ((2n-1)N(n-1,k-1) + (2n-1)N(n-1,k)$$
$$- (n-2)N(n-2,k-2) + 2(n-2)N(n-2,k-1)$$
$$- (n-2)N(n-2,k))x^k = \sum_k (n+1)N(n,k)x^k$$
$$= (n+1)N_n(x).$$

☐

Note that a similar result can be obtained for elements of Pascal's triangle. In fact, the generating function $C_n(x)$ of the sequence $\binom{n}{0}, \binom{n}{1}, \ldots, \binom{n}{n}$ has the form $C_n(x) = (1+x)^n$, i.e., it is an integer polynomial of degree n. For the sequence $C_0(x)$, $C_1(x), \ldots, C_n(x), \ldots$, the following recurrent relation holds:

$$C_n(x) = (1+x)C_{n-1}(x).$$

It can be obtained using a well-known recurrent relation for binomial coefficients: $\binom{n}{k} = \binom{n-1}{k} + \binom{n-1}{k-1}$:

$$(1+x)C_n(x) = (1+x)\sum_k \binom{n-1}{k} x^k$$

$$= \sum_k \binom{n-1}{k} x^k + \sum_k \binom{n-1}{k-1} x^k$$

$$= \sum_k \left(\binom{n-1}{k} + \binom{n-1}{k-1}\right) x^k$$

$$= \sum_k \binom{n}{k} x^k = C_n(x).$$

4.4.17. It is easy to find a connection between Narayana polynomials and Catalan numbers:

- The value $N_n(1)$ is equal to the nth Catalan number C_n: $N_n(1) = C_n$.

□ It follows from the definition $N_n(1) = \sum_k N(n,k) = C_n$. □

Moreover, a connection between Narayana polynomials and Schröder numbers can be established:

- The value $N_n(2)$ is equal to the nth Schröder number S_n: $N_n(2) = S_n$.

□ In fact, $N_1(2) = 2 = S_1$, $N_2(2) = 2^2 + 2 = 6$, $N_3(2) = 2^3 + 3 \cdot 2^2 + 2 = 22 = S_3$, etc. The general proof can be obtained by considering $N_n(2)$ as the number of plane trees having n edges with leaves colored by one of two colors. □

Properties of Narayana numbers

4.4.18. The main property of Narayana numbers is the earlier obtained closed formula

$$N(n,k) = \frac{1}{n}\binom{n}{k}\binom{n}{k-1} = \frac{1}{k}\binom{n}{k-1}\binom{n-1}{k-1}$$

$$= \frac{1}{n+1}\binom{n+1}{k+1}\binom{n-1}{k}.$$

Using this formula, we can prove that:

- *Narayana's triangle is symmetrical with respect to the principal diagonal:*

$$N(n,k) = N(n, n-k+1).$$

☐ In fact, it holds that

$$N(n,k) = \frac{1}{n}\binom{n}{k}\binom{n}{k-1} = \frac{1}{n}\binom{n}{n-k}\binom{n}{n-k+1}$$
$$= \frac{1}{n}\binom{n}{n-k+1}\binom{n}{(n-k+1)-1} = N(n, n-k+1). \;\square$$

Another important property is connected with Catalan numbers:

- *The sum of the elements of the nth row of Narayana's triangle is equal to the nth Catalan number C_n:*

$$N(n,1) + N(n,2) + N(n,3) + \cdots + N(n,n) = C_n.$$

☐ It follows from the definitions of these two number sets. ☐

4.4.19. It turns out that there exist other connections between Narayana and Catalan numbers.

- *The Narayana number $N(n,k)$ can be found using Catalan numbers:*

$$N(n,k) = \sum_{j \geq 0} \binom{n-1}{2j}\binom{n-2j-1}{n-j-k}C_j.$$

☐ For example,

$$N(7,4) = 175 = 1 \cdot 20 \cdot 1 + 15 \cdot 6 \cdot 1 + 15 \cdot 2 \cdot 2 + 1 \cdot 1 \cdot 1 \cdot 5$$
$$= \binom{6}{0}\binom{6}{3}C_0 + \binom{6}{2}\binom{4}{2}C_1 + \binom{6}{4}\binom{2}{1}C_2 + \binom{6}{6}\binom{0}{0}C_3$$
$$= \sum_{j \geq 0}\binom{6}{2j}\binom{6-2j}{3-j}C_j.$$

A general proof can be obtained through a combinatorial approach, using the representation of $N(n,k)$ as the number of rhyme schemes of an n-line poem having exactly k symbols. □

4.4.20. There are some divisibility properties of Narayana's triangle.

- *For a given prime p, the pth row of Narayana's triangle is divisible by p:*

$$p \in P \Rightarrow p | N(p,k) \quad \text{for } 2 \le k \le n-1.$$

☐ In fact, for small n, it is true: the third row $1, 3, 1$ is divisible by 3, the fifth row $1, 10, 20, 10, 1$ is divisible by 5, and the seventh row $1, 21, 105, 175, 21, 1$ is divisible by 7. A general proof follows from the well-known property of binomial coefficients: for a prime p, the number $\binom{p}{k}$ is divisible by p for any $1 \le k \le p-1$. □

It is possible to prove the following more general fact:

- *If p is a prime, $n = p^m$, and $m \in \mathbb{N}$, then $p | N(n,k)$ for any $1 \le k \le n-1$.*

☐ It is true for small values of parameters. For example, the ninth row

$$1, 36, 336, 1176, 1764, 1176, 336, 36, 1$$

of the triangle is divisible by 3. For a general proof, see a similar construction in Chapter 2. □

On the other hand:

- *if p is a prime, $n = p^m - 1$, and $m \in \mathbb{N}$, then the nth row of Narayana'a triangle is not divisible by p:*

$$p \in P, n = p^m - 1, m \in \mathbb{N} \Rightarrow p \nmid N(n,k), \quad 1 \le k \le n.$$

☐ It is true for small values of the parameters. For example, if $p = 2$, then we can see that the first, third, and seventh rows contain only odd numbers. If $p = 3$, we can check that the second and eighth rows have no numbers divisible by 3, etc. □

4.4.21. Finally, it can be proven that:

- *for a fixed positive integer n, the sequence $N(n, k)$, $k = 0, 1, 2, \ldots, n$, has a single maximum, which is achieved for an odd n at one point, $k_n = \lfloor \frac{n+1}{2} \rfloor = \frac{n+1}{2}$, and for an even n at two consecutive points, $k_n = \lfloor \frac{n}{2} \rfloor = \frac{n}{2}$ and $k_n + 1 = \lfloor \frac{n}{2} \rfloor + 1 = \frac{n+2}{2}$.*

□ We can obtain a proof using similar considerations followed for Pascal's triangle, Stirling's triangles, and Delannoy's triangle (see, for example, [Deza24]). □

Narayana number in the family of special numbers

4.4.22. We define Narayana numbers as the numbers of partitions of a special form. Thus, the Narayana number $N(n, k)$ counts the number of non-crossing partitions of an n-set into k parts.

Since the Stirling number of the second kind $S(n, k)$ counts the number of all partitions of an n-set into k parts, the Bell number $B(n)$ gives the number of all partitions of an n-set, and the Catalan number C_n counts the number of all non-crossing partitions of an n-set, then:

- *the following relations hold:*

$$N(n, k) \leq S(n, k) \leq B(n), \quad \text{and} \quad N(n, k) \leq C_n \leq B(n).$$

4.4.23. There are interesting connections between Narayana numbers and triangular numbers. It is easy to note that

$$N(n, 2) = N(n, n-1) = S_3(n-1), \quad n \geq 2.$$

This result can be obtained by direct consideration:

$$N(n, 2) = N(n, n-1) = \frac{1}{n}\binom{n}{2}\binom{n}{1} = \frac{n(n-1)}{2} = S_3(n-1).$$

So, any triangular number is a Narayana number.

Similarly, the recurrent relation $N(n+1, k+1) = \frac{n(n+1)}{k(k+1)} N(n,k)$ can be rewritten as

$$N(n+1, k+1) = \frac{S_3(n)}{S_3(k)} N(n,k).$$

So, we obtain that

$$N(n+1, k+1)$$
$$= \frac{S_3(n)}{S_3(k)} N(n,k) = \frac{S_3(n) S_3(n-1)}{S_3(k) S_3(k-1)} N(n-1, k-1)$$
$$= \cdots = \frac{S_3(n) S_3(n-1) \cdots S_3(n-k+1)}{S_3(k) S_3(k-1) \cdots S_3(1)} N(n-k+1, 1)$$
$$= \frac{S_3(n) S_3(n-1) \cdots S_3(n-k+1)}{S_3(k) S_3(k-1) \cdots S_3(1)}.$$

For a given triangular number $S_3(n)$, define $S_3(n)! = S_3(n) S_3(n-1) \cdots S_3(1)$. Then, the following equality holds:

$$N(n+1, k+1) = \frac{S_3(n)!}{S_3(k)! S_3(n-k)!}.$$

4.4.24. Consider now the numbers $N(2n+1, n+1)$. They form the principal diagonal of the isosceles Narayana's triangle and are called *central Narayana numbers*. The direct computation implies that

$$N(2n+1, n+1) = (2n+1)(C_n)^2.$$

Therefore, we can state that the central Narayana number is a square number if and only if $2n+1$ is a square. Moreover, $N(2n+1, n+1)$ is odd if and only if C_n is odd, that is, if and only if n is a Mersenne number, $2^k - 1$.

4.4.25. There is another integer sequence associated with the name of Narayana. (In this case, we speak about an Indian mathematician of the 16th century with the same name.)

In particular, *Narayana's cow sequence* is the sequence $cn(n)$, $n = 0, 1, 2, 3, \ldots$, defined recursively:

$$cn(0) = cn(1) = cn(2) = 1, \quad \text{and}$$
$$cn(n) = cn(n-1) + cn(n-3), \quad n \geq 3.$$

The first few elements of this sequence are 1, 1, 1, 2, 3, 4, 6, 9, 13, 19, ... (sequence A000930 in the OEIS).

The number $cn(n)$ gives the number of compositions of n into parts 1 and 3; the number of tilings of a $3 \times n$ rectangle with straight trominoes; the number of ways to arrange $n - 1$ tatami mats in a $2 \times (n-1)$ room, such that no four mats meet at a point (for example, there are six ways to cover a 2×5 room: 11111, 2111, 1211, 1121, 1112, and 212).

This number set belongs to a family of sequences which satisfy a recurrence of the form

$$a(n) = a(n-1) + a(n-m), \quad \text{with } a(n) = 1 \quad \text{for } n = 0, \ldots, m-1.$$

The generating function of such sequence is $\frac{1}{1-x-x^m}$. Moreover, $a(n) = \sum_{k=0}^{\lfloor \frac{n}{m} \rfloor} \binom{n-(m-1)k}{i}$.

This family of binomial summations gives the number of ways to cover (without overlapping) a linear lattice of n sites with molecules that are m sites wide.

A special case of $m = 1$ gives the sequence of powers of two: 1, 2, 4, 8, 16, 32, 64, 128, 256, 512, ... (sequence A000079 in the OEIS).

Exercises

1. Find all partitions of the set $\{1, 2, 3, \ldots, n\}$ if $n = 1, 2, 3, 4, 5, 6$. Check that: the number of such partitions into k parts, $k = 1, \ldots, n$, is equal to the Stirling number of the second kind $S(n, k)$; the number of non-crossing partitions into k parts, $k = 1, \ldots, n$, is equal to the Narayana number $N(n, k)$; the total number of partitions is the Bell number $B(n)$; and the total number of non-crossing partitions is the Catalan number C_n.
2. Using the formula $N(n, k) = \frac{1}{n} C_n^k C_n^{k-1}$, build the first 10 rows of Narayana's triangle. Formulate a rule for the construction of an arbitrary element of Narayana's triangle, using Pascal's triangle. Build the first 10 rows of Narayana's isosceles triangle.
3. For $n = 1, 2, 3, 4, 5, 6, 7, 8, 9, 10$, build the Narayana polynomials $N_n(x)$ using Narayana's triangle. Construct these polynomials using the recurrent relation. Compare the obtained results.

4. For $n = 1, 2, 3, 4, 5, 6, 7, 8, 9, 10$, find the values of the Narayana polynomial $N_n(x)$ at $x = 1$ and at $x = 2$. Check that we obtain the nth Catalan number and the nth Schröder number, respectively: $N_n(1) = C_n$, $N_n(2) = S_n$. Prove this relation for any positive integer n.
5. For $n = 1, 2, 3, 4, 5, 6$ and $k = 1, \ldots, n$, check that the Narayana number $N(n, k)$ is the number of polynomials of perimeter $2k+2$, having exactly k columns.
6. Prove that for any non-negative integers n and k, the number $\frac{1}{n} C_n^k C_n^{k+1}$ is divisible by n. Prove that $t C_n^k C_n^{k+t}$ is divisible by n for any non-negative integer t.
7. Find the first 10 entries of *Narayana's cow sequence* $cn(n)$. Prove that the generation function of this sequence has the form $\frac{1}{1-x-x^3}$. Prove that $cn(n) = \sum_{k=0}^{\lfloor \frac{n}{3} \rfloor} \binom{n-2k}{k}$.

4.5. Stirling Numbers

History of the question

4.5.1. James Stirling (1692–1770) was a Scottish mathematician. He was born in Garden (Stirling, Scotland) and educated in Oxford (England), and his scientific activities began in Venice (Italy).

In 1722, Stirling had returned to Glasgow, and late in 1724 or early in 1725, he went to London.

It was during this very mathematically productive period in London that Stirling published his most important work, *Methodus Differentialis sive Tractatus de Summatione et Interpolatione Serierum Infinitarum* (*Differential Method with a Tract on Summation and Interpolation of Infinite Series*, 1730), a treatise on infinite series, summation, interpolation, etc. It contains an asymptotic formula for $n!$, now known as *Stirling's formula*.

From 1734, J. Stirling was temporarily employed by the Scotch Mines Company, Leadhills (Scotland), and in 1737, he took up a permanent position with the company as chief agent. In this position, he remained until the end of his life.

In addition to *Stirling's series* and *Stirling's asymptotic formula* for $n!$, two classes of integers of the special form are named after J. Stirling: the *Stirling numbers of the first kind* and the *Stirling numbers of the second kind*.

These two classes of special numbers arise in a variety of analytic and combinatorial problems.

J. Stirling introduced them in a purely algebraic setting in his book *Methodus differentialis* in 1730. They were rediscovered and given a combinatorial meaning by M. Saka in 1782 ([MaSc15]).

Recurrent construction of Stirling numbers

4.5.2. A *Stirling number of the second kind* (or *Stirling set number*, *Stirling partition number*) $S(n,k)$ is the number of partitions of a given n-set into k non-empty subsets, for $n = 1, 2, \ldots, k = 1, 2, \ldots, n$.

This definition leads to the well-known recurrent relation.

- *For the numbers $S(n,k)$, the following recurrent relation holds:*

$$S(n,k) = kS(n-1,k) + S(n-1,k-1), \quad n > 0.$$

☐ Let there be an n-set, $n > 0$. We are going to find a decomposition of this set into k non-empty parts. In this situation, we can place, on the one hand, the last (nth) element into a new part. There are $S(n-1, k-1)$ of such decompositions since, by constructing one part from the nth element, we can simply construct additional $k-1$ parts from the first $n-1$ elements of the considered n-set; the number of such decompositions is, by definition, $S(n-1, k-1)$. On the other hand, we can place the last (nth) element into some non-empty subset of the $(n-1)$-set of the first $n-1$ elements. In this case, we have $kS(n-1,k)$ possibilities, as every partition from all the $S(n-1,k)$ decompositions of the first $n-1$ elements into k non-empty parts gives k subsets with which we can combine the last (nth) element. ☐

Using the conditions $S(0,0) = 1$, $S(n,0) = 0, n > 0$, and $S(n,k) = 0, k > n$, we can obtain all the values of the Stirling numbers of the second kind, i.e., we can construct the *triangle of Stirling numbers of the second kind* (sequence A008277 in the OEIS):

n\k	0	1	2	3	4	5	6	7	8	9
0	1									
1	0	1								
2	0	1	1							
3	0	1	3	1						
4	0	1	7	6	1					
5	0	1	15	25	10	1				
6	0	1	31	90	65	15	1			
7	0	1	63	301	350	140	21	1		
8	0	1	127	966	1701	1050	266	28	1	
9	0	1	255	3025	7770	6951	2646	462	36	1

A *Bell number* $B(n)$ is defined as the number of all possible partitions of a given n-set into non-empty parts. So, by definition:

- *the sum of the elements of the nth row of the triangle of Stirling numbers of the second kind is the nth Bell number $B(n)$, $n \geq 0$.*

4.5.3. A *Stirling number of the first kind* $s(n, k)$ is defined by the equality

$$s(n, k) = (-1)^{n+k}|s(n, k)|,$$

where $|s(n, k)|$ is the number of ways of decomposing a given n-set into k cycles, for $n = 1, 2, \ldots$ and $k = 1, 2, \ldots, n$. The numbers $|s(n, k)|$ are called *unsigned Stirling numbers of the first kind*.

This combinatorial definition leads to a well-known recurrent formula for $s(n, k)$.

- *For the Stirling numbers of the first kind, the following recurrent formula holds:*

$$s(n, k) = s(n-1, k-1) - (n-1)s(n-1, k), \quad n > 0, k > 0.$$

☐ Each representation of n objects in the form of k cycles either places the last object in a separate cycle (in $|s(n-1, k-1)|$ ways) or inserts this object into one of the $|s(n-1, k)|$ cyclic representations of the first $n-1$ objects. In the second case, there exists $n-1$ ways of such insertion. In fact, it is easy to check that there are j ways to put a new element in a j-cycle in order to get a $(j+1)$-cycle. For example, if $j = 3$, then the cycle $[A, B, C]$ leads to the cycles $[A, B, C, D]$, $[A, B, D, C]$, or $[A, D, B, C]$ if we insert a new element D, and no other possibilities exist. Summation over all j gives $n-1$ ways to insert the nth (last) object into a circular partition of the first $n-1$ objects. Thus, it holds that

$$|s(n,k)| = |s(n-1, k-1)| + (n-1)|s(n-1, k)|, \quad n > 0, \ k > 0,$$

and, therefore,

$$s(n,k) = s(n-1, k-1) - (n-1)s(n-1, k), \quad n > 0, \ k > 0. \quad \square$$

Using the obtained recurrent relation, we can build the *triangle of Stirling numbers of the first kind* (sequence A048994 in the OEIS). At the vertex of the triangle is the number 1 ($s(0,0) = 1$), the left-hand side is formed by zeros ($s(n,0) = 0$, $n > 0$), and the right-hand side consists of unities ($s(n,n) = 1$, $n \geq 1$):

n/k	0	1	2	3	4	5	6	7	8	9
0	1									
1	0	1								
2	0	−1	1							
3	0	2	−3	1						
4	0	−6	11	−6	1					
5	0	24	−50	35	−10	1				
6	0	−120	274	−225	85	−15	1			
7	0	720	−1764	1624	−735	175	−21	1		
8	0	−5040	13268	−13132	6769	−1960	322	−28	1	
9	0	40320	−109584	118124	−67284	22449	−4536	546	−36	1

Examining the sum of the elements of the nth row of the constructed triangle, we can see that:

- the sum of the elements of the nth row, $n \geq 2$, of the triangle of Stirling numbers of the first kind is equal to zero; the alternating sum of the elements of the nth row, $n \geq 2$, of the triangle of Stirling numbers of the first kind is equal to $(-1)^n n!$.

□ This statement follows from the used recurrent construction and can be proved by induction. However, it is merely a consequence of the well-known fact: the number of all permutations of an n-set is equal to $n!$. □

Stirling numbers and falling factorials

4.5.4. For a positive integer n, define the *falling factorial* (or *falling power*) $x^{\underline{n}}$ of x, $x \in \mathbb{R}$, as the product $x(x-1)\cdots(x-n+1)$; moreover, assume, that $x^{\underline{0}} = 1$. Then, the main property of the Stirling numbers of the second kind will take the following form:

- The *Stirling numbers of the second kind* $S(n,k)$ satisfy the relationship

$$x^n = S(n,0)x^{\underline{0}} + S(n,1)x^{\underline{1}} + S(n,2)x^{\underline{2}} + \cdots S(n,3)x^{\underline{3}} + \cdots$$
$$+ S(n,n)x^{\underline{n}},$$

i.e., using the Stirling numbers of the second kind, we can express the ordinary powers x^n of x through the falling powers (falling factorials) $x^{\underline{k}}$ of x.

☐ For $n = 0$ and $n = 1$, the formulas are true: $x^0 = S(0,0)$, as $S(0,0) = 1$, and $x^1 = S(1,0) + S(1,1)x$, as $S(1,0) = 0$, $S(1,1) = 1$.

Let the formula be true for $n-1$; we prove it for n, $n \geq 2$, using the recurrent relation for $S(n,k)$:

$$x^n = x \cdot x^{n-1} = x \sum_{k=0}^{n-1} S(n-1,k)x(x-1)\cdots(x-k+1)$$

$$= (x-k)\sum_{k=0}^{n-1} S(n-1,k)x(x-1)\cdots(x-k+1)$$

$$+ k\sum_{k=0}^{n-1} S(n-1,k)x(x-1)\cdots(x-k+1)$$

$$= \sum_{k=0}^{n-1} S(n-1,k)x(x-1)\cdots(x-k+1)(x-k)$$

$$+ k\sum_{k=0}^{n-1} S(n-1,k)x(x-1)\cdots(x-k+1)$$

$$= \sum_{m=0}^{n} S(n-1, m-1)x(x-1)\cdots(x-m+1)$$

$$+ k \sum_{k=0}^{n} S(n-1, k)x(x-1)\cdots(x-k+1)$$

$$= \sum_{k=0}^{n} (S(n-1, k-1)$$

$$+ kS(n-1, k))x(x-1)\cdots(x-k+1)$$

$$= \sum_{k=0}^{n} S(n, k)x(x-1)\cdots(x-k+1).$$

Here, we use the facts $S(n-1, 0) = 0$ and $S(n-1, n) = 0$, and we introduce a change of the variable of summation: $m = k+1$. The statement is proven. □

4.5.5. In terms of the *falling factorial* $x^{\underline{n}} = x(x-1)\cdots(x-n+1)$, a similar property for the Stirling numbers of the first kind holds.

• The *Stirling numbers of the first kind* $s(n, k)$ satisfy the equality

$$x^{\underline{n}} = s(n, 0) + s(n, 1)x + s(n, 2)x^2 + \cdots + s(n, n-1)x^{n-1}$$
$$+ s(n, n)x^n,$$

i.e., using the Stirling numbers of the first kind, we can express the falling powers (falling factorials) $x^{\underline{n}}$ of x through the ordinary powers $x^0, x^1, \ldots, x^n$ of x.

□ In order to prove this relationship, let us to use induction by n.

For $n = 0$ and $n = 1$, we have the identities $1 = s(0, 0) \cdot 1$ and $x = s(1, 0) \cdot 1 + s(1, 1) \cdot x$, correspondingly.

Let $n \geq 2$. Suppose that the statement is true for $n-1$ (i.e., the equality $x(x-1)\cdots(x-n+2) = \sum_{k=0}^{n-1} s(n-1, k)x^k$ holds), and we prove it for n using the recurrent equation $s(n, k) = s(n-1, k-1) - (n-1)s(n-1, k)$ for the numbers $s(n, k)$, $n, k > 0$, obtained above.

In this case, it holds that

$$\sum_{k=0}^{n} s(n,k) = \sum_{k=1}^{n} s(n,k)x^k$$

$$= x^k = \sum_{k=1}^{n}(s(n-1,k-1) - (n-1)s(n-1,k))x^k$$

$$= \sum_{k=1}^{n} s(n-1,k-1)x^k - (n-1)\sum_{k=1}^{n} s(n-1,k)x^k$$

$$= x \sum_{m=0}^{n-1} s(n-1,m)x^m - (n-1)\sum_{k=0}^{n-1} s(n-1,k)x^k$$

$$= (x-n+1)\sum_{k=0}^{n-1} s(n-1,k)x^k$$

$$= (x-n+1)x(x-1)(x-2)\cdots(x-n+2).$$

Here, we change the limits of the summation using the following obvious results: $s(n,0) = 0$, $s(n-1,0) = 0$, and $s(n-1,n) = 0$. The property is proven. □

Closed formulas for Stirling numbers

4.5.6. There are no convenient closed formulas for the Stirling numbers. However, we have several well-known possibilities to find $S(n,k)$ and $s(n,k)$ as functions of the parameters n, k.

- The *Stirling numbers of the second kind* $S(n,k)$ satisfy the equality

$$S(n,k) = \frac{k^{n-1}}{(k-1)!0!} - \frac{(k-1)^{n-1}}{(k-1)!1!}$$

$$+ \frac{(k-2)^{n-1}}{(k-3)!3!} + \cdots + (-1)^{k-1}\frac{1^{n-1}}{0!(k-1)!}.$$

☐ In fact, it holds that

$$S(n,1) = \frac{1}{1} \cdot 1^n = 1; \quad S(n,2) = \frac{1}{2}(1 \cdot 2^n - 2 \cdot 1^n) = 2^{n-1} - 1;$$

$$S(n,3) = \frac{1}{6}(3^n - 3 \cdot 2^n + 3); \quad S(n,4) = \frac{1}{24}(4^n - 4 \cdot 3^n + 6 \cdot 2^n - 4);$$

$$S(n,5) = \frac{1}{120}(5^n - 5 \cdot 4^n + 10 \cdot 3^n - 10 \cdot 2^n + 5);$$

$$S(n,6) = \frac{1}{720}(6^n - 6 \cdot 5^n + 15 \cdot 4^n - 20 \cdot 3^n + 15 \cdot 2^n - 6).$$

The general proof can be found in [Deza24]. □

4.5.7. Unfortunately, a compact closed formula for the Stirling numbers of the first kind does not exist. However, it is possible to prove ([Male11]) that

$$s(n, n-m) = \frac{1}{(n-m-1)!} \sum_{(k_1, k_2, \ldots, k_m)} (-1)^{k_1 + k_2 + \cdots + k_m}$$

$$\times \frac{(n + k_1 + k_2 + \cdots + k_m - 1)!}{k_1! k_2! \cdots k_m! (1!)^{k_1} (2!)^{k_1} \cdots ((p+1)!)^{k_m}},$$

where $0 \le k_1, \ldots, k_m \le m$, and $k_1 + 2k_2 + 3k_3 + \cdots + mk_m = m$.

Generating functions of the sequences of Stirling numbers

4.5.8. Using the recurrent approach, it is easy to get the generating function $D_k(x) = \sum_{n=0}^{\infty} S(n,k)x^n$ of the sequence $S(n,k)$, $n = 0, 1, 2, \ldots$, with fixed k, $k \ge 0$. In fact, the following statement holds:

- For a fixed $k \ge 0$, the generating function $D_k(x) = \sum_{n=0}^{\infty} S(n,k)x^n$ of the sequence $S(n,k)$, $n = 0, 1, 2, \ldots$, has the form $D_0(x) \equiv 1$ if $k = 0$; it has the form $D_k(x) = \frac{x^k}{(1-x)(1-2x)\cdots(1-kx)}$ if $k > 0$, i.e., it holds that

$$\frac{x^k}{(1-x)(1-2x)\cdots(1-kx)} = S(0,k) + S(1,k)x + S(2,k)x^2$$

$$+ S(n,k)x^n + \cdots, \quad |x| \le \frac{1}{k}.$$

□ A proof can be obtained by the ordinary recurrent algorithm. See for details Chapter 3. See also [Deza24]. □

4.5.9. Using the results of the previous sections, it is not difficult to get the generating function of the sequence of Stirling numbers of the first kind $s(n,k)$, $k = 0, 1, 2, \ldots$, for a fixed n.

- The *generating function $K_n(x)$ of the sequence of Stirling numbers of the first kind $s(n,k)$, $k = 0, 1, 2, \ldots$, for a fixed $n = 0, 1, 2, \ldots$, has the form $K_n(x) = x(x-1)\cdots(x-k+1)$.*

☐ This statement follows from the property

$$x(x-1)\cdots(x-n+1) = s(n,0) + s(n,1)x + s(n,2)x^2$$
$$+ \cdots + s(n, n-1)x^{n-1} + s(n,n)x^n$$

and the definition of the generating function. As $s(n, n+1) = s(n, n+2) = \cdots = 0$, we get that

$$K_n(x) = s(n,0) + s(n,1)x + s(n,2)x^2 + \cdots + s(n, n-1)x^{n-1}$$
$$+ s(n,n)x^n + s(n, n+1)x^{n+1} + \cdots, \quad x \in \mathbb{R}. \qquad \square$$

Properties of Stirling numbers

4.5.10. There are many combinatorial, recurrent, and number-theoretic properties of Stirling numbers. A large list can be found in [Deza24]. A few examples are as follows:

- The *Stirling numbers of the second kind $S(n,k)$ satisfy the equality*

$$S(n,k) = \sum_{\substack{k_1 + k_2 + \cdots + k_n = k, \\ k_1 + 2k_2 + \cdots + nk_n = n}} \frac{n!}{k_1! k_2! \cdots k_k! (1!)^{k_1} (2!)^{k_2} \cdots (n!)^{k_n}}.$$

☐ The proof follows from the definition of Stirling numbers of the second kind and the properties of ordered partitions (see Chapter 1). ☐

For a given positive integer k, an asymptotic estimation of the behavior of the Stirling numbers of the second kind, as $n \to \infty$, has

the form
$$S(n,k) \sim \frac{k^n}{k!}.$$

Using Stirling's formula $k! \sim \sqrt{2\pi k}(\frac{k}{e})^n$ for factorials, we get that $S(n,k) \sim \frac{e^n}{\sqrt{2\pi k}}$, i.e., $\lim_{n\to\infty} \frac{S(n,k)}{\frac{e^n}{\sqrt{2\pi k}}} = 1$. Therefore, for big n and a fixed k, the value $S(n,k)$ as a function of n behaves approximately as the exponential function e^n.

4.5.11. As for the Stirling numbers of the first kind, they are closely connected with *harmonic numbers* $H_n = 1 + \frac{1}{2} + \cdots + \frac{1}{n}$. In fact, we can state the following:

- The *Stirling numbers of the first kind satisfy the identities*

$$s(n,2) = (-1)^n (n-1)! H_{n-1},$$

$$s(n,3) = \frac{1}{2}(-1)^{n-1}(n-1)!((H_{n-1})^2 - H_{n-1}^{(2)}),$$

$$s(n,4) = \frac{1}{6}(-1)^n (n-1)!((H_{n-1})^3 - 3H_{n-1}H_{n-1}^{(2)} + 2H_{n-1}^{(3)}),$$

where $H_n = 1 + \frac{1}{2} + \cdots + \frac{1}{n}$ is the nth harmonic number and $H_n^{(m)} = 1 + \frac{1}{2^m} + \cdots + \frac{1}{n^m}$ is the nth generalized harmonic number.

□ The proof can be obtained using a recurrent approach. See [Deza24] for details. □

There are some estimations of the asymptotic behavior of the Stirling numbers of the first kind.

The first estimate, uniformly for $k = o(\log n)$, is given in terms of the Euler–Mascheroni constant $\gamma = 0.5772156649\ldots$:

$$|s(n+1, k+1)| \sim \frac{n!}{k!}(\gamma + \log n)^k.$$

For fixed n, as $k \to \infty$, we get the estimation

$$|s(n+k, k)| \sim \frac{k^{2n}}{2^n n!}.$$

For more information, see [AbSt72], [OLBC10], and [Temm93].

Stirling numbers in the family of special numbers

4.5.12. There are many natural connections of Stirling numbers with other classes of special numbers.

As was noted above, the sum of the elements of the nth row of the triangle of Stirling numbers of the second kind gives the nth Bell number $B(n)$. A similar summation for the triangle of Stirling numbers of the first kind is connected with *factorial numbers*.

There are interesting connections of Stirling numbers with *triangular numbers*:

- Any triangular number, $S_3(n) = \frac{n(n+1)}{2}$, is a Stirling number of the second kind and an (unsigned) Stirling number of the first kind:

$$S_3(n) = S(n+1, n), \quad S_3(n) = |s(n+1, n)| = -s(n+1, n).$$

☐ It can be obtained using the definition of Stirling numbers. ☐

Similarly, we have natural connections of Stirling numbers with Mersenne numbers.

- Any Mersenne number, $\mathcal{M}_n = 2^n - 1$, is a Stirling number of the second kind:

$$\mathcal{M}_n = 2^n - 1 = S(n+1, 2).$$

☐ It also follows from the combinatorial definition of the Stirling numbers of the second kind. ☐

As was shown above,

- the *Stirling numbers of the first kind are connected with harmonic numbers*, $H_n = 1 + \frac{1}{2} + \cdots + \frac{1}{n}$:

$$s(n+1, 2) = (-1)^{n+1} n! H_n.$$

4.5.13. In the context of our consideration, the most important are the connections of Stirling numbers with Pascal's triangle. For example, the following property can be proven.

- Any Stirling number can be obtained by the summation of pairwise products of the entries of a row of Pascal's triangle and a column

of the triangle of Stirling numbers of the second kind:
$$S(n+1, k+1) = C_n^0 S(0, k) + C_n^1 S(1, k) + \cdots + C_n^k S(k, k)$$
$$+ \cdots + C_n^n S(n, k), \quad n \geq 0, \; k \geq 0.$$

□ For a proof, see Chapter 3 in [Deza24]. □

On the other hand, the closed formula for $S(n, k)$ also contains elements of Pascal's triangle.

- *Any Stirling number of the second kind can be obtained using the alternating summation of pairwise products of the entries of a row of Pascal's triangle and a string of given powers of consecutive non-negative integers:*

$$S(n, k) = \frac{1}{k!}(C_k^0 \cdot k^n - C_k^1 \cdot (k-1)^n + C_k^2 \cdot (k-2)^n$$
$$+ C_k^3 \cdot (k-3)^n + \cdots + (-1)^k C_k^k \cdot 0^n).$$

□ See Chapter 3 in [Deza24]. □

Similar to the Stirling numbers of the second kind, the Stirling numbers of the first kind have many connections with the elements of Pascal's triangle. For example, the following property holds:

- *There are infinite many Stirling numbers of the first kind which can be represented as a (signed) product of the elements of Pascal's triangle:*

$$|s(n, n-3)| = C_n^2 C_n^4; \quad s(n, n-3) = -C_n^2 C_n^4.$$

In other words, *all the entries of the third descending diagonal of the triangle of Stirling numbers of the first kind can be obtained as products of two binomial coefficients with the "minus" sign.*
□ This property was proven in Chapter 4 of [Deza24]. □

Moreover, two classes of the Stirling numbers of the first kind can be represented in terms of the binomial coefficients:

$$|s(n, n-1)| = C_n^2, \quad s(n, n-1) = -C_n^2, \quad |s(n, n-2)|$$
$$= s(n, n-2) = \frac{1}{4}(3n-1)C_n^3.$$

Exercises

1. Using definition of the Stirling numbers of the second kind, find $S(2,1)$, $S(2,2)$, $S(5,1)$, $S(5,2)$, $S(5,3)$, $S(5,4)$, and $S(5,5)$. Write the corresponding partitions, and draw the appropriate diagrams. Using the recurrent relation for the Stirling numbers of the second kind, check the values of $S(n,k)$, represented in the rows 0–10 of the triangle above. Find $S(10,k)$, $k = 0,1,2,3,\ldots,10$; find $S(11,k)$, $k = 0,1,2,3,\ldots,11$.

2. Prove that $S(n+1,2) = C_n^1 + \cdots + C_n^n$, $n \in \mathbb{N}$; prove that $S(n,n-2) = C_2^2 + 2C_3^2 + 3C_4^2 + \cdots + (n-2)C_{n-1}^2$.

3. Using the recurrent relation for the Stirling numbers of the first kind $s(n,k)$, find the numbers $s(7,k)$, $k = 0,1,2,3,\ldots,7$; $s(8,k)$, $k = 0,1,2,3,\ldots,8$; $s(9,k)$, $k = 0,1,2,3,\ldots,9$; $s(10,k)$, $k = 0,1,2,3,\ldots,10$; $s(11,k)$, $k = 0,1,2,3,\ldots,11$.

4. Calculate $s(4,2)$, $s(5,3)$, $s(6,4)$, $s(7,5)$, and $s(8,6)$ using the closed formula. Prove that $s(n,n-2) = \frac{n(n-1)(n-2)(3n-1)}{24}$. Check the result by performing direct calculations using the closed formula for the values $s(4,2)$, $s(5,3)$, $s(6,4)$, $s(7,5)$, $s(8,6)$, and $s(9,7)$.

5. Calculate $s(5,2)$, $s(6,3)$, $s(7,4)$, $s(8,5)$, and $s(9,6)$ using the closed formula above. Prove that $s(n,n-3) = -\frac{n^2(n-1)^2(n-2)(n-3)}{48}$. Check the result by performing direct calculations using the closed formula for the values $s(5,2)$, $s(6,3)$, $s(7,4)$, $s(8,5)$, and $s(9,6)$.

6. Check that the pth row of the triangle of Stirling numbers of the second kind is divisible by p for $p = 3,5,7,11,13,17,19$.

7. Prove the relation $|s(n+1,k)| = \sum_{j=0}^n j! C_n^j s(n-j, k-1)$ in two ways: directly, using the formula $|s(n+1, m+1)| = n! \sum_{k=0}^n \frac{(-1)^{m+k}|s(k,m)|}{k!}$, and from the formula $s(n+1,k) = \sum_{j=0}^n (-1)^j \cdot j! C_n^j s(n-j, k-1)$. Using these results, calculate the Stirling numbers of the first kind for $n = 1,2,3,4,5,6$, $m = 1,2,\ldots,n$.

8. Using the formula $s(n,k) = \sum_{i=k}^n n^{i-k} s(n+1, i+1)$, calculate the Stirling numbers of the first kind $s(n,k)$ for $n = 1,2,3,4,5,6$, $k = 1,2,\ldots,n$.

4.6. Bell Numbers

A history of the question

4.6.1. In mathematics, *Bell numbers* count the possible partitions of a set. These numbers have been studied by mathematicians since the 19th century, and their roots go back to medieval Japan. In an example of Stigler's law of eponyms, they are named after Eric Temple Bell, who wrote about them in the 1930s.

E. T. Bell (1883–1960) was a Scottish-born mathematician and science fiction writer who lived in the United States for most of his life. He was born in Peterhead, Aberdeen, Scotland. He was educated at Bedford Modern School, England. He received degrees from Stanford University (1904), the University of Washington (1908), and Columbia University (1912). E. T. Bell was part of the faculty first at the University of Washington and later at the California Institute of Technology.

In 1924, he was awarded the Bôcher Memorial Prize for his work in mathematical analysis. In 1927, he was elected to the National Academy of Sciences. He was elected to the American Philosophical Society in 1937. He died in 1960 in Watsonville, California.

E. T. Bell researched number theory and did much work using generating functions, treated as formal power series. He is the eponym of the *Bell series* (number theory), the *Bell polynomials*, and the *Bell numbers* (combinatorics).

Construction of Bell numbers

4.6.2. In combinatorics, the nth *Bell number* $B(n)$ is defined as the number of partitions of n different elements.

Starting with $B(0) = B(1) = 1$, the first few Bell numbers are 1, 1, 2, 5, 15, 52, 203, 877, 4140, 21247,... (sequence A000110 in the OEIS).

It is not difficult to get the following recurrent relation for Bell numbers.

- Bell numbers satisfy the recurrent relation

$$B(n+1) = C_n^0 B(n) + C_n^1 B(n-1) + C_n^2 B(n-2)$$
$$+ \cdots + C_n^k B(n-k) + \cdots + C_n^n B(0).$$

□ Consider any partition S of the set $\{a_1, a_2, \ldots, a_n, a_{n+1}\}$, and assume that the subset, containing a_{n+1}, also contains k other elements, $0 \leq k \leq n$. If k is fixed, then the number of such partitions is equal to $C_n^k B(n-k)$. In fact, the k elements can be chosen from the set $\{a_1, a_2, \ldots, a_n\}$ in C_n^k ways, while the set of the remaining $n-k$ elements gives $B(n-k)$ partitions. Given all possible values of k, starting with zero (a subset containing a_{n+1} has no other elements) and ending at n (for $k = n$, there are no remaining elements, but the convention $B(0) = 1$ includes this case), we consider all possible partitions of the set $\{a_1, a_2, \ldots, a_n, a_{n+1}\}$. So, starting with $B(0) = 1$, we get that $B(1) = C_0^0 B(0) = 1$, $B(2) = C_1^0 B(1) + C_1^1 B(0) = 2$, $B(3) = C_2^0 B(2) + C_2^1 B(1) + C_2^2 B(0) = 5$, etc. □

Bell numbers can be easily calculated using the *Bell triangle*, also called *Aitken matrix*. We can realize this construction as follows:

- Start with the number 1. It forms the starting row of the triangle.
- Start a new row from the right element of the previous row.
- Get the next element of a row as the sum of two numbers that are located on the left and on the top left diagonally.
- Repeat the previous step until we get a row by one number longer than the previous one.

The first rows of the Bell triangle are

$$
\begin{array}{cccccc}
1 & & & & & \\
1 & 2 & & & & \\
2 & 3 & 5 & & & \\
5 & 7 & 10 & 15 & & \\
15 & 20 & 27 & 37 & 52 &
\end{array}
$$

The Bell numbers $B(n)$ form the edges of the Bell triangle (see for details [Deza24]).

Combinatorial problems related to Bell numbers

4.6.3. Consider the following several important combinatorial problems related to Bell numbers:

- *The Bell number $B(n)$ give the number of all possible equivalence relations built on a given n-set.*

☐ Indeed, it is well known that the introduction of an equivalence relation on a given set S results in a partition of this set S into equivalence classes that are non-empty disjoint subsets of the set S, whose union gives the entire set S. Thus, setting the equivalence relation by S leads to a partition of the set S. On the other hand, any partition of the set S gives an equivalence relation on S by the following law: two elements a and b in S are equivalent, $a \sim b$, if and only if they belong to the same subset of the considered partition. ☐

- *The Bell number $B(n)$ gives the number of ways to place n different balls in one or more indistinguishable boxes.*

☐ For example, suppose that $n = 3$. So, we have three different balls, say, $a, b,$ and c, and several boxes. If the boxes cannot be distinguished from each other, then there are five possible options for placing balls into the boxes: each ball is placed in its own box; all three balls are placed in one box; the ball a is placed in one box, the balls b and c are placed in the other box; the ball b is placed in one box, the balls a and c are placed in the other box; the ball c is placed in one box, the balls a and b are placed in the other. In general, any partition of a given n-set can be represented in this way: in the same box, those elements of the set S that belong to the same subset of a given partition of S are placed. ☐

Bell numbers are also connected with rhyme schemes.

- *The Bell number $B(n)$ counts the rhyme schemes of an n-line poem or stanza.*

☐ A *rhyme scheme* describes which lines of the poem rhyme with each other and so may be interpreted as a partition of the set of lines into rhyming subsets. Rhyme schemes are usually written as sequences of Roman letters, one per line, with rhyming lines marked by the same letter and with the first line in each rhyming set labeled in alphabetical order. Thus, the 15 possible four-line rhyme schemes are *aaaa, aaab, aaba, aabb, aabc, abaa, abab, abac, abba, abbb, abba, abca, abcb, abcc,* and *abcd*. One of them, *abcd*, indicates an absence of any rhyme.

For example, the following piece of dialogue at the moment when Romeo and Juliet (Shakespeare) meet can be illustrated by the rhyme scheme *abab*:

> "If I profane with my unworthiest hand
> This holy shrine, the gentle fine is this:
> My lips, two blushing pilgrims, ready stand
> To smooth that rough touch with a tender kiss."

A transition from partitions to rhymes is easily possible if, instead of each element k of the set $V_n = \{1, 2, \ldots, n\}$, we write the symbol of the class S_i, into which the element k falls under a given partition $V_n = S_1 \cup \cdots \cup S_q$ of the set V_n: symbol a for S_1, symbol b for S_2, etc.

It is easy to check that:

- the Bell number $B(n)$ gives the number of distinct multiplicative partitions of squareless number N with n prime divisors.

That is, we consider the representations of the number N as products of natural numbers greater than unity; two representations are considered the same partition if they have the same factors, regardless of their order.

For example, a positive integer 30 is a product of three primes, 2, 3, and 5, and has five such representations: $30 = 2 \cdot 15 = 3 \cdot 10 = 5 \cdot 6 = 2 \cdot 3 \cdot 5$.

Closed formula for Bell numbers

4.6.4. Since the Bell number $B(n)$ is the number of all partitions of a given n-set, and the Stirling number of the second kind $S(n, k)$ gives the number of all partitions of a given n-set, consisting of k subsets, then the Bell number $B(n)$ can be obtained by a summation of values of the Stirling numbers of the second kind $S(n, k)$ over all possible values of k:

$$B(n) = S(n,0) + S(n,1) + S(n,2) + \cdots + S(n,n).$$

Therefore, as we know a closed formula for the Stirling numbers of the second kind,

$$S(n,k) = \frac{k^n}{k!} - \frac{(k-1)^n}{1!(k-1)!} + \frac{(k-2)^n}{2!(k-2)!} - \cdots + (-1)^{k-1}\frac{1^n}{(k-1)!1!},$$

we can also get a closed formula for the Bell number $B(n)$.

- Bell numbers satisfy the following closed formula:

$$B(n) = \left(1 - \frac{0}{0!}\right)\frac{n^n}{n!} + \left(1 - \frac{1}{1!}\right)\frac{(n-1)^n}{(n-1)!}$$
$$+ \left(1 - \frac{1}{1!} + \frac{1}{2!}\right)\frac{(n-2)^n}{(n-2)!}$$
$$+ \cdots + \left(1 - \frac{1}{1!} + \frac{1}{2!} - \cdots + (-1)^{n-1}\frac{1}{(n-1)!}\right)\frac{1^n}{1!}.$$

□ The result follows from direct transformations. In other words, it holds that

$$B(n) = \sum_{k=0}^{n-1}\left(1 - \frac{1}{1!} + \frac{1}{2!} - \cdots + (-1)^k\frac{1}{k!}\right)\frac{(n-k)^n}{(n-k)!}.$$

The coefficients in brackets are the sums of partial series for the decomposition of $\frac{1}{e}$. □

Generating function of the sequence of Bell numbers

4.6.5. It is relatively easy to obtain the exponential generating function for the sequence of Bell numbers.

- *The exponential generating function of the sequence of Bell numbers has the form e^{e^x-1}, i.e., it holds that*

$$e^{e^x-1} = B(0) + \frac{B(1)}{1!}x + \frac{B(2)}{2!}x^2 + \frac{B(3)}{3!}x^3 + \cdots + \frac{B(n)}{n!}x^n + \cdots.$$

□ In order to prove this property, we once again use the fact that the nth Bell number is the sum of the Stirling numbers of the second kind. As was proven earlier,

$$S(n,k) = \sum_{(k_1,k_2,\ldots,k_n)} \frac{n!}{(1!)^{k_1}(2!)^{k_2}\cdots(n!)^{k_n}k_1!\cdots k_n!},$$

where $k_1, k_2, \ldots, k_n$ are non-negative integers, such that $k_1 + 2k_2 + \cdots + nk_n = n$ and $k_1 + k_2 + \cdots + k_n = k$. Then, the nth Bell number can be represented as

$$B(n) = \sum_{(k_1,k_2,\ldots,k_n)} \frac{n!}{(1!)^{k_1}(2!)^{k_2}\cdots(n!)^{k_n}k_1!\cdots k_n!},$$

where $k_1, k_2, \ldots, k_n$ are non-negative integers, such that $k_1 + 2k_2 + \cdots + nk_n = n$.

Now, we can get the result using the classical decomposition

$$e^x = 1 + \frac{1}{1!}x + \frac{1}{2!}x^2 + \frac{1}{3!}x^3 + \cdots + \frac{1}{n!}x^n + \cdots, \quad |x| < 1. \quad □$$

Properties of Bell numbers

4.6.6. There are several formulas connecting Bell numbers and the elements of Pascal's triangle (binomial coefficients), including the classical recurrent relation

$$B(n+1) = C_n^0 B(n) + C_n^1 B(n-1) + \cdots + C_n^n B(n).$$

Another natural connection exists between Bell numbers and the Stirling numbers of the second kind. Let us write the corresponding formula again:

- *For the $n$the Bell number $B(n)$, it holds that*

$$B(n) = S(n,0) + S(n,1) + S(n,2) + \cdots + S(n,n-1) + S(n,n).$$

☐ This property immediately follows from the definitions of Bell numbers and the Stirling numbers of the second kind. ☐

On the other hand, we can obtain another formula that glues together Bell numbers, the Stirling numbers of the second kind, and binomial coefficients. Thus, the following property holds:

- Bell numbers satisfy the relation

$$B(n+m) = \sum_{k=0}^{n}\sum_{j=0}^{m} j^{n-k} S(m,j) C_n^k B(k).$$

☐ For a proof of the considered property, let us use some simple combinatorial reasons. For two given sets, an n-set and an m-set, one can count the number of partitions of their union, that is, the corresponding $(n+m)$-set, as follows. Break down the m-set into exactly j subsets. This can be done in $S(m,j)$ ways. Choose k elements from the n-set in order to split them into several new subsets, and distribute the remaining $n-k$ objects among j subsets, already formed from the m-set. There are C_n^k ways to choose k objects from n objects, $B(k)$ ways to break them down into several non-empty subsets, and j^{n-k} ways to distribute the remaining $n-k$ objects among j existing subsets of a given partition of the m-set. Thus, there are $j^{n-k} S(m,j) C_n^k B(k)$ ways to realize the considered operation. Summation over all possible values of j and k gives the result. ☐

The considered formula generalizes two properties, already discussed above. Namely, in order to obtain the formula $B(n) = \sum_{k=0}^{n} S(n,k)$, expressing Bell numbers as the sum of Stirling numbers of the second kind, we should simply take $n=0$ in the new formula. In order to obtain the recurrent relation $B(n) = \sum_{k=0}^{n} C_n^k B(k)$ for Bell numbers, we should take $m=0$ in the new formula.

- Bell numbers can be represented in the form

$$B(n) = \sum_{(k_1,\ldots,k_n)} \frac{n!}{((1!)^{k_1}(2!)^{k_2}\cdots(n!)^{k_n} k_1!\cdots k_n!},$$

where $k_1, k_2, \ldots, k_n$ are non-negative integers, such that $k_1 + 2k_2 + \cdots + nk_n = n$.

□ In order to prove this property, we should consider and study the $(r_1, r_2, \ldots, r_k)$-partitions of a given n-set M: ordered strings of k non-empty subsets of the set M, such that $M_i \cap M_j = \emptyset$, if $i \neq j$, $M_1 \cup M_2 \cup \cdots \cup M_k = M$, and $|M_1| = |r_1|, |M_2| = r_2, \ldots, |M_k| = r_k$, where $|M|$ is the cardinality of the set M. For details, see [Deza24]. □

- Bell numbers and the Stirling numbers of the first kind are connected by the following relation:

$$s(n,0)B(0) + s(n,1)B(1) + s(n,2)B(2)$$
$$+ \cdots + s(n, n-1)B(n-1) + s(n,n)B(n) = 1.$$

□ In fact, the well-known *Stirling's transform* states that two sequences, $a_1, a_2, \ldots$ and $A_1, A_2, \ldots$, are connected by the relation $A_n = \sum_{k=1}^{n} s(n,k) a_k$, $n = 1, 2, \ldots$, if and only if $a_n = \sum_{k=1}^{n} S(n,k) A_n$, $n = 1, 2, \ldots$. Therefore, it holds that $\sum_{k=0}^{n} s(n,k) B(k) = 1$, i.e., the sequence $B(0), B(1), B(2), \ldots$ of Bell numbers is the Stirling's transform of the sequence $1, 1, 1, \ldots$, and vice versa. □

4.6.7. Using the property $x^n = \sum_{k=1}^{n} S(n,k) x(x-1)\cdots(x-k+1)$ of the Stirling numbers of the second kind, it is possible to prove the *Dobinskii formula*.

- For Bell numbers, there exists the Dobinskii formula

$$B(n) = \frac{1}{e}\left(\frac{0^n}{0!} + \frac{1^n}{1!} + \frac{2^n}{2!} + \cdots + \frac{k^n}{k!} + \cdots\right).$$

□ As $m^n = \sum_{k=1}^{n} S(n,k) m(m-1)\cdots(m-k+1)$, $\frac{m^n}{m!} = \sum_{k=1}^{n} \frac{S(n,k)}{(m-k)!}$, and

$$\sum_{m=1}^{\infty} \frac{m^n}{m!} x^m = \sum_{m=1}^{\infty} \sum_{k=1}^{n} \frac{S(n,k) x^m}{(m-k)!} = \sum_{m=1}^{\infty} \sum_{k=1}^{n} S(n,k) x^k \cdot \frac{x^{m-k}}{(m-k)!}$$

$$= \left(\sum_{k=1}^{n} S(n,k) x^k\right) \cdot \left(\sum_{j=0}^{\infty} \frac{x^j}{j!}\right) = \left(\sum_{k=1}^{n} S(n,k) x^k\right) \cdot e^x.$$

Then,

$$\sum_{k=1}^{n} S(n,k)x^k = e^{-x} \sum_{m=1}^{\infty} \frac{m^n}{m!} x^m, \quad \text{and}$$

$$\sum_{k=1}^{n} S(n,k) = B(n) = e^{-1} \sum_{m=1}^{\infty} \frac{m^n}{m!}. \qquad \square$$

It is easy to check that the sequence of Bell numbers grows faster than the exponential function.

More precisely, for the Bell numbers with large indices, there exists an asymptotic expression:

$$\frac{\log B(n)}{n} = \log n - \log \log n + O(1).$$

Bell numbers in the family of special numbers

4.6.8. Many different connections of Bell numbers with other special numbers are already considered in this Chapter. We noted the close connection between Bell numbers and the Stirling numbers of the second kind, the relations between Bell numbers and the Stirling numbers of the first kind, a recurrent relation for Bell numbers, including binomial coefficients, etc. A problem of rhyme schemes made it possible to connect Bell numbers and Catalan numbers.

In this section, we consider some generalizations of Bell numbers.

4.6.9. *Complementary Bell numbers* (or *Uppuluri–Carpenter numbers*) $\widetilde{B}(n)$ are determined by the formula

$$\widetilde{B}(n) = S(n,0) - S(n,1) + S(n,2) - \cdots + (-1)^n S(n,n),$$

where $S(n,k)$ are the Stirling numbers of the second kind (see, for example, [Deza18], [Weis24], and [Sloa24]).

Direct consideration shows that for $n = 0, 1, 2, 3, \ldots$, these numbers are 1, -1, 0, 1, 1, -2, -9, -9, 50, 267, 413,... (sequence A000587 in the OEIS).

Simple consideration gives us *Dobinskii's identity* for complimentary Bell numbers; it has the form

$$\widetilde{B}(n) = e \sum_{k=0}^{\infty} (-1)^k \frac{k^n}{n!}.$$

4.6.10. *Ordered Bell numbers* (or *Fubini numbers*) $OB(n)$ count the number of weak orderings on a given n-set. For $n = 0, 1, 2, 3, \ldots$, these numbers are 1, 1, 3, 13, 75, 541, 4683, 47293, 545835, 7087261, 102247563, ... (sequence A000670 in the OEIS).

It is easy to get the following property:

- *Ordered Bell numbers $OB(n)$ can be calculated using the following linear combination of the Stirling numbers of the second kind:*

$$OB(n) = S(n,0) + 1!S(n,1) + 2!S(n,2) + \cdots + k!S(n,k) + \cdots$$
$$+ n!S(n,n).$$

☐ This result follows from our previous considerations. If a partition of a given n-set has exactly k parts, then we have exactly $k!$ linear orders of these parts. ☐

In other words, it is possible to obtain ordered Bell numbers from the triangle of Stirling numbers of the second kind, but, summing the elements of the corresponding row of the triangle, we should multiply each element by the weight factor $k!$, corresponding to the index of this element in the row.

Exercises

1. Using the definition, find the first five Bell numbers. Using the recurrent relation, find the first 15 Bell numbers. Construct a rule for the calculation of the nth Bell number using the nth row of Pascal's triangle.
2. Construct the Bell triangle modulo 2: indicate by white color its even elements; indicate by black color its odd elements. What properties does the resulting triangle possess? What is the ratio of the number of white points of the triangle to the number of its black points?

3. Find the first 15 Bell numbers using the obtained closed formula. Compare the complexity of different methods of calculation of Bell numbers.
4. Prove that $B(n) = \frac{1}{k!}(C_k^0 k^n + C_k^1(k-1)^n + C_k^2(k-2)^n + \cdots + C_k^{k-2} 2^n + C_k^{k-1} + 1)$. Formulate a rule for the calculation of the nth Bell number using Pascal's triangle.
5. Find the smallest prime divisor of $B(n)$, $n = 2, 3, 4, 5, 6, 7, 8, 9, 10$. Check that $B(p) \equiv 2 (mod\ p)$, for $p = 2, 3, 5, 7$.
6. Using the triangle of Stirling numbers of the second kind, find the first 10 elements of the sequence of complimentary Bell numbers.
7. Find the first seven ordered Bell numbers using their combinatorial definition. Find the first seven ordered Bell numbers using the corresponding recurrent relation. Find the first seven ordered Bell numbers using the triangle of Stirling numbers of the second kind.

4.7. Factorial Numbers

A history of the question

4.7.1. In mathematics, the *factorial* of a non-negative integer n, denoted by $n!$, is the product of all positive integers less than or equal to n.

The concept of factorials emerged independently in many cultures.

In India, one of the earliest known descriptions of factorials comes from the Anuyogadvāra-sūtra (300 BC–400 AD). The product rule for permutations was also described in the 6th century by Jain monk Jinabhadra (520–623). Hindu scholars have been using factorial formulas since at least 1150, with Bhāskara II (circa 1114–circa 1185) mentioning factorials in his work *Līlāvatī*.

In the mathematics of the Middle East, the Hebrew mystic book of creation *Sefer Yetzirah*, from the Talmudic period (circa 200–circa 500), lists factorials up to 7! as part of an investigation into the number of words that can be formed using the Hebrew alphabet. Factorials were also studied for similar reasons in the 8th century by an Arab grammarian Al-Khalilibn Ahmad al-Farahidi (circa 718–circa 786).

The Arab mathematician Ibn al-Haytham (circa 965–circa 1040) was the first to formulate *Wilson's theorem*, connecting the factorials with the prime numbers.

In Europe, although Plato famously used 5040 = 7! as the population of an ideal community, in part because of its divisibility properties, the first works on factorials were by Jewish scholars, such as Shabbethai Donnolo (circa 913 – circa 982). From the 15th century, factorials became the subject of study by Western mathematicians, including Luca Pacioli (circa 1447 – 1517), Christopher Clavius (1538–1612), and Marin Mersenne (1588–1648). The power series for the exponential function with the reciprocals of factorials for its coefficients was first formulated in 1676 by Isaac Newton (1642–1726/1727). Other important works on factorials include a 1685 treatise by John Wallis (1616–1703), a study of their approximate values by Abraham de Moivre (1667–1754), asymptotic results derived by James Stirling (1692–1770), the works of Daniel Bernoulli (1700–1782) and Leonhard Euler (1707–1783) that presented the continuous extension of the factorial function to the *gamma function*.

The notation $n!$ was introduced by the French mathematician Christian Kramp (1760–1826) in 1808. The word *factorial* was first used in 1800 by Louis François Antoine Arbogast (1759–1803) but referring to a more general concept of products of arithmetic progressions.

Definition of factorial

4.7.2. The *factorial* of a positive integer n, denoted by $n!$, is the product of all positive integers less than or equal to n:

$$n! = 1 \cdot 2 \cdot \cdots \cdot (n-1) \cdot n, \quad n \in \mathbb{N}.$$

The value of $0!$ is 1, according to the convention for an empty product.

The earliest uses of the factorial function involve counting permutations: there are $n!$ different ways of arranging n distinct objects into a sequence. So, we have

$$P_n = n!.$$

From this point of view, it is important to remember that the unsigned Stirling numbers of the first kind $|s(n,k)|$ count the number of permutations of n elements, containing exactly k cycles. So, the sum of such numbers over all possible k gives the number of all permutations of n elements, i.e., $n!$:

$$n! = |s(n,0)| + |s(n,1)| + |s(n,2)| + \cdots + |s(n,n)|.$$

The factorial of n also equals the product of n with the next smaller factorial: $n! = n \cdot (n-1)!$.

So, we can obtain the following simple recurrent relation for the sequence 1, 1, 2, 6, 24, 120, 720, 5040, 40320, 362880, ... (sequence A000142 in the OEIS) of factorials:

$$n! = n \cdot (n-1)!, \quad 0! = 1.$$

Problems connected with factorials

4.7.3. Factorials appear in many formulas of combinatorics to account for different orderings of objects.

As was considered above, the factorial function involve counting permutations: there are $n!$ permutations of a given n-set.

The number A_n^k of partial k-permutations of a given n-set can be given by the formula

$$A_n^k = \frac{n!}{(n-k)!}.$$

Similarly, the number C_n^k of k-combinations (subsets of k elements) from a set with n elements can be computed from factorials using the formula

$$C_n^k = \frac{n!}{k!(n-k)!}.$$

The unsigned Stirling numbers of the first kind sum to the factorials and count the permutations of n grouped into subsets with the same numbers of cycles. Another combinatorial application is in counting *derangements*, permutations that do not leave any element in its original position; the number of derangements of n items is the nearest integer to $n!/e$.

4.7.4. In algebra, the factorials originate from the binomial theorem, which uses binomial coefficients to expand the powers of sums. In fact, the binomial coefficients $\binom{n}{k}$, i.e., the coefficients of the decomposition

$$(x+y)^n = \binom{n}{0}x^n + \binom{n}{1}x^{n-1}y + \binom{n}{2}x^{n-2}y^2$$
$$+ \cdots + \binom{n}{n-1}xy^{n-1} + \binom{n}{n}y^n,$$

coincide with the numbers of k-element combinations of a set with n elements and can be computed from factorials using the formula

$$\binom{n}{k} = \frac{n!}{k!(n-k)!}.$$

Their use in counting permutations can also be restated algebraically: the factorials are the orders of finite symmetric groups.

4.7.5. In calculus, factorials occur in *Faà di Bruno's formula* for chaining higher derivatives:

$$\frac{d^n}{dx^n} f(g(x)) = \sum \frac{n!}{m_1!\, 1!^{m_1}\, m_2!\, 2!^{m_2} \cdots m_n!\, n!^{m_n}}$$
$$\cdot f^{(m_1+\cdots+m_n)}(g(x)) \cdot \prod_{j=1}^{n}(g^{(j)}(x))^{m_j},$$

where the sum is over all n-tuples of non-negative integers $(m_1, \ldots, m_n)$, satisfying the constraint $1 \cdot m_1 + 2 \cdot m_2 + 3 \cdot m_3 + \cdots + n \cdot m_n = n$.

In mathematical analysis, factorials frequently appear in the denominators of power series, most notably in the series for the exponential function

$$e^x = 1 + \frac{x}{1!} + \frac{x^2}{2!} + \frac{x^3}{3!} + \cdots = \sum_{i=0}^{\infty} \frac{x^i}{i!}$$

(this usage of factorials in power series leads to analytic combinatorics), and in the coefficients of other Taylor series (in particular, those of the trigonometric and hyperbolic functions).

Closed formula for factorials

4.7.6. Formally, the definition of factorial is given by its closed formula. For a given non-negative integer n, we can find the nth factorial number $n!$ as the product of consecutive integers:
$$0! = 1, \quad n! = n \cdot (n-1) \cdots 2 \cdot 1.$$
But, for large n, it is difficult to realize this step-by-step multiplication of consecutive positive integers from 1 up to n.

However, there exist other formulas that allow us to calculate large factorials, or, more exactly, to study their arithmetic nature.

In fact, the most salient property of factorials is the divisibility of $n!$ by all positive integers up to n and, hence, by all prime numbers less than or equal to n. This fact for prime factors of n is described more precisely by *Legendre's formula*. This formula gives an expression for the exponent of the largest power of a prime p that divides the factorial $n!$. It is named after A.-M. Legendre but is also known as *de Polignac's formula*, after A. de Polignac.

- For any prime number p and any positive integer n, let $\nu_p(n)$ be the exponent of the largest power of p that divides n. Then,
$$\nu_p(n!) = \sum_{i=1}^{\infty} \left\lfloor \frac{n}{p^i} \right\rfloor.$$

While the sum on the right-hand side is an infinite sum, for any particular values of n and p, it has only finitely many non-zero terms: for every i large enough that $p^i > n$, one has $\lfloor \frac{n}{p^i} \rfloor = 0$. This reduces the infinite sum above to
$$\nu_p(n!) = \sum_{i=1}^{L=\lfloor \log_p n \rfloor} \left\lfloor \frac{n}{p^i} \right\rfloor.$$

So, going through all primes up to n, we obtain that the integer factorization of the factorial $n!$, $n \geq 2$, has the following form:

- For any positive integer $n \geq 2$, it holds that $n! = p_1^{\alpha_1} \cdots p_k^{\alpha_k}$, where p_i are all prime numbers less than or equal to n, and
$$\alpha_i = \left\lfloor \frac{n}{p_i} \right\rfloor + \left\lfloor \frac{n}{p_i^2} \right\rfloor + \left\lfloor \frac{n}{p_i^3} \right\rfloor + \cdots.$$

For example, if $n = 6$, one has $6! = 720 = 2^4 \cdot 3^2 \cdot 5^1$. The exponents $\alpha_1 = \nu_2(6!) = 4$, $\alpha_2 = \nu_3(6!) = 2$, and $\alpha_3 = \nu_5(6!) = 1$ can be computed using Legendre's formula as follows:

$$\nu_2(6!) = \sum_{i=1}^{\infty} \left\lfloor \frac{6}{2^i} \right\rfloor = \left\lfloor \frac{6}{2} \right\rfloor + \left\lfloor \frac{6}{4} \right\rfloor = 3 + 1 = 4,$$

$$\nu_3(6!) = \sum_{i=1}^{\infty} \left\lfloor \frac{6}{3^i} \right\rfloor = \left\lfloor \frac{6}{3} \right\rfloor = 2, \quad \nu_5(6!) = \sum_{i=1}^{\infty} \left\lfloor \frac{6}{5^i} \right\rfloor = \left\lfloor \frac{6}{5} \right\rfloor = 1.$$

☐ Since $n!$ is the product of the integers 1 through n, we obtain at least one factor of p in $n!$ for each multiple of p in $\{1, 2, \ldots, n\}$, of which there are $\lfloor \frac{n}{p} \rfloor$. Each multiple of p^2 contributes an additional factor of p, each multiple of p^3 contributes yet another factor of p, etc. Adding up the number of these factors gives the infinite sum for $\nu_p(n!)$. ☐

Properties of factorials

4.7.7. In the following, we consider some simple but important arithmetical properties of factorials:

- For a given positive integer n, any positive integer k, $k \leq n$, divides $n!$.

☐ This fact is obvious, as $n! = 1 \cdot 2 \cdot \cdots \cdot k \cdot \cdots \cdot n$. ☐

The product formula for the factorial implies that $n!$ is divisible by all prime numbers that are at most n and by no larger prime numbers.

- For a given positive integer n, any prime number p, $p \leq n$, divides $n!$, and any prime number q, $q > n$, does not divide $n!$.

☐ The first statement is obvious, as any $p \leq n$ is a factor of $n!$. The second statement follows from the basic property of primes: if $p|ab$, then $p|a$ or $p|b$. So, if $p|1 \cdot 2 \cdot \cdots \cdot n$, then $p|k$ for some $k \leq n$ and, hence, $p \leq k \leq n$. ☐

It means that factorials have only relatively small prime divisors. On the other hand, the numbers $n! \pm 1$ have only large prime divisors.

Another result concerning the divisibility of factorials is *Wilson's theorem*:

- *For any positive integer $n > 1$, $(n-1)! + 1$ is divisible by n if and only if n is a prime number.*

☐ 1. If n is a composite number, then there exists a positive integer a, such that $a|n$ and $1 < a < n$. In this case, it holds that $a|(n-1)!$. So, if n divides $(n-1)! + 1$, then a divides $(n-1)! + 1$, and a divides $(n-1)!$. It implies that $a|1$, but $a > 1$ – a contradiction.

2. Let $n = p$ be a prime. The result is trivial for $p = 2$: $(2-1)! + 1 = 2$ is divisible by 2. Let p be an odd prime, $p \geq 3$. Since the residue classes modulo p form a field, every non-zero class a has an unique multiplicative inverse, a^{-1}. The only values of a for which $a \equiv a^{-1} (\bmod\ p)$ are $a \equiv \pm 1 (\bmod\ p)$ because the congruence $a^2 \equiv 1 (\bmod\ p)$ can have at most two roots. Therefore, with the exception of ± 1, the factors of $(p-1)!$ can be arranged in disjoint pairs, such that the product of each pair is congruent to 1 modulo p. For example, for $p = 11$, we have

$$10! = [(1 \cdot 10)] \cdot [(2 \cdot 6)(3 \cdot 4)(5 \cdot 9)(7 \cdot 8)]$$
$$\equiv [-1] \cdot [1 \cdot 1 \cdot 1 \cdot 1] \equiv -1 \pmod{11}.$$

This proves Wilson's theorem. ☐

As a function of n, the factorial has faster-than-exponential growth, but grows more slowly than a double exponential function. Its growth rate is similar to n^n but slower by an exponential factor. In fact, the formula

$$\ln n! = \sum_{x=1}^{n} \ln x \approx \int_{1}^{n} \ln x\, dx = n \ln n - n + 1$$

allows us to approximate $n!$ as $(n/e)^n$.

The classical *Stirling's approximation* (see Chapter 1) has the form

$$n! \sim \sqrt{2\pi n} \left(\frac{n}{e}\right)^n.$$

The following version of the bound holds for all $n \geq 1$ rather than only asymptotically ([Robb55]):

$$\sqrt{2\pi n}\left(\frac{n}{e}\right)^n e^{\frac{1}{12n+1}} < n! < \sqrt{2\pi n}\left(\frac{n}{e}\right)^n e^{\frac{1}{12n}}.$$

Factorials in the family of their relatives

4.7.8. The *falling factorial* $x^{\underline{n}}$ (or *descending factorial, falling power, falling sequential product, lower factorial*) is defined as the polynomial

$$x^{\underline{n}} = \overbrace{x(x-1)(x-2)\cdots(x-n+1)}^{n \text{ factors}} = \prod_{k=1}^{n}(x-k+1) = \prod_{k=0}^{n-1}(x-k).$$

The *rising factorial* $x^{\overline{n}}$ (or *Pochhammer function, Pochhammer polynomial, rising power, ascending factorial, rising sequential product, upper factorial*) is defined as

$$x^{\overline{n}} = \overbrace{x(x+1)(x+2)\cdots(x+n-1)}^{n \text{ factors}} = \prod_{k=1}^{n}(x+k-1) = \prod_{k=0}^{n-1}(x+k).$$

The value of each product is taken to be 1 (an empty product) when $n = 0$: $x^{\underline{0}} = x^{\overline{0}} = 1$.

These symbols are collectively called *factorial powers* (see [Knut97], [GKP94], [Deza18], and [Wiki24]).

We consider several simple properties of the falling and the rising factorials in the following:

- The rising and falling factorials are related to each other by the following formulas:

$$x^{\overline{n}} = (x+n-1)^{\underline{n}} = (-1)^n(-x)^{\underline{n}};$$
$$x^{\underline{n}} = (x-n+1)^{\overline{n}} = (-1)^n(-x)^{\overline{n}}.$$

□ It is easy to check by definition. For example,

$$x^{\overline{n}} = x(x+1)\cdots(x+n-1)$$
$$= (-1)^n(-x)(-x-1)\cdots(-x-n+1) = (-1)^n(-x)^{\underline{n}}. \quad \square$$

- The rising and falling factorials are related to the ordinary factorials:

$$n! = 1^{\overline{n}} = n^{\underline{n}}; \quad m^{\underline{n}} = \frac{m!}{(m-n)!}; \quad m^{\overline{n}} = \frac{(m+n-1)!}{(m-1)!}.$$

☐ It is obvious; for example, $m^{\underline{n}} = m(m-1)\cdots(m-n+1) = \frac{m!}{(m-n)!}$. ☐

- The rising and falling factorials can be used to express a generalized binomial coefficient: for any real x, it holds that

$$\frac{x^{\overline{n}}}{n!} = \binom{x+n-1}{n}; \quad \frac{x^{\underline{n}}}{n!} = \binom{x}{n}.$$

☐ The statement follows from the definition of $x^{\overline{n}}$, $x^{\underline{n}}$, and $\binom{x}{n}$. ☐

4.7.9. The *double factorial* (or *semifactorial*) of a positive number n, denoted by $n!!$, is the product of all the integers from 1 up to n that have the same parity (odd or even) as n (see [Call09]). For even n, the double factorial is

$$n!! = \prod_{k=1}^{\frac{n}{2}}(2k) = n(n-2)(n-4)\cdots 4 \cdot 2,$$

and for odd n, it is

$$n!! = \prod_{k=1}^{\frac{n+1}{2}}(2k-1) = n(n-2)(n-4)\cdots 3 \cdot 1.$$

The double factorial $0!! = 1$ as an empty product.

The sequence of double factorials for even $n = 0, 2, 4, 6, 8, \ldots$ starts as 1, 2, 8, 48, 384, 3840, 46080, 645120,... (sequence A000165 in the OEIS). The sequence of double factorials for odd $n = 1, 3, 5, 7, 9, \ldots$ starts as 1, 3, 15, 105, 945, 10395, 135135,... (sequence A001147 in the OEIS).

The factorial of a positive integer n can be written as the product of two double factorials:

$$n! = n!! \cdot (n-1)!!.$$

For an even non-negative integer $n = 2k$ with $k \geq 0$, the double factorial may be expressed as

$$n!! = 2^k k!.$$

For odd $n = 2k - 1$ with $k \geq 1$, combining the two formulas above yields

$$n!! = \frac{(2k)!}{2^k k!} = \frac{(2k-1)!}{2^{k-1}(k-1)!}.$$

Stirling's approximation for the factorial, $n! \sim \sqrt{2\pi n}(\frac{n}{e})^n$, can be used to derive the following asymptotic equivalence as n tends to infinity:

$$n!! \sim \sqrt{2n^{n+1}e^{-n}} \sim \begin{cases} \sqrt{\pi n}\left(\dfrac{n}{e}\right)^{n/2}, & \text{if } n \text{ is even,} \\ \sqrt{2n}\left(\dfrac{n}{e}\right)^{n/2}, & \text{if } n \text{ is odd.} \end{cases}$$

An approximation for the ratio of the double factorial of two consecutive integers, which is quite accurate for large n, is

$$\frac{(2n)!!}{(2n-1)!!} \approx \sqrt{\pi n}.$$

The double factorials were originally introduced in order to simplify the expression of certain trigonometric integrals that arise in the derivation of the *Wallis product*. They also arise when expressing the volume of a hypersphere, and they have many applications in enumerative combinatorics.

For instance, $n!!$ for odd values of n counts the *perfect matchings* of the complete graph K_{n+1} (see, for example, [Ore80], [Hara03], [HaPa77], and [PeSk03]).

On the other hand, the numbers of matchings in the complete graph K_n without constraining the matchings to be perfect are instead given by the *telephone numbers* (or *involution numbers*) $T(n)$, which also count the ways n people can be connected by person-to-person telephone calls, or the number of permutations on n elements that are involutions, and may be expressed as a summation, involving double factorials:

$$T(n) = \sum_{k=0}^{\lfloor n/2 \rfloor} \binom{n}{2k}(2k-1)!! = \sum_{k=0}^{\lfloor n/2 \rfloor} \frac{n!}{2^k(n-2k)!k!}.$$

The double factorials are used in *Stirling permutations*. Here, the set of all the *Stirling permutations* of order k is defined as a set of all the permutations of the multiset $\{1,1,2,2,\ldots,k,k\}$, in which each pair of equal numbers is separated only by larger numbers.

- *The number of Stirling permutations of order k is given by the double factorial $(2k-1)!!$.*

☐ It is easy to see that the two copies of k must be adjacent. Removing them from the permutation leaves a permutation in which the maximum element is $k-1$ with n positions into which the adjacent pair of k values may be placed. From this recursive construction, a proof that the Stirling permutations are counted by the double permutations follows by induction.

Alternatively, instead of the restriction that a values between a pair may be larger than it, one may also consider the permutations of this multiset in which the first copies of each pair appear in sorted order; such a permutation defines a matching on the $2k$ positions of the permutation; therefore, again, the number of permutations may be counted by the double factorial. ☐

4.7.10. In the same way that the double factorial generalizes the notion of the single factorial, the following definition of the integer-valued multiple factorial functions, called *multifactorials* (or α-*factorials*), extends the notion of the double factorial function for $\alpha \in \mathbb{Z}^+$:

$$n!_\alpha = \begin{cases} n \cdot (n-\alpha)!_\alpha, & \text{if } n > 0; \\ 1, & \text{if } -\alpha < n \leq 0; \\ 0, & \text{otherwise.} \end{cases}$$

Alternatively, the multifactorial $n!_\alpha$ can be extended to most real and complex numbers n by noting that, when n is one more than a positive multiple of α, then

$$n!_\alpha = n(n-\alpha)\cdots(\alpha+1) = \alpha^{\frac{n-1}{\alpha}}\left(\frac{n}{\alpha}\right)\left(\frac{n-\alpha}{\alpha}\right)\cdots\left(\frac{\alpha+1}{\alpha}\right)$$

$$= \alpha^{\frac{n-1}{\alpha}}\frac{\Gamma\left(\frac{n}{\alpha}+1\right)}{\Gamma\left(\frac{1}{\alpha}+1\right)}.$$

This last expression is defined much more broadly than the original. In addition to extending $n!_\alpha$ to most complex numbers n, this definition has the feature of working for all positive real values of α. Furthermore, when $\alpha = 1$, this definition is mathematically equivalent to the classical factorial function. Also, when $\alpha = 2$, this definition is mathematically equivalent to the alternative extension of the double factorial.

For more information see, for example, [Schm10] and [Wiki24].

Exercises

1. Prove that the exponential generating function of the *Pochhammer polynomials* $x^{\underline{n}}$, $n = 0, 1, 2, \ldots$, has the form $(1+t)^x$, i.e.,

$$\sum_{n=0}^{\infty} x^{\underline{n}} \frac{t^n}{n!} = (1+t)^x.$$

2. A symmetric generalization of falling and rising factorials is defined as $x^{\overline{\underline{m}}} = \frac{x^{\overline{m}}x^{\underline{m}}}{x}$. Give several examples of such constrictions. Find several properties.

3. The *hyperfactorial* of a positive integer n is the product of the numbers $1^1, 2^2, \ldots, n^n$. Check that the sequence of hyperfactorials, beginning with $hf(0) = 1$, is 1, 1, 4, 108, 27648, 86400000, 4031078400000, 3319766398771200000, $\ldots$ (sequence A002109 in the OEIS).

4. The *primorial* $n\#$ is the product of prime numbers less than or equal to n. Check that the first primorials $p_n\#$, $n = 0, 1, 2, \ldots$, including $p_0\# = 1$ as an empty product, are 1, 2, 6, 30, 210, 2310, $\ldots$ (sequence A002110 in the OEIS).

5. The *n-compositorial* of a composite number n is the product of all composite numbers up to and including n. Prove that the

n-compositorial is equal to $n!$ divided by $n\#$. Check that the first compositorials are 1, 4, 24, 192, 1728, 17280, 207360, 2903040, 43545600, 696729600, ... (sequence A036691 in the OEIS).

References

[Abra74], [AbSt72], [Adam97], [Aign98], [Andr98], [Apos86], [BaCo87], [Bell34], [Bell38], [Bern99], [BeSl95], [Bhar00], [Bond93], [BPS91], [Bron01], [Brua97], [Buch09], [Cata44], [ChRe14], [CoGu96], [Comt74], [CoRo96], [Cofm75], [CWZ20], [Dede63], [DeDe12], [Deza17], [Deza18], [Deza21], [Deza24], [Dick05], [DoRe69], [DoSh77], [Dutk91], [Eule48], [Ficht01], [FlSe09], [Gard61], [Gard88], [GKP94], [Glai77], [Goul85], [GPP01], [Hogg69], [Kara83], [Knut76], [Knut97], [Kord95], [Lah54], [Lege79], [Male11], [Mazu10], [Motz48], [Ore48], [OsJe15], [Pasc54], [Rose18], [Schw97], [Sier64], [SlPl95], [Sloa24], [Stan97], [Stru87], [StWa78], [Uspe76], [WaZh15], [Weis24], [Wiki24], [ZhQi17].

Chapter 5

Catalan Numbers and Their Relatives on Integer Lattice

5.1. Integer Lattice: Basic Notions

Lattices

5.1.1. In geometry and group theory, a *lattice* in the real coordinate space $\mathbb{R}^n$ is an infinite set of points in this space with the properties that coordinate-wise addition or subtraction of two points in the lattice produces another lattice point, that the lattice points are all separated by some minimum distance, and that every point in the space is within some maximum distance of a lattice point.

For any basis of $\mathbb{R}^n$, the subgroup of all linear combinations with integer coefficients of the basis vectors forms a lattice, and every lattice can be formed from a basis in this way.

A lattice may be viewed as a regular tiling of a space by a primitive cell.

A typical lattice Λ in $\mathbb{R}^n$ has the form

$$\Lambda = \left\{ \sum_{i=1}^{n} a_i v_i \,\middle|\, a_i \in \mathbb{Z} \right\},$$

where $\{v_1, \ldots, v_n\}$ is a basis for $\mathbb{R}^n$. Different bases can generate the same lattice, but the absolute value of the determinant of the vectors v_i is uniquely determined by Λ and denoted by $d(\Lambda)$. If one thinks of a lattice as dividing $\mathbb{R}^n$ into equal polyhedra (copies of an n-dimensional parallelepiped, known as the *fundamental region* of the lattice), then $d(\Lambda)$ is equal to the n-dimensional volume of this polyhedron. This is why $d(\Lambda)$ is called the *covolume* of the lattice. If this equals 1, the lattice is called *unimodular*.

A lattice in $\mathbb{C}^n$ is a discrete subgroup of $\mathbb{C}^n$ which spans $\mathbb{C}^n$ as a real vector space.

For example, the Gaussian integers

$$\mathbb{Z}[i] = \{a + bi \mid a, b \in \mathbb{Z}\}$$

form a lattice in $\mathbb{C} = \mathbb{C}^1$, as $(\{1, i\}$ is a basis of $\mathbb{C}$ over $\mathbb{R}$.

Lattices have many significant applications in pure mathematics, particularly in connection to Lie algebras, number theory, and group theory. They also arise in applied mathematics in connection with coding theory and cryptography, and they are used in various ways in the physical sciences.

As a group (dropping its geometric structure), a lattice is a finitely generated free abelian group and thus isomorphic to $\mathbb{Z}^n$.

A lattice in the sense of a three-dimensional array of regularly spaced points coinciding with, for example, the atomic or molecular positions in a crystal or, more generally, the orbit of a group action under translational symmetry, is a translation of the translation lattice: a coset, which need not contain the origin and, therefore, need not be a lattice in the previous sense.

A simple example of a lattice in $\mathbb{R}^n$ is the subgroup $\mathbb{Z}^n$. More complicated examples include the *E8 lattice* and the *Leech lattice*.

The *E8 lattice* is a discrete subgroup of $\mathbb{R}^8$ of full rank. It can be given explicitly by the set Γ_8 of points in $\mathbb{R}^8$ such that:

- all the coordinates are integers, or all the coordinates are half-integers (a mixture of integers and half-integers is not allowed);
- the sum of all eight coordinates is an even integer.

More precisely,

$$\Gamma_8 = \{(x_i) \in \mathbb{Z}^8 \cup (\mathbb{Z}+\tfrac{1}{2})^8 \mid \sum_i x_i \equiv 0 \pmod{2}\}.$$

The *Leech lattice* $\Lambda 24$ is a unique lattice in the 24-dimensional Euclidean space, $\mathbb{E}^{24}$, with the following list of properties:

- it is unimodular, i.e., it can be generated by the columns of a certain 24×24 matrix with determinant 1;
- it is even, i.e., the square of the length of each vector in $\Lambda 24$ is an even integer;
- the length of every non-zero vector in $\Lambda 24$ is at least 2.

Integer lattices

5.1.2. In mathematics, the n-dimensional *integer lattice* (or *cubic lattice*), denoted $\mathbb{Z}^n$, is the lattice in the Euclidean space $\mathbb{R}^n$ whose lattice points are n-tuples of integers.

The two-dimensional integer lattice $\mathbb{Z}^2 = \{(x,y) \mid x, y \in \mathbb{Z}\}$ is also called the *square lattice* (or *grid lattice*).

In the study of *Diophantine geometry*, the square lattice of points with integer coordinates is often referred to as the *Diophantine plane*.

In algebraic terms, the Diophantine plane is the Cartesian product $\mathbb{Z} \times \mathbb{Z}$ of the set $\mathbb{Z}$ of integers. The study of Diophantine figures focuses on the selection of nodes in the Diophantine plane such that all pairwise distances are integers.

In geometry, the square lattice is considered the *square grid* (or *square tiling, square tessellation*): a regular tiling of the Euclidean plane. It has the *Schläfli symbol* of $\{4,4\}$, meaning it has four squares around every vertex. J. H. Conway called it a *quadrille*.

The integer lattice is an important mathematical object. Carl Friedrich Gauss (1777–1855) systematically used the integer lattice to solve some number-theoretical problems. This in turn led to the creation by Hermann Minkowski (1864–1909) of the geometry of numbers. At the moment, the integer lattice has found its application in combinatorics and in the theory of functions of the complex variable. The study of the symmetry of the integer lattice helps in the classification of crystal systems.

Lattice graphs

5.1.3. In graph theory, a *lattice graph* (or *mesh graph* or *grid graph*) is a graph whose drawing, embedded in some Euclidean space $\mathbb{E}^n$, forms a regular tiling. This implies that the group of bijective transformations that send the graph to itself is a lattice in the group-theoretical sense.

Typically, no clear distinction is made between such a graph in the more abstract sense of graph theory and its drawing in space (often the plane or 3D space). This type of graph may be more succinctly called a *lattice*, *mesh*, or *grid*. Moreover, these terms are also commonly used for a finite section of the infinite graph.

The term *lattice graph* has also been given in the literature to various other kinds of graphs with some regular structure, such as the Cartesian product of a number of complete graphs.

For example, a *triangular grid graph* is a graph that corresponds to a triangular grid, i.e., the triangular tiling of the Euclidean plane. A *Hanan grid graph* for a finite set of points on a plane is produced by the grid obtained by the intersections of all vertical and horizontal lines through each point of the set.

A common type of lattice graph is a *square grid graph*, corresponding to a square grid, i.e., the square tiling of the Euclidean plane. It is a graph whose vertices correspond to the points on the plane with integer coordinates, with x-coordinates being in the range $0, \ldots, n$, y-coordinates being in the range $0, \ldots, m$, and two vertices are connected by an edge whenever the corresponding points are at distance 1. In other words, it is a *unit distance graph* for the described point set.

A square grid graph is a *Cartesian product* of graphs, namely, of two path graphs with n and m edges.

It is well known that any path graph is a *median graph*. Every three vertices, a, b, and c, have a unique median: a vertex $m(a, b, c)$ that belongs to the shortest paths between each pair of a, b, and c. The latter fact implies that the square grid graph is also a median graph.

All square grid graphs are *bipartite*, which is easily verified by the fact that one can color the vertices in a checkerboard fashion.

A path graph P_n may also be considered to be a grid graph on the grid $(n-1) \times 0$. An 1×1 grid graph is a 4-cycle.

Lattice paths

5.1.4. In combinatorics, a *lattice path* L (in the d-dimensional integer lattice $\mathbb{Z}^d$) of length k, with steps in the set S is a sequence of vectors $v_0, v_1, \ldots, v_k \in \mathbb{Z}^d$, such that each consecutive difference $v_i - v_{i-1}$ lies in S.

A lattice path may lie in any lattice in $\mathbb{R}^d$, but the integer lattice $\mathbb{Z}^d$ is most commonly used.

An example of a lattice path in $\mathbb{Z}^2$ of length 5, with steps in $S = \{(2,0), (1,1), (0,-1)\}$ is $L = \{(-1,-2), (0,-1), (2,-1), (2,-2), (2,-3), (4,-3)\}$.

5.1.5. A *north-east lattice path* (or *NE lattice path*) is a lattice path in $\mathbb{Z}^2$, with steps in $S = \{(0,1), (1,0)\}$. The $(0,1)$ steps are called *north steps* (or *up steps*) and denoted by n's; the $(1,0)$ steps are called *east steps* (or *right steps*) and denoted by e's.

NE lattice paths most commonly begin at the origin $(0,0)$. This convention allows us to encode all the information about a NE lattice path L in a single permutation word, containing only the symbols e and n. The length of the word gives us the number k of steps of the lattice path. The order of the n's and e's communicates the sequence of L. Furthermore, the numbers of n's and e's in the word determine the end point of L. In fact, if the permutation word for a NE lattice path contains n n-steps and m e-steps, and if the path begins at the origin $(0,0)$, then the path necessarily ends at (m,n). This follows because you have "walked" exactly n steps north and m steps east from $(0,0)$.

5.1.6. Lattice paths are often used to count certain combinatorial objects. Similarly, there are many combinatorial objects that count the number of lattice paths of a certain kind. This occurs when the lattice paths are in a bijection with the objects under consideration. Let us give several examples:

- Catalan numbers C_n count the number of *Dyck paths* (or *Catalan's paths*). A *Dyck path* is a NE lattice path from $(0,0)$ to (n,n) that lies strictly below (but may touch) the diagonal $y = x$.
- Schröder numbers S_n count the number of lattice paths from $(0,0)$ to (n,n), with steps in $S = \{(1,0),(0,1),(1,1)\}$, that never rise above the diagonal $y = x$.
- Elements of Pascal's triangle $\binom{n+m}{n}$ count he number of all NE lattice paths from $(0,0)$ to (m,n).

5.1.7. The graphical representation of lattice paths leads to many bijective proofs involving combinatorial identities.

For example, in order to prove the identity

$$\sum_{k=0}^{n} \binom{n}{k}^2 = \binom{2n}{n},$$

we can note that the right-hand side is equal to the number of NE lattice paths from $(0,0)$ to (n,n). Each of these NE lattice paths intersects exactly one of the lattice points in the square array with coordinates $(x, n-x)$ for $x \in \{0, 1, \ldots, n\}$ (the points of the descending diagonal $y = n - x$).

On the left-hand side, the squared binomial coefficient, $\binom{n}{k}^2$, represents two copies of the set of NE lattice paths from $(0,0)$ to $(k, n-k)$, with their endpoints attached to their start points. Rotate the second copy 90° clockwise. This does not change the combinatorics of the object: $\binom{n}{k} = \binom{n}{n-k}$. So, the total number of lattice paths remains the same.

Superimpose the squared NE lattice paths onto the same rectangular array. We see that all NE lattice paths from $(0,0)$ to (n,n) are accounted for. In particular, note that any lattice path passing through a fixed lattice point on the descending diagonal is counted by the fixed squared set of lattice paths.

5.1.8. In this chapter, we construct a mini-theory of several classes of special numbers (Catalan numbers, Schröder numbers, Delannoy numbers, Narayana numbers, elements of Pascal's triangle, etc.), starting with the definition of a given number set in terms of the number of certain lattice paths.

Exercises

1. For an integer lattice $\Lambda = \mathbb{Z}^n$, $n = 2, 3, 4$, find its covolume $d(\Lambda)$.
2. Prove that the complete graph K_3 is a lattice graph. Prove that the complete bipartite graph $K_{2,2}$ is a lattice graph. Give other examples of lattice graphs.
3. For $n = 2, 3, 4, 5$ and $m = 0, 1, 2, \ldots, n$, check that there are exactly $\binom{n+m}{n}$ NE lattice paths from $(0,0)$ to (m, n).
4. For $n = 2, 3, 4, 5$, check that there are exactly $D(n) = D(n, n)$ lattice paths from $(0,0)$ to (n, n), with steps in $\{(1, 0), (0, 1), (1, 1)\}$, where $D(n)$ is the nth central Delannoy number.
5. For $n = 2, 3, 4, 5$, check that there are exactly S_n lattice paths from $(0,0)$ to (n, n), with steps in $\{(1, 0), (0, 1), (1, 1)\}$, that never rise above the diagonal $y = x$, where S_n is the nth Schröder number.
6. For $n = 2, 3, 4, 5$, check that there are exactly $\binom{2n}{n}$ NE lattice paths from $(0,0)$ to (n, n); check that there are exactly C_n Dyck paths from $(0,0)$ to (n, n). Using this fact, obtain a graphical proof of the closed formula $C_n = \frac{1}{n+1}\binom{2n}{n}$.
7. For $n = 3, 4, 5$, give the illustrations of the graphical proof of the combinatorial identity $\sum_{k=0}^{n} \binom{n}{k}^2 = \binom{2n}{n}$.

5.2. Catalan Numbers on Integer Lattice

Path-related construction of Catalan numbers

5.2.1. Define the nth *Catalan number* C_n as the number of lattice paths from the point $(0, 0)$ to the point (n, n), with steps in $S = \{(1, 0), (0, 1)\}$, that never rise above the diagonal $y = x$. Such paths are called *Dyck paths* (or *Catalan's paths*).

Denoting a right (east) step $(1, 0)$ by r and an up (north) step $(1, 0)$ by u, we represent each Dyck path from $(0, 0)$ to (n, n) (i.e., in the square $n \times n$ grid) as a word of length $2n$, consisting of the symbols r and u.

As there is exactly one such lattice path, ru, from $(0, 0)$ to $(1, 1)$, we obtain that $C_1 = 1$.

As there are exactly two such lattice paths, $rruu$ and $ruru$, from $(0,0)$ to $(2,2)$, we obtain that $C_2 = 2$.

As there are exactly five such lattice paths, $rurruu$, $rururu$, $rrurru$, $rrruur$, and $rrurur$, from $(0,0)$ to $(3,3)$, we obtain that $C_2 = 5$.

Moreover, it is natural to assume that there exists exactly one lattice path from the point $(0,0)$ to the same point $(0,0)$. So, $C_0 = 1$.

5.2.2. There exists an equivalent representation of the nth Catalan number as the number of certain lattice paths.

- *The nth Catalan number C_n is equal to the number of lattice paths from the point $(0,0)$ to the point $(2n,0)$, with steps in $S = \{(1,1),(1,-1)\}$, that never pass below the OX-axis.*

□ In fact, these two approaches are equivalent. We should just replace any right step $(1,0)$ with the right-up step $(1,1)$ and replace any up step $(0,1)$ with the right-down step $(1,-1)$. □

Thus, instead of the path ru from $(0,0)$ to $(1,1)$, we obtain the path ⟋⟍ from $(0,0)$ to $(2,0)$. Instead of the paths $rruu$ and $ruru$ from $(0,0)$ to $(1,1)$, we obtain the paths ⟋⟋⟍⟍ and ⟋⟍⟋⟍ from $(0,0)$ to $(4,0)$. Instead of the paths $rurruu$, $rururu$, $rruuru$, $rrruuu$, and $rruruu$ from $(0,0)$ to $(3,3)$, we obtain the paths ⟋⟍⟋⟋⟍⟍, ⟋⟍⟋⟍⟋⟍, ⟋⟋⟍⟍⟋⟍, ⟋⟋⟋⟍⟍⟍, and ⟋⟋⟍⟋⟍⟍ from $(0,0)$ to $(6,0)$.

Recurrent relation for Catalan numbers

5.2.3. Using the representation of Catalan numbers as the number of lattice paths from the point $(0,0)$ to the point $(2n,0)$, with steps in $S = \{(1,1),(1,-1)\}$, that never pass below the OX-axis, we can obtain the classical recurrent relation for Catalan numbers.

- *Catalan numbers can be calculated using the following recurrent formula:*

$$C_0 = C_1 = 1; C_n = \sum_{k=0}^{n-1} C_k C_{n-1-k}.$$

□ It is already checked that, for $n = 0,1,2,3$, the statement is true. In fact, counting Dyck paths from $(0,0)$ to $(2n,0)$, $n = 0,1$, we obtain that $C_0 = C_1 = 1$. For $n = 2$, the number of Dyck paths from $(0,0)$ to $(4,0)$ is equal to 2. As $2 = 1 \cdot 1 + 1 \cdot 1$, we have that $C_2 = C_0 C_1 + C_1 C_0$. For $n = 3$, the number of Dyck paths from $(0,0)$ to $(6,0)$ is equal to 5. As $5 = 1 \cdot 2 + 1 \cdot 1 + 2 \cdot 1$, we have that $C_3 = C_0 C_2 + C_1 C_1 + C_2 C_0$.

Consider now a positive integer $n \geq 4$. Let us count all Dyck paths from $(0,0)$ to $(2n,0)$ using a certain recurrent procedure.

It is obvious that any Dyck path starting at $(0,0)$ and ending at $(2n,0)$ has exactly n right-up steps ↗ and exactly n right-down steps ↘. Moreover, it always starts with ↗ and ends with ↘. At last, in any of its subpaths starting at $(0,0)$, the number of ↗ is greater or equal to the number of ↘.

For any Dyck path from $(0,0)$ to $(2n,0)$, fix its first step; by construction, this is always a right-up step ↗. Then, let us find and fix its "pair," i.e., the first possible right-down step ↘ such that the substring, starting with the first step ↗ and ending with its pair-step ↘, is a Dyck path; so, the chosen pair of steps defines the shortest Dyck path, which can be obtained by walking along the considered path.

The fixed right-down step ↘ ends at some point $(2s+2, 0)$, where $0 \leq s \leq n-1$. In this case, all other right-up and right-down steps are divided into two groups. The first one, having s pairs of steps, is placed between ("in") two fixed steps. The second one, having $n - s - 1$ pairs of steps, is placed to the right ("out") of two fixed steps.

The first group can be arranged in some Dyck path (from $(1,1)$ to $(2s+2, 0)$ that never passes below the horizontal line $y = 1$) in C_s ways. The second group can be arranged in some Dyck path (from $(2s+2, 0)$ to $(2n, 0)$ that never passes below the OX-axis) in C_{n-s-1} ways. So, for a fixed $0 \leq s \leq n-1$, we have $C_s C_{n-s-1}$ possibilities. Therefore, in total, there are exactly $C_0 C_{n-1} + C_1 C_{n-2} + \cdots + C_s C_{n-s-1} + \cdots + C_{n-1} C_0$ possibilities.

In other words, we have proven that the number of Dyck paths from $(0,0)$ to $(2n,0)$ can be represented as the sum $C_0 C_{n-1} + C_1 C_{n-2} + \cdots + C_s C_{n-s-1} + \cdots + C_{n-1} C_0$. On the other hand, the number of such paths is, by definition, the nth Catalan number C_n.

So, we have proven that

$$C_n = C_0 C_{n-1} + C_1 C_{n-2} + \cdots + C_s C_{n-s-1} + \cdots + C_{n-1} C_0, \ n \geq 1. \ \square$$

For example, consider the algorithm of the proof for the number C_4. If $s = 0$, then there is $C_0 = 1$ possibility to construct a Dyck path from $(1, 1)$ to $(1, 1)$ between the fixed paths $\nearrow$ and $\searrow$, while there are $C_3 = 5$ possibilities to construct a Dyck path from $(2, 0)$ to $(8, 0)$ "outside" of the fixed paths $\nearrow$ and $\searrow$. Altogether, there are $C_0 \cdot C_3 = 1 \cdot 5 = 5$ possibilities.

If $s = 1$, then there is $C_1 = 1$ possibility to construct a Dyck path from $(1, 1)$ to $(3, 1)$ between the fixed paths $\nearrow$ and $\searrow$, while there are $C_2 = 2$ possibilities to construct a Dyck path from $(4, 0)$ to $(8, 0)$ "outside" of the fixed paths $\nearrow$ and $\searrow$. Altogether, there are $C_1 \cdot C_2 = 1 \cdot 2 = 2$ possibilities.

If $s = 2$, then there are $C_2 = 2$ possibilities to construct a Dyck path from $(1, 1)$ to $(5, 1)$ between the fixed paths $\nearrow$ and $\searrow$, while there is $C_1 = 1$ possibility to construct a Dyck path from $(6, 0)$ to $(8, 0)$ "outside" of the fixed paths $\nearrow$ and $\searrow$. Altogether, there are $C_2 \cdot C_1 = 2 \cdot 1 = 2$ possibilities.

If $s = 3$, then there are $C_3 = 5$ possibilities to construct a Dyck path from $(1, 1)$ to $(7, 1)$ between the fixed paths $\nearrow$ and $\searrow$, while there is $C_0 = 1$ possibility to construct a Dyck path from $(8, 0)$ to $(8, 0)$ "outside" of the fixed paths $\nearrow$ and $\searrow$. Altogether, there are $C_3 \cdot C_0 = 5 \cdot 1 = 5$ possibilities. So, we have shown that $C_4 = C_0 C_3 + C_1 C_2 + C_2 C_1 + C_3 C_0 = 5 + 2 + 2 + 5 = 14$.

A lot of other representations of Catalan numbers as the numbers of certain (often exotic) paths on the integer lattice are listed in [Stan13] and [Stan15].

Properties of Catalan numbers and lattice paths

5.2.4. Once we obtain the recurrent relation for Catalan numbers, we can construct the theory of this number set using a classical approach, which was described in detail in Chapter 3.

However, some considerations become really simple if we define Catalan numbers as the numbers of corresponding lattice paths.

Thus, many counting problems related to Catalan numbers can be solved by the construction of a simple one-to-one correspondence between the set of all Dyck paths from $(0,0)$ to $(2n,0)$ and the set under consideration. Let us give several examples:

- *The number of mountains with n ascents and n descents is equal to the nth Catalan number C_n.*

□ In this case, the one-to-one correspondence is obvious: $ascent \leftrightarrow \nearrow$, $descent \leftrightarrow \searrow$. □

- *The number of correct arrangements of n pairs of brackets is equal to the nth Catalan number C_n.*

□ The one-to-one correspondence is thus $(\leftrightarrow \nearrow,) \leftrightarrow \searrow$. □

- *The number of Dyck words of length $2n$ is equal to the nth Catalan number C_n.*

□ We have the one-to-one correspondence $X \leftrightarrow \nearrow, Y \leftrightarrow \searrow$ (see Chapter 3). □

- *The number of sequences, consisting of n "1" and n "−1", all partial sums of which are non-negative, is equal to the nth Catalan number C_n.*

□ The one-to-one correspondence is thus $1 \leftrightarrow \nearrow$, $-1 \leftrightarrow \searrow$. □

- *The number of plane rooted trees with n edges is equal to the nth Catalan number C_n.*

□ The one-to-one correspondence can be constructed as follows. For a given (planted) plane rooted tree with n edges, let us mark each of its edges with one pair of right-up and right-down steps, $\nearrow$ and $\searrow$, respectively, putting $\nearrow$ to the left and $\searrow$ to the right. Now, starting with the root and moving from the left to the right along the edges of the tree, we can collect all n pairs of steps, obtaining a corresponding Dyck path from $(0,0)$ to $(2n,0)$. □

5.2.5. Lattice path representations are also used in certain proofs of the closed formula for Catalan numbers:

$$C_n = \frac{1}{n+1}\binom{2n}{n}.$$

In this model, we consider the nth Catalan number as the number of Dyck paths from the point $(0,0)$ to the point (n,n) in the square $n \times n$ grid. They are paths with the set of steps $S = \{(1,0),(0,1)\}$, which lies strictly below (but may touch) the diagonal $y = x$.

If we now consider all possible paths from the point $(0,0)$ to the point (n,n) with the set of steps $S = \{(1,0),(0,1)\}$, i.e., the set of all NE paths in the square $n \times n$ grid, we can prove that there are exactly $\binom{2n}{n}$ such paths. In fact, any path has exactly n east steps and exactly n north steps; ordering these two sets (or equivalently choosing from $2n$ possible places n places for the symbol e), we obtain $P_{(n,n)} = \binom{2n}{n}$ variants.

Noting that all constructed NE lattice paths can be divided into $n+1$ groups of the same cardinality such that every group contains exactly one NE lattice path, which lies strictly below (but may touch) the diagonal $y = x$, we obtain the following result:

$$C_n = \frac{1}{n+1} P_{(n,n)} = \frac{1}{n+1}\binom{2n}{n}.$$

Exercises

1. For $n = 1, 2, 3, 4, 5, 6$, construct all Dyck paths from $(0,0)$ to $(2n, 0)$. Check that there are exactly C_n such paths.
2. For $n = 1, 2, 3, 4, 5$, construct all correct bracket sequences with n pairs of brackets. Establish the one-to-one correspondence between the obtained Dyck paths and the correct bracket sequences. Check that there are C_n such sequences.
3. For $n = 1, 2, 3, 4, 5$, construct all Dyck words of length $2n$. Establish the one-to-one correspondence between Dyck paths and Dyck words. Check that there are C_n such words.
4. For $n = 1, 2, 3, 4, 5$, construct all plane rooted trees with n edges. Establish the one-to-one correspondence between Dyck paths of

length $2n$ and plane rooted trees with n edges. Check that there are C_n such trees.
5. For $n = 1, 2, 3, 4, 5$, construct all correct $(1, -1)$-sequences of length $2n$. Establish the one-to-one correspondence between the Dyck paths of length $2n$ and the correct $(1, -1)$-sequences of length $2n$. Check that there are C_n such sequences.
6. Find a one-to-one correspondence between planted binary trees with n internal vertices and Dyck paths of length $2n$.

5.3. Motzkin Numbers on Integer Lattice

Path-related construction of Motzkin numbers

5.3.1. Define the nth *Motzkin number* M_n as the number of lattice paths from the point $(0, 0)$ to the point (n, n), with steps in $S = \{(2, 0), (0, 1), (1, 1)\}$, that never rise above the diagonal $y = x$. Let us call such paths *Motzkin's paths*.

Denoting a double right (east) step $(2, 0)$ by R, a double up (north) step $(2, 0)$ by U, and a diagonal (east-north) step $(1, 1)$ by d, we represent each Motzkin's path from $(0, 0)$ to (n, n) (i.e., in the square $n \times n$ grid) as a word, consisting of the symbols R, U, and d.

As there is exactly one such lattice path, d, from $(0, 0)$ to $(1, 1)$, we obtain that $M_1 = 1$.

As there are exactly two such lattice paths, dd and RU, from $(0, 0)$ to $(2, 2)$, we obtain that $M_2 = 2$.

As there are exactly three such lattice paths, ddd, dRU, and RUd, from $(0, 0)$ to $(3, 3)$, we obtain that $M_2 = 3$.

Moreover, it is natural to assume that there exists exactly one Motzkin's path from the point $(0, 0)$ to the same point $(0, 0)$. So, $M_0 = 1$.

5.3.2. There exists an equivalent representation of the nth Motzkin number as the number of certain lattice paths.

- The nth Motzkin number M_n is equal to the number of lattice paths from the point $(0,0)$ to the point $(n,0)$, with steps in $S = \{(1,0),(1,1),(1,-1)\}$, that never pass below the OX-axis.

□ In fact, these two approaches are equivalent. We should just replace any double right step $(2,0)$ with the right-up step $(1,1)$, replace any double up step $(0,2)$ with the right-down step $(1,-1)$, and replace any diagonal step $(1,1)$ with the right step $(1,0)$. This correspondence is quite clearly visible under the simple transformation of the considered square $n \times n$ grid, consisting of the rotation by $45°$ and scaling. □

Thus, instead of the path d from $(0,0)$ to $(1,1)$, we obtain the path → from $(0,0)$ to $(1,0)$. Instead of the paths dd and RU from $(0,0)$ to $(2,2)$, we obtain the paths →→ and ↗↘ from $(0,0)$ to $(2,0)$. Instead of the paths ddd, dRU, and RUd from $(0,0)$ to $(3,3)$, we obtain the paths →→→, →↗↘, and ↗↘→ from $(0,0)$ to $(3,0)$.

Recurrent relation for Motzkin numbers

5.3.3. Using the representation of Motzkin numbers as the number of lattice paths from the point $(0,0)$ to the point $(n,0)$, with steps in $S = \{(1,0),(1,1),(1,-1)\}$, that never pass below the OX-axis, we can obtain the classical recurrent relation for Motzkin numbers.

- Motzkin numbers can be calculated using the following recurrent formula:

$$M_0 = M_1 = 1;\ M_n = M_{n-1} + \sum_{k=0}^{n-2} M_k M_{n-2-k}.$$

□ It is already checked that, for $n = 0, 1, 2, 3$, the statement is true. In fact, counting Motzkin's paths from $(0,0)$ to $(n,0)$, $n = 0, 1$, we obtain that $M_0 = M_1 = 1$. For $n = 2$, the number of Motzkin's paths from $(0,0)$ to $(2,0)$ is equal to 2; as $2 = 1 + 1 \cdot 1$, we have that $M_2 = M_1 + M_0 M_0$. For $n = 3$, the number of Motzkin's paths from $(0,0)$ to $(3,0)$ is equal to 4; as $4 = 2 + 1 \cdot 1 + 1 \cdot 1$, we have that $M_3 = M_2 + M_0 M_1 + M_1 M_0$.

Consider now a positive integer $n \geq 4$. Let us count all Motzkin's paths from $(0,0)$ to $(n,0)$ using a certain recurrent procedure.

It is obvious that any Motzkin's path starting at $(0,0)$ and ending at $(n,0)$ has exactly k right-up steps ↗ and exactly k right-down steps ↘, where $0 \leq k \leq \lfloor \frac{n}{2} \rfloor$, and, for a fixed k, exactly $n - 2k$ horizontal steps. Moreover, it always starts with → or ↗ and always ends with → or ↘. At last, in any of its subpaths, starting at $(0,0)$, the number of ↗ is greater or equal to the number of ↘.

For any Motzkin's path from $(0,0)$ to $(n,0)$, let us fix its first step; by construction, this is a horizontal step →, or a right-up step ↗.

If the path starts with →, it is possible to construct M_{n-1} Motzlin's paths from $(1,0)$ to $(n,0)$; so, there are exactly M_{n-1} Motzkin's paths from $(0,0)$ to $(n,0)$, starting with →.

If the path starts with ↗, fix it, and find and fix its "pair," i.e., the first possible right-down step ↘, such that the substring, starting with the first step ↗ and ending with its pair-step ↘, is a Motzkin's path; so, the chosen pair of steps defines the shortest Motzkin's path, which can be obtained by walking along the considered path.

The fixed right-down step ↘ ends at the point $(s+2, 0)$, where $0 \leq s \leq n-2$. In this case, all other right, right-up, and right-down steps are divided into two groups. The first one is placed between ("in") two fixed steps. The second one is placed to the right ("out") of two fixed steps.

The first group can be arranged in some Motzkin's path (from $(1,1)$ to $(s+1,1)$ that never pass below the horizontal line $y = 1$) in M_s ways. The second group can be arranged in some Motzkin's path (from $(s+2, 0)$ to $(n,0)$, that never pass below the OX-axis) in M_{n-s-2} ways. So, for a fixed $0 \leq s \leq n-2$, we have $M_s M_{n-s-2}$ possibilities. Therefore, in total, there are exactly $M_0 M_{n-2} + M_1 M_{n-3} + \cdots + M_s M_{n-s-2} + \cdots + M_{n-2} M_0$ Motzkin's paths from $(0,0)$ to $(n,0)$, starting with ↗.

In other words, we have proven that the number of all Motzkin's paths from $(0,0)$ to $(n,0)$ can be represented as the sum $M_{n-1} + M_0 M_{n-2} + M_1 M_{n-3} + \cdots + M_s M_{n-s-2} + \cdots + M_{n-2} M_0$. On the other hand, the number of such paths is, by definition, the nth Motzkin number M_n. So, we have proven that

$$M_n = M_{n-1} + M_0 M_{n-2} + M_1 M_{n-3} + \cdots + M_s M_{n-s-2}$$
$$+ \cdots + M_{n-2} M_0, \ n \geq 2. \qquad \square$$

For example, consider the algorithm of the proof for the number M_4.

If a Motzkin's path starts with $\rightarrow$, we have $M_3 = 4$ possibilities to continue it from the point $(1, 0)$ to the point $(4, 0)$.

If the path starts with $\nearrow$, consider its "pair" $\searrow$, ending at $(s+2, 0)$.

If $s = 0$, then there is $M_0 = 1$ possibility to construct a Motzkin's path from $(1, 1)$ to $(1, 1)$ between the fixed steps $\nearrow$ and $\searrow$, while there are $M_2 = 2$ possibilities to construct a Dyck path from $(2, 0)$ to $(4, 0)$ "outside" of the fixed steps $\nearrow$ and $\searrow$. Altogether, there are $M_0 \cdot C_2 = 1 \cdot 2 = 2$ possibilities.

If $s = 1$, then there is $M_1 = 1$ possibility to construct a Motzkin's path from $(1, 1)$ to $(2, 1)$ between the fixed steps $\nearrow$ and $\searrow$, while there is $M_1 = 1$ possibility to construct a Motzkin's path from $(3, 0)$ to $(4, 0)$ "outside" of the fixed steps $\nearrow$ and $\searrow$. Altogether, there is $M_1 \cdot M_1 = 1 \cdot 1 = 1$ possibility.

If $s = 2$, then there are $M_2 = 2$ possibilities to construct a Motzkin's path from $(1, 1)$ to $(3, 1)$ between the fixed steps $\nearrow$ and $\searrow$, while there is $M_0 = 1$ possibility to construct a Motzkin's path from $(4, 0)$ to $(4, 0)$ "outside" of the fixed steps $\nearrow$ and $\searrow$. Altogether, there are $M_2 \cdot M_0 = 2 \cdot 1 = 2$ possibilities.

So, we have shown that $M_4 = M_3 + M_0 C_2 + M_1 M_1 + M_2 M_0 = 4 + 1 \cdot 2 + 1 \cdot 1 + 2 \cdot 1 = 9$.

Properties of Motzkin numbers and lattice paths

5.3.4. Once we obtain the recurrent relation for Motzkin numbers, we can construct the theory of this number set using a classical approach, which was described in detail in Chapter 4.

However, some considerations become really simple if we define Motzkin numbers as the numbers of corresponding lattice paths.

Thus, many counting problems related to Motzkin numbers can be solved by the construction of a simple one-to-one correspondence between the set of all Motzkin's paths from $(0,0)$ to $(n,0)$ and the set under consideration. Let us give several examples:

- *The number of mountains with plateaus of length n is equal to the nth Motzkin number M_n.*

☐ In this case, the one-to-one correspondence is obvious: *ascent* $\leftrightarrow \nearrow$, *descent* $\leftrightarrow \searrow$, *plateau* $\leftrightarrow \rightarrow$. ☐

- *The number of regular bracket sequences separated by zeros of length n is equal to the nth Motzkin number M_n.*

☐ The one-to-one correspondence is thus $(\leftrightarrow \nearrow,\ ) \leftrightarrow \searrow$, $0 \leftrightarrow \rightarrow$. ☐

- *The number of Dyck words with zeros of length n is equal to the nth Motzkin number M_n.*

☐ We have the one-to-one correspondence $X \leftrightarrow \nearrow$, $Y \leftrightarrow \searrow$, $0 \leftrightarrow \rightarrow$ (see Chapter 3). ☐

- *The number of correct $(1,-1)$-sequences, separated by zeros of length n, is equal to the nth Motzkin number M_n.*

☐ The one-to-one correspondence is thus $1 \leftrightarrow \nearrow$, $-1 \leftrightarrow \searrow$, $0 \leftrightarrow \rightarrow$. ☐

- *The number of plane rooted quasi-trees of length n is equal to the nth Motzkin number M_n.*

☐ The one-to-one correspondence can be constructed as follows. For a given (planted) plane rooted tree of length n, let us mark each of its edges with one pair of right-up and right-down steps, $\nearrow$ and $\searrow$, respectively, putting $\nearrow$ to the left and $\searrow$ to the right. Moreover, let us mark any loop with zero. Now, starting with the root and moving from the left to the right along the edges of the tree, we can collect all marks, obtaining a corresponding Motzkin's path from $(0,0)$ to $(n,0)$. ☐

5.3.5. Lattice path representations are also used in the classical proof of the closed formula for Motzkin numbers:

$$M_n = \sum_{k=0}^{\lfloor \frac{n}{2} \rfloor} \binom{n}{2k} C_k = \sum_{k=0}^{\lfloor \frac{n}{2} \rfloor} \frac{1}{k+1} \binom{n}{2k} \binom{2k}{k}.$$

In this model, we consider the nth Motzkin number as the number of Motzkin's paths from the point $(0,0)$ to the point $(n,0)$.

Any such path has k pairs of right-up $\nearrow$ and right-down $\searrow$ steps, $0 \le k \le \lfloor \frac{n}{2} \rfloor$, as well as $n - 2k$ horizontal $\rightarrow$ steps. For a fixed k, in order to obtain a Motzkin's path, we should just choose from n possible places exactly $2k$ places (it can be done in $\binom{n}{2k}$ ways) and construct on the chosen places any Dyck path (it can be done in C_k ways); on the remaining $n - 2k$ places, the horizontal steps will be used (the only possibility).

So, for a fixed k, we obtain $\binom{n}{2k} C_k$ possibilities. Thus, in total, there are $\sum_{k=0}^{\lfloor \frac{n}{2} \rfloor} \binom{n}{2k} C_k$ possibilities, i.e., there are exactly $\sum_{k=0}^{\lfloor \frac{n}{2} \rfloor} \binom{n}{2k} C_k$ Motzkin's paths from $(0,0)$ to $(n,0)$. As, by definition, this number is equal to the nth Motzkin number M_n, the statement is proven.

A similar combinatorial approach can be used for obtaining other equalities related to Motzkin numbers, including the recurrent relation $(n + 3)M_{n+1} = (2n + 3)M_n + 3nM_{n-1}$ ([Sula01]; see also Chapters 3 and 4).

Exercises

1. For $n = 1, 2, 3, 4, 5, 6, 7, 8$, construct all Motzkin's paths from $(0, 0)$ to $(n, 0)$. Check that there are exactly M_n such paths.
2. For $n = 1, 2, 3, 4, 5, 6, 7, 8$, construct all correct bracket sequences separated by zeros of length n. Establish the one-to-one correspondence between the obtained Motzkin's paths and the correct bracket sequences separated by zeros. Check that there are M_n such sequences.
3. For $n = 1, 2, 3, 4, 5, 6, 7, 8$, construct all Dyck words with zeros of length n. Establish the one-to-one correspondence between

Motzkin's paths and Dyck words with zeros. Check that there are M_n such words.
4. For $n = 1, 2, 3, 4, 5, 6, 7, 8$, construct all plane rooted quasi-trees of length n. Establish the one-to-one correspondence between Motzkin's paths of length n and plane rooted quasi-trees with n edges. Check that there are M_n such trees.
5. For $n = 1, 2, 3, 4, 5, 6, 7, 8$, construct all correct $(1, -1)$-sequences with zeros of length n. Establish the one-to-one correspondence between Motzkin's paths of length n and the correct $(1, -1)$-sequences with zeros of length n. Check that there are M_n such sequences.
6. Find a one-to-one correspondence between planted binary quasi-trees and Motzkin's paths.

5.4. Schröder Numbers on Integer Lattice

Path-related construction of Schröder numbers

5.4.1. Define the nth *Schröder number* S_n as the number of lattice paths from the point $(0,0)$ to the point (n,n), with steps in $S = \{(1,0), (0,1), (1,1)\}$, that never rise above the diagonal $y = x$. Let us call such paths *Schröder's paths*.

Denoting a right (east) step $(1,0)$ by r, an up (north) step $(1,0)$ by u, and a diagonal (east-north) step $(1,1)$ by d, we represent each Schröder's path from $(0,0)$ to (n,n) (i.e., in the square $n \times n$ grid) as a word, consisting of the symbols r, u, and d.

As there are exactly two such lattice paths, d and ru, from $(0,0)$ to $(1,1)$, we obtain that $S_1 = 1$.

As there are exactly six such lattice paths, dd, dru, rud, $ruru$, rdu, and $rruu$, from $(0,0)$ to $(2,2)$, we obtain that $S_2 = 6$.

As there are exactly 22 such lattice paths, ddd, $ddru$, $drud$, $druru$, $drdu$, $drruu$, $rudd$, $rudru$, $rurud$, $rururu$, $rurdu$, $rurruu$, $rdud$, $rduru$, $rruud$, $rruru$, $rrdu$, $rdruu$, $rrudu$, $rrruu$, $rrduu$, and $rrruuu$, from $(0,0)$ to $(3,3)$, we obtain that $S_2 = 3$.

Moreover, it is natural to assume that there exists exactly one Schröder's path from the point $(0,0)$ to the same point $(0,0)$. So, $S_0 = 1$.

5.4.2. There exists an equivalent representation of the nth Schröder number as the number of certain lattice paths.

- The nth Schröder number S_n is equal to the number of lattice paths from the point $(0,0)$ to the point $(2n,0)$, with steps in $S = \{(2,0),(1,1),(1,-1)\}$, that never pass below the OX-axis.

☐ In fact, these two approaches are equivalent. We should just replace any right step $(1,0)$ with the right-up step $(1,1)$, replace any up step $(0,1)$ with the right-down step $(1,-1)$, and replace any diagonal step $(1,1)$ with the double right step $(2,0)$. This correspondence is quite clearly visible under the simple transformation of the considered square $n \times n$ grid, consisting of the rotation by 45° and scaling: if we deal with the paths from $(0,0)$ to $(2n,0)$, the pair ↗↘ of right-up and right-down steps correspond to two right steps $(1,0)$; if we deal with the paths from $(0,0)$ to (n,n), the pair →↑ of right and up steps correspond to one diagonal step $(1,1)$. ☐

Thus, instead of the paths d and ru from $(0,0)$ to $(1,1)$, we obtain the paths →→ and ↗↘ from $(0,0)$ to $(2,0)$. Instead of the paths dd, dru, rud, $ruru$, rdu, and $rruu$ from $(0,0)$ to $(2,2)$, we obtain the paths →→→→, →→↗↘, ↗↘→→, ↗↘↗↘, ↗→→↘, and ↗↗↘↘ from $(0,0)$ to $(4,0)$. Instead of the paths ddd, $ddru$,..., $rrduu$, and $rrruuu$ from $(0,0)$ to $(3,3)$, we obtain the paths →→→→→→, →→→→↗↘,..., ↗↗→→↘↘, and ↗↗↗↘↘↘ from $(0,0)$ to $(6,0)$.

Recurrent relation for Schröder numbers

5.4.3. Using the representation of Schröder numbers as the number of lattice paths from the point $(0,0)$ to the point $(2n,0)$, with steps in $S = \{(2,0),(1,1),(1,-1)\}$, that never pass below the OX-axis, we can obtain the classical recurrent relation for Schröder numbers.

- Schröder numbers can be calculated using the following recurrent formula:

$$S_0 = 1, S_1 = 2; S_n = S_{n-1} + \sum_{k=0}^{n-1} S_k S_{n-k-1}.$$

☐ It is already checked that, for $n = 0, 1, 2, 3$, the statement is true. In fact, counting Schröder's paths from $(0,0)$ to $(2n,0)$, $n = 0, 1$, we obtain that $S_0 = 1$ and $S_1 = 2$. For $n = 2$, the number of Schröder's paths from $(0,0)$ to $(4,0)$ is equal to 6; as $6 = 2 + 1 \cdot 2 + 2 \cdot 1$, we have that $S_2 = S_1 + S_0 S_1 + S_1 S_0$. For $n = 3$, the number of Schröder's paths from $(0,0)$ to $(3,0)$ is equal to 22; as $22 = 6 + 1 \cdot 6 + 2 \cdot 2 + 6 \cdot 1$, we have that $S_3 = S_2 + S_0 S_2 + S_1 S_1 + S_2 S_0$.

Consider now a positive integer $n \geq 4$. Let us count all Schröder's paths from $(0,0)$ to $(2n,0)$ using a certain recurrent procedure.

It is obvious that any Schröder's path starting at $(0,0)$ and ending at $(2n,0)$ has exactly k right-up steps ↗ and exactly k right-down steps ↘, where $0 \leq k \leq n$, and, for a fixed k, exactly $n - k$ double horizontal steps. Moreover, it always starts with →→ or ↗ and always ends with →→ or ↘. At last, in any of its subpaths starting with $(0,0)$, the number of ↗ is greater or equal to the number of ↘.

For any Schröder's path from $(0,0)$ to $(2n,0)$, let us fix its first step; by construction, this is a double horizontal step →→, or a right-up step ↗.

If the path starts with →→, it is possible to construct S_{n-1} Schröder's paths from $(2,0)$ to $(2n,0)$; so, there are exactly S_{n-1} Schröder's paths from $(0,0)$ to $(2n,0)$, starting with →→.

If the path starts with ↗, fix it, and find and fix its "pair," i.e., the first possible right-down step ↘, such that the substring, starting with the first step ↗ and ending in its pair-step ↘, is a Schröder's path; so, the chosen pair of steps defines the shortest Schröder's path, which can be obtained by walking along the considered path.

The fixed right-down step ↘ ends at the point $(2s + 2, 0)$, where $0 \leq s \leq n - 1$. In this case, all other double right, right-up, and right-down steps are divided into two groups. The first one is placed between ("in") two fixed steps. The second one is placed to the right ("out") of two fixed steps.

382   Catalan Numbers

The first group can be arranged in some Schröder's path (from $(1,1)$ to $(2s+1, 1)$ that never pass below the horizontal line $y = 1$) in S_s ways. The second group can be arranged in some Schröder's path (from $(2s + 2, 0)$ to $(2n, 0)$, that never pass below the OX-axis) in S_{n-s-1} ways. So, for a fixed $0 \leq s \leq n - 1$, we have $S_s S_{n-s-1}$ possibilities. Therefore, in total, there are exactly $S_0 S_{n-1} + S_1 S_{n-2} + \cdots + S_s S_{n-s-1} + \cdots + S_{n-1} S_0$ Schröder's paths from $(0, 0)$ to $(2n, 0)$, starting with ↗.

In other words, we have proven that the number of all Schröder's paths from $(0, 0)$ to $(2n, 0)$ can be represented as the sum $S_{n-1} + S_0 S_{n-1} + S_1 S_{n-2} + \cdots + S_s S_{n-s-1} + \cdots + S_{n-1} S_0$. On the other hand, the number of such paths is, by definition, the nth Schröder number S_n. So, we have proven that

$$S_n = S_{n-1} + S_0 S_{n-S} + S_1 S_{n-2} + \cdots + S_s S_{n-s-1} + \cdots + S_{n-2} S_0,$$

$n \geq 2$. □

For example, consider the algorithm of the proof for the number S_4.

If Schröder's path starts with →→, we have $S_3 = 22$ possibilities to continue it from the point $(2, 0)$ to the point $(8, 0)$.

If the path starts with ↗, consider its "pair" ↘, ending at $(2s+2, 0)$.

If $s = 0$, then there is $S_0 = 1$ possibility to construct a Schröder's path from $(1, 1)$ to $(1, 1)$ between the fixed steps ↗ and ↘, while there are $S_3 = 22$ possibilities to construct a Schröder's path from $(2, 0)$ to $(8, 0)$ "outside" of the fixed steps ↗ and ↘. Altogether, there are $S_0 \cdot S_3 = 1 \cdot 22 = 22$ possibilities.

If $s = 1$, then there are $S_1 = 2$ possibilities to construct a Schröder's path from $(1, 1)$ to $(3, 1)$ between the fixed steps ↗ and ↘, while there are $S_2 = 6$ possibilities to construct a Schröder's path from $(4, 0)$ to $(8, 0)$ "outside" of the fixed steps ↗ and ↘. Altogether, there are $S_1 \cdot S_2 = 2 \cdot 6 = 12$ possibilities.

If $s = 2$, then there are $S_2 = 6$ possibilities to construct a Schröder's path from $(1, 1)$ to $(5, 1)$ between the fixed steps ↗ and ↘, while there are $S_1 = 2$ possibilities to construct a Schröder's path from $(6,0)$ to $(8,0)$ "outside" of the fixed steps $\nearrow$ and $\searrow$. Altogether, there are $S_2 \cdot S_1 = 6 \cdot 2 = 12$ possibilities.

If $s = 3$, then there are $S_3 = 22$ possibilities to construct a Schröder's path from $(1, 1)$ to $(7, 1)$ between the fixed steps $\nearrow$ and $\searrow$, while there is $S_0 = 1$ possibility to construct a Schröder's path from $(8, 0)$ to $(8, 0)$ "outside" of the fixed steps $\nearrow$ and $\searrow$. Altogether, there are $S_3 \cdot S_0 = 22 \cdot 2 = 22$ possibilities.

So, we have shown that $S_4 = S_3 + S_0 S_3 + S_1 S_2 + S_2 S_1 + S_3 S_0 = 22 + 1 \cdot 22 + 2 \cdot 6 + 6 \cdot 2 + 22 \cdot 1 = 90$.

Properties of Schröder numbers and lattice paths

5.4.4. Once we obtain a recurrent relation for Schröder numbers, we can construct the theory of this number set using a classical approach, which was described in detail in Chapter 4.

However, some considerations become really simple if we define Schröder numbers as the numbers of corresponding lattice paths.

Thus, many counting problems related to Schröder numbers can be solved by the construction of a simple one-to-one correspondence between the set of all Schröder's paths from $(0, 0)$ to $(2n, 0)$ and the set under consideration. Let us give several examples:

- *The number of mountains with superplateaus of length $2n$ is equal to the nth Schröder number S_n.*

□ In this case, the one-to-one correspondence is obvious: $ascent \leftrightarrow \nearrow$, $descent \leftrightarrow \searrow$, $superplateau \leftrightarrow \rightarrow\rightarrow$. □

- *The number of regular bracket sequences separated by pairs of zeros of length $2n$ is equal to the nth Schröder number S_n.*

□ The one-to-one correspondence is thus $(\leftrightarrow \nearrow,\ ) \leftrightarrow \searrow$, $00 \leftrightarrow \rightarrow\rightarrow$.□

- *The number of Dyck words with pairs of zeros of length $2n$ is equal to the nth Schröder number S_n.*

□ We have the one-to-one correspondence $X \leftrightarrow \nearrow$, $Y \leftrightarrow \searrow$, $00 \leftrightarrow \rightarrow\rightarrow$. (see Chapter 3). □

- *The number of correct $(1,-1)$-sequences, separated by pairs of zeros of length $2n$, is equal to the nth Schröder number S_n.*

□ The one-to-one correspondence is thus $1 \leftrightarrow \nearrow$, $-1 \leftrightarrow \searrow$, $00 \leftrightarrow \rightarrow\rightarrow$. □

- *The number of plane rooted quasi-trees with bows of length $2n$ is equal to the nth Schröder number S_n.*

□ The one-to-one correspondence can be constructed as follows. For a given (planted) plane rooted tree of length n, let us mark each of its edges with one pair of right-up and right-down steps, $\nearrow$ and $\searrow$, respectively, putting $\nearrow$ to the left and $\searrow$ to the right. Moreover, let us mark any pair of loops with $\rightarrow\rightarrow$. Now, starting with the root and moving from the left to the right along the edges of the quasi-tree, we can collect all marks, obtaining a corresponding Schröder's path from $(0,0)$ to $(2n,0)$. □

5.4.5. Lattice path representations are also used in the classical proof of the closed formula for Schröder numbers:

$$S_n = \sum_{k=0}^{n} \binom{n+k}{2k} C_k,$$

where $\binom{n}{m}$ are binomial coefficients and C_k are Catalan numbers.

□ Any Schröder's path from $(0,0)$ to $(2n,0)$ has several pairs of right-up and right-down steps, which, in turn, are correctly placed in the sequence of steps, i.e., in any initial subsequence, the number of $\nearrow$ is equal to or greater than the number of $\searrow$. In other words, the subsequence, formed only from right-up and right-down steps, forms a Dyck (Catalan's) path. It is easy to see that the number k of such pairs in a Schröder's path from $(0,0)$ to $(2n,0)$ satisfies the conditions $0 \le k \le n$. If there are k pairs of right-up and right-down steps, $0 \le k \le n$, then there are exactly $n-k$ double right steps. In order to obtain a Schröder's path from k pairs of right-up and right-down steps and $n-k$ pairs of $\rightarrow\rightarrow$, we should choose some places for $\rightarrow\rightarrow$. In can be done in $\binom{n+k}{n-k} = \binom{n+k}{2k}$ ways (see Chapter 4). The remaining $2k$ places will be used for a Dyck (Catalan's) path, consisting of k right-up and k right-down steps. On $2k$ places, we can order k right-up and k right-down steps in C_k ways, where C_k is the kth Catalan number. So, for a fixed $0 \leq k \leq n$, we have $\binom{n+k}{2k} C_k$ possible Schröder's paths. Therefore, the total number of Schröder's paths from $(0,0)$ to $(2n, 0)$ is $\sum_{k=0}^{n} \binom{n+k}{2k} C_k$. On the other hand, it was proven that the number of such sequences is equal to the nth Schröder number. So, we have proven that $S_n = \sum_{k=0}^{n} \binom{n+k}{2k} C_k$. □

A similar combinatorial approach can be used for obtaining other equalities related to Schröder numbers, including the recurrent relation $(n+2)S_{n+1} = (6n+3)S_n - (n-1)S_{n-1}$, $n \geq 1$ ([Sula01]; see also Chapters 3 and 4).

Exercises

1. For $n = 1, 2, 3, 4, 5$, construct all Schröder's paths from $(0,0)$ to $(2n, 0)$. Check that there are exactly S_n such paths.
2. For $n = 1, 2, 3, 4, 5$, construct all correct bracket sequences separated by pairs of zeros of length $2n$. Establish the one-to-one correspondence between the obtained Schröder's paths and the correct bracket sequences separated by pairs of zeros. Check that there are S_n such sequences.
3. For $n = 1, 2, 3, 4, 5$, construct all Dyck words with pairs of zeros of length $2n$. Establish the one-to-one correspondence between Schröder's paths and Dyck words with pairs of zeros. Check that there are S_n such words.
4. For $n = 1, 2, 3, 4, 5$, construct all plane rooted quasi-trees with bows of length $2n$. Establish the one-to-one correspondence between Schröder's paths of length $2n$ and plane rooted quasi-trees with bows of length $2n$. Check that there are S_n such trees.
5. For $n = 1, 2, 3, 4, 5$, construct all correct $(1, -1)$-sequences with pairs of zeros of length $2n$. Establish the one-to-one correspondence between Schröder's paths of length $2n$ and the correct $(1, -1)$-sequences with pairs of zeros of length $2n$. Check that there are S_n such sequences.
6. Find a one-to-one correspondence between planted binary quasi-trees with bows and Schröder's paths.

5.5. Narayana Numbers on Integer Lattice

Path-related construction of Narayana numbers

5.5.1. Define the *Narayana number* $N(n,k)$ as the number of lattice paths from the point $(0,0)$ to the point (n,n), with steps in $S = \{(1,0),(0,1)\}$, that never rise above the diagonal $y = x$ and have exactly k "lower right angles," consisting of two consecutive right and up steps. Let us call such paths *Dyck paths* (or *Catalan's paths*) *with k peaks*.

Denoting a right (east) step $(1,0)$ by r and an up (north) step $(1,0)$ by u, we represent each (n,k)-Narayana's path from $(0,0)$ to (n,n) (i.e., in the square $n \times n$ grid) as a word of length $2n$, consisting of the symbols r and u and having exactly k entries, ru.

As there is exactly one Dyck path, ru, from $(0,0)$ to $(1,1)$ and it has exactly one ru entry, we obtain that $N(1,1) = 1$.

As there are exactly two Dyck paths, $rruu$ and $ruru$, from $(0,0)$ to $(2,2)$, the first with one ru entry and the second with two ru entries, we obtain that $N(2,1) = 1, N(2,2) = 1$.

As there are exactly five Dyck paths, $rurruu$, $rururu$, $rrurru$, $rrruuu$, and $rruruu$, from $(0,0)$ to $(3,3)$, among which we have one ($rrruuu$) with one ru entry, three ($rurruu, rrurru,$ and $rruruu$) with two ru entries, and one ($rururu$) with three ru entries, we obtain that $N(3,1) = 1, N(3,2) = 3, N(3,3) = 1$.

Moreover, it is natural to assume that there exists exactly one lattice path from the point $(0,0)$ to the same point $(0,0)$, having zero ru entries. So, $N(0,0) = 1$, while $N(n,0) = 0$ for any positive integer n.

5.5.2. There exists an equivalent representation of the Narayana number $N(n,k)$ as the number of certain lattice paths.

- The Narayana number $N(n,k)$ is equal to the number of lattice paths from the point $(0,0)$ to the point $(2n,0)$, with steps in

$S = \{(1,1), (1,-1)\}$, *that never pass below the OX-axis and have exactly k peaks* ⟋⟍.

☐ In fact, these two approaches are equivalent. We should just replace any right step $(1,0)$ with the right-up step $(1,1)$ and replace any up step $(0,1)$ with the right-down step $(1,-1)$. The lower right angles, formed by two consecutive right and up steps, will be represented as the peaks ⟋⟍. ☐

Thus, instead of the path ru from $(0,0)$ to $(1,1)$, we obtain the path ⟋⟍ from $(0,0)$ to $(2,0)$. Instead of the paths $rruu$ and $ruru$ from $(0,0)$ to $(1,1)$, we obtain the paths ⟋⟋⟍⟍ and ⟋⟍⟋⟍ from $(0,0)$ to $(4,0)$. Instead of the paths $rurruu$, $rururu$, $rruuru$, $rrruuu$, and $rruruu$ from $(0,0)$ to $(3,3)$, we obtain the paths ⟋⟍⟋⟋⟍⟍, ⟋⟍⟋⟍⟋⟍, ⟋⟋⟍⟋⟍⟍, ⟋⟋⟋⟍⟍⟍, and ⟋⟋⟍⟋⟍⟍ from $(0,0)$ to $(6,0)$.

Recurrent relation for Narayana numbers

5.5.3. Using the representation of Narayana numbers as the number of lattice paths from the point $(0,0)$ to the point $(2n,0)$, with steps in $S = \{(1,1),(1,-1)\}$, that never pass below the OX-axis and have exactly k peaks, we can obtain a recurrent relation for Narayana numbers. However, the additional restriction, connected with parameter k, makes the considerations more difficult.

- *Narayana numbers can be calculated using the following recurrent formula:*

$$N(n+1,k) = N(n,k-1) + \sum_{t=1}^{n} \sum_{\substack{l+m=k,\\ l\geq 0, m\geq 0}} N(t,l)N(n-t,m),$$

$$N(0,0) = 1; N(n,0) = N(0,n) = 0 \text{ for } n \geq 1.$$

☐ Let us give the general proof, considering the Narayana number $N(n,k)$ as the number of Catalan's paths from the point $(0,0)$ to the point $(2n,0)$, having exactly k peaks.

388   Catalan Numbers

It is easy to check that the statement is true for small values of n. Assuming that the statement is true for all positive integers $k \leq n$, try to prove it for a given positive integer $n+1$:

I. If a path from the point $(0,0)$ to the point $(2n+2,0)$ starts with the first right-up and the second right-down steps, i.e., with the construction $\nearrow\searrow$, then we already have one peak, and the remaining part of the considered path should be a Catalan's path from $(2,0)$ to $(2n+2,0)$, containing exactly $k-1$ peaks. There are exactly $N(n, k-1)$ such possibilities.

II. If the path always starts with an up step $\nearrow$ and has the down step associated with the first up step, and it ends at the point $(2s, 0)$, $2 \leq s \leq n+1$, let us fix both of these steps. Then, we obtain between the points $(0,0)$ and $(2s,0)$ exactly $2s-1$ points, while for the "out" construction, we have $2(n-s+1)+1$ points, starting with the point $(2s, 0)$.

If we want to count the number of possible paths of this configuration, we should consider the number of possible "between" and "out" constructions. If a "between" construction has l peaks, $l \geq 1$, then the corresponding "out" construction should have $m = k-l$ peaks, $m \geq 0$ (zero is possible if $s = n+1$). For a fixed $1 \leq l \leq k$, we obtain $N(s-1, l)$ "between" possibilities and $N(n-(s-1), k-l)$ "out" possibilities. Denoting $t = s-1$ and $m = k-l$, we obtain the result.   □

For example, consider the algorithm of the proof for the number $N(4,3)$.

For any path counted by $N(4,3)$, starts with $\nearrow$, find its pair-step $\searrow$. It ends at the point $(2s+2,0)$, $s = 0, 1, 3$.

If $s = 0$, then we already have one peak $\nearrow\searrow$, and there are exactly $N(3,2) = 3$ possibilities to construct a Dick path with two peaks (from $(2,0)$ to $(8,0)$) "outside" of the fixed peak $\nearrow\searrow$.

If $s = 1$, then there is $N(1,1) = 1$ possibility to construct a Dyck path (from $(1,1)$ to $(3,1)$) between the fixed steps $\nearrow$ and $\searrow$ (it has one peak), while there are $N(2,2) = 3$ possibilities to construct a Dyck path (from $(4,0)$ to $(8,0)$) with two peaks "outside" of the fixed steps ↗ and ↘. Altogether, there is $N(1,1) \cdot N(2,2) = 1 \cdot 1 = 1$ possibility.

If $s = 2$, then there is $N(2,1) = 1$ possibility to construct a Dyck path (from $(1,1)$ to $(5,1)$) with one peak between the fixed paths ↗ and ↘, while there are no possibilities to construct a Dyck path (from $(6,0)$ to $(8,0)$) with two peaks "outside" of the fixed steps ↗ and ↘. Similarly, there are $N(2,1) = 1$ possibilities to construct a Dyck path (from $(1,1)$ to $(5,1)$) with two peaks between the fixed steps ↗ and ↘, while there are no possibilities to construct a Dyck path (from $(6,0)$ to $(8,0)$) without peaks "outside" of the fixed steps ↗ and ↘.

If $s = 3$, then there are $N(3,2) = 3$ possibilities to construct a Dyck path (from $(1,1)$ to $(7,1)$) with two peaks between the fixed steps ↗ and ↘, while there is $N(0,0) = 1$ possibility to construct a Dyck path (from $(8,0)$ to $(8,0)$) with zero peaks "outside" of the fixed steps ↗ and ↘. Altogether, there are $N(3,2) \cdot N(0,0) = 3 \cdot 1 = 3$ possibilities.

So, we have shown that $N(4,3) = N(3,2) + N(1,1)N(2,2) + N(3,2)N(0,0) = 3 + 1 \cdot 1 + 3 \cdot 1 = 6$.

Properties of Narayana numbers and lattice paths

5.5.4. Unfortunately, the obtained recurrent relation for Narayana numbers is complicated. However, we can construct the theory of this number set using a classical approach, based on path or bracket representations of Narayana numbers, which was described in detail in Chapter 4.

In fact, most of the results of the theory of Narayana numbers can be obtained using their lattice path representations.

Thus, many counting problems related to Narayana numbers can be solved by the construction of a simple one-to-one correspondence between the set of all Dyck paths from $(0,0)$ to $(2n,0)$ with k peaks and the set under consideration. Let us give several examples:

- *The number of mountains with n ascents, n descents, and k peaks is equal to the Narayana number $N(n,k)$.*

□ In this case, the one-to-one correspondence is obvious: ascent ↔ ↗, descent ↔ ↘; the peaks correspond to the peaks. □

- *The number of correct arrangements of n pairs of brackets with k nestings is equal to the Narayana number $N(n,k)$.*

□ The one-to-one correspondence is thus (↔ ↗,) ↔ ↘. The nestings () correspond to the peaks ↗↘. □

- *The number of Dyck words of length 2n with k entries XY is equal to the Narayana number $N(n,k)$.*

□ We have the one-to-one correspondence X ↔ ↗, Y ↔ ↘; the entries XY correspond to the peaks ↗↘. (see Chapter 3). □

- *The number of sequences, consisting of n "1" and n "−1" and having k 1 − 1 entries, all partial sums of which are non-negative, is equal to the Narayana number $N(n,k)$.*

□ The one-to-one correspondence is thus 1 ↔ ↗, −1 ↔ ↘. The entries 1 − 1 correspond to the peaks ↗↘. □

- *The number of plane rooted trees with n edges and k leaves is equal to the Narayana number $N(n,k)$.*

□ The one-to-one correspondence can be constructed as follows. For a given (planted) plane rooted tree with n edges, let us mark each of its edges with one pair of right-up and right-down steps, ↗ and ↘, respectively, putting ↗ to the left and ↘ to the right. Now, starting with the root and moving from the left to the right along the edges of the tree, we can collect all n pairs of steps, obtaining a corresponding Dyck path from $(0,0)$ to $(2n,0)$ with k peaks. The leaves correspond to the peaks ↗↘. □

5.5.5. Lattice path representations (or equivalently, the correct bracket arrangement representations) are also sufficiently used in the proof of the closed formula for Narayana numbers:

$$N(n,k) = \frac{1}{n}\binom{n}{k}\binom{n}{k-1} = \frac{1}{k}\binom{n}{k-1}\binom{n-1}{k-1}.$$

In this model, we consider the set of cardinality $\binom{n}{k-1}\binom{n-1}{k-1}$, consisting of certain infinite periodical paths related to the Dyck paths between $(0,0)$ and $(2n,0)$ with k peaks; each constructed sequence contains exactly one Dyck path of length $2n$, and this path has exactly k peaks. This set can be divided into k groups with equal cardinalities such that all sequences from a given group are equivalent, i.e., any group produces exactly one Dyck path of length $2n$ with k peaks. Therefore, the number of such paths is equal to $\frac{1}{k}\binom{n}{k-1}\binom{n-1}{k-1}$.

Exercises

1. For $n = 1, 2, 3, 4, 5, 6$ and $k = 1, 2, \ldots, n$, construct all Dyck paths from $(0,0)$ to $(2n,0)$, having k peaks. Check that there are exactly $N(n,k)$ such paths.
2. For $n = 1, 2, 3, 4, 5$ and $k = 1, 2, \ldots, n$, construct all correct bracket sequences with n pairs of brackets and k nestings. Establish the one-to-one correspondence between the obtained Dyck paths and the correct bracket sequences. Check that there are $N(n,k)$ such sequences.
3. For $n = 1, 2, 3, 4, 5$ and $k = 1, 2, \ldots, n$, construct all Dyck words of length $2n$ with k entries XY. Establish the one-to-one correspondence between Dyck paths and Dyck words. Check that there are $N(n,k)$ such words.
4. For $n = 1, 2, 3, 4, 5$ and $k = 1, 2, \ldots n$, construct all plane rooted trees with n edges and k leaves. Establish the one-to-one correspondence between Dyck paths of length $2n$ with k peaks and plane rooted trees with n edges and k leaves. Check that there are $N(n,k)$ such trees.
5. For $n = 1, 2, 3, 4, 5$ and $k = 1, 2, \ldots, n$, construct all correct $(1,-1)$-sequences of length $2n$ with k $1-1$ entries. Establish the one-to-one correspondence between Dyck paths of length $2n$ with k peaks, and the correct $(1,-1)$-sequences of length $2n$ with k $1-1$ entries. Check that there are $N(n,k)$ such trees.

6. Find a one-to-one correspondence between planted binary trees with n internal vertices and k "left" leaves and Dyck paths of length $2n$ with k peaks.

5.6. Delannoy Numbers on Integer Lattice

Path-related construction of Delannoy numbers

5.6.1. Define the *Delannoy number* $D(m, n)$ as the number of lattice paths from the point $(0, 0)$ to the point (m, n), with steps in $S = \{(1, 0), (0, 1), (1, 1)\}$. Let us call such paths *Delannoy's paths*.

If $m = n$, we obtain *central Delannoy numbers* $D(n) = D(n, n)$ that count Delannoy's paths from $(0, 0)$ to (n, n), i.e., in the square $n \times n$ grid.

Denoting a right (east) step $(1, 0)$ by r, an up (north) step $(1, 0)$ by u, and a right-up (northeast) step by d, we represent each Delannoy's path from $(0, 0)$ to (m, n) (i.e., in the rectangular $m \times n$ grid) as a word, consisting of the symbols r, u, and d.

It is easy to see that there exists exactly one Delannoy's path $rrr \cdots rrr$ from $(0, 0)$ to $(m, 0)$, i.e., $D(m, 0) = 1$, for any positive integer m. Similarly, there exists exactly one Delannoy's path $uuu \cdots uuu$ from $(0, 0)$ to $(0, n)$, i.e., $D(0, n) = 1$, for any positive integer n. Moreover, it is natural to assume that there exists exactly one lattice path from the point $(0, 0)$ to the same point $(0, 0)$. So, $D(0, 0) = 1$.

It is easy to check that there are exactly three Delannoy paths, ru, ur, and d, from $(0, 0)$ to $(1, 1)$; exactly five Delannoy paths, rru, rur, urr, dr, and rd, from $(0, 0)$ to $(2, 1)$; exactly five Delannoy paths, uur, uru, ruu, du, and ud, from $(0, 0)$ to $(1, 2)$; and there are exactly 13 Delannoy paths, rud, urd, dd, $rruu$, $ruru$, $urru$, dru, rdu, $uurr$, $urur$, $ruur$, dur, and udr, from $(0, 0)$ to $(2, 2)$.

5.6.2. There exists an equivalent representation of the Delannoy number $D(m, n)$ as the number of certain lattice paths.

- The Delannoy number $D(m,n)$ is equal to the number of lattice paths from the point $(0,0)$ to the point $(m+n, m-n)$, with steps in $S = \{(1,1), (2,0), (1,-1)\}$.

☐ In fact, these two approaches are equivalent. We should just replace any right step $(1,0)$ with the right-up step $(1,1)$, replace any up step $(0,1)$ with the right-down step $(1,-1)$, and replace any diagonal step $(1,1)$ with the double step $(2,0)$. ☐

If $m = n$, a Delannoy's path in a square $n \times n$ grid corresponds to a lattice path from $(0,0)$ to $(2n, 0)$.

- The central Delannoy number $D(n) = D(n,n)$ is equal to the number of lattice paths from the point $(0,0)$ to the point $(2n, 0)$, with steps in $S = \{(1,1), (2,0), (1,-1)\}$.

☐ The proof is similar to the corresponding proofs for Motzkin and Schröder numbers. ☐

Thus, instead of the paths ru, ur, and d from $(0,0)$ to $(1,1)$, we obtain the paths ↗↘, ↘↗, and →→ from $(0,0)$ to $(2,0)$. Instead of the paths rru, rur, urr, dr, and rd from $(0,0)$ to $(2,1)$, we obtain the paths ↗↗↘, ↗↘↗, ↘↗↗, →→↗, and ↗→→ from $(0,0)$ to $(3,1)$. Instead of the paths uur, uru, ruu, du, and ud from $(0,0)$ to $(1,2)$, we obtain the paths ↘↘↗, ↘↗↘, ↗↘↘, →→↘, and ↘→→ from $(0,0)$ to $(3,-1)$.

In the following figure, all 13 Delannoy's paths from the point $(0,0)$ to the point $(4,0)$ are represented: $D(2) = 13$.

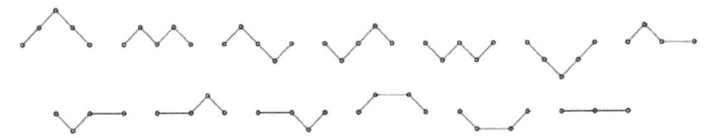

5.6.3. The nth central Delannoy number $D(n)$ can be interpreted as the number of *weighted paths* from $(0,0)$ to $(n,0)$ using the step $(1,1)$ with the weight 2, the step $(1,-1)$ with the weight 0, and the step $(1,0)$ with the weight 3. For example, $D(2) = 2 + 2 + 3 \cdot 3 = 13$ (see the following figure).

The nth central Delannoy number $D(n)$ can also be interpreted as a weighted path from $(0,0)$ to $(2n,0)$ using the steps $(1,1)$ and $(1,-1)$ with the weight 2.

For example, $D(2) = 2+4+2+2+2+1 = 13$ (see the following figure).

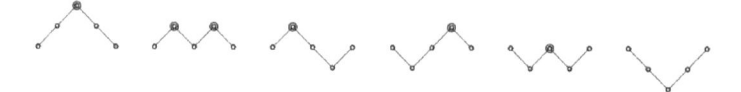

Recurrent relation for Delannoy numbers

5.6.4. Using the representation of Delannoy numbers as the number of lattice paths from the point $(0,0)$ to the point (m,n), with steps in $S = \{(1,1),(1,-1)\}$, we can obtain the classical recurrent relation for Delannoy numbers.

- Delannoy numbers can be calculated using the following recurrent formula:

$$D(m+1, n+1) = D(m+1, n) + D(m, n+1) + D(m, n),$$
$$D(0,0) = 1, D(m,0) = D(0,n) = 0 \text{ for } m, n \geq 1.$$

□ In fact, moving along a Delannoy's path from $(0,0)$ to $(m+1, m+1)$, we can reach the last step to the point $(m+1, n+1)$ from any of three possible positions: $(m+1, n), (m, n+1)$, or (m, n). Thus, the number $D(m+1, n+1)$ of Delannoy's paths from $(0,0)$ to $(m+1, n+1)$ is equal to the sum of the number $D(m+1, n)$ of Delannoy's paths leading to the point $(m+1, n)$, the number $D(m, n+1)$ of Delannoy's paths leading to the point $(m, n+1)$, and the number $D(m, n)$ of Delannoy's paths leading to the point (m, n). The initial conditions of this recurrent structure were discussed earlier: $D(0,0) = D(0,n) = D(m,0) = 1$, $m, n \in \mathbb{N}$. □

For example, consider the algorithm of the proof for the number $D(4,3)$.

For any path counted by $D(4,3)$, we should do the last step to the point $(4,3)$ from the point $(4,2)$, or from the point $(3,3)$, or from the point $(3,2)$. As there are exactly $D(4,2) = 41$ Delannoy's paths from $(0,0)$ to $(4,2)$, exactly $D(3,3) = 63$ Delannoy's paths from $(0,0)$ to $(3,3)$, and exactly $D(3,2) = 25$ Delannoy's paths from $(0,0)$ to $(3,2)$, we obtain that $D(4,3) = D(4,2) + D(3,3) + D(3,2) = 41 + 63 + 25 = 129$.

Properties of Delannoy numbers and lattice paths

5.6.5. The classical definition of Delannoy numbers is based on the lattice path representation; so, the theory of this number set uses the lattice path model as the classical approach, and almost all known properties of Delannoy numbers are proven using this model and with simple combinatorial reasoning, which were described in detail in Chapter 4.

Thus, many counting problems related to Delannoy numbers can be solved by the construction of a simple one-to-one correspondence between the set of all Delannoy paths from $(0,0)$ to (m,n) and the set under consideration.

In fact, the Delannoy bumber $D(m,n)$ is equal to the number of alignments of two sequences of lengths m and n; the number of points in an m-dimensional integer lattice located no further than n steps from the origin; the number of cells in an m-dimensional von Neumann neighborhood of radius (rank) n, etc. (see Chapter 4).

5.6.6. Lattice path representations are also sufficiently used in the proof of the closed formula for Delannoy numbers:

$$D(m,n) = \sum_{k=0}^{m} P_{(k,m-k,n-k)} = \sum_{k=0}^{m} \frac{(m+n-k)!}{k!(m-k)!(n-k)!}.$$

□ The proof is very simple. In order to obtain this statement, it is enough to note that, for a fixed number k of diagonal steps, any path between the points $(0,0)$ and (m,n) will contain exactly $m-k$ horizontal steps and exactly $n-k$ vertical steps.

A sequence, containing exactly k diagonal, $m-k$ horizontal, and $n-k$ vertical steps will uniquely define a Delannoy's path from $(0,0)$ to (m,n). The number of such sequences is the number $P_{(k,m-k,n-k)}$ of permutations with repetitions, in which the first element (diagonal step's symbol) is repeated k times, the second element (horizontal step's symbol) is repeated $m - k$ times, and the third element (vertical step's symbol) is repeated $n - k$ times. As $P_{(k, m-k, n-k)} = \frac{(m+n-k)!}{k!(m-k)!(n-k)!}$, the Delannoy number $D(m, n)$ can be calculated using the formula

$$D(m, n) = \sum_{k=0}^{m} P_{(k, m-k, n-k)} = \sum_{k=0}^{m} \frac{m+n-k)!}{k!(m-k)!(n-k)!}. \qquad \square$$

Similar considerations allow us to obtain closed formulas for generalized Delannoy numbers $D^k(m, n)$ and $D_k(m, n)$ (see Chapter 4):

$$D^k(m, n) = \sum_{j=0}^{n} C_n^j \cdot C_{m+n-j+k}^{n+k}, \quad D_k(m, n) = \sum_{j=0}^{n} C_n^j \cdot C_{m+n-j-k}^{n-k}.$$

A representation of the nth central Delannoy number $D(n)$ as a weighted path from $(0, 0)$ to $(2n, 0)$ using the steps $(1, 1)$ and $(1, -1)$ with weight 2 provides a combinatorial proof of the formula

$$D(n) = \sum_{k=0}^{n} 2^k \cdot (C_n^k)^2.$$

Exercises

1. For $n = 1, 2, 3, 4, 5, 6$ and $m = 1, 2, \ldots, n$, construct all Delannoy paths from $(0, 0)$ to (m, n). Check that there are exactly $D(m, n)$ such paths.
2. For $n = 1, 2, 3, 4, 5$ and $m = 1, 2, \ldots, n$, check that the Delannoy number $D(m, n)$ can be calculated using the formula $D(m, n) = \sum_{k=0}^{m} 2^k \cdot \binom{m}{k} \cdot \binom{n}{k}$, $m \geq n$. Prove this formula.
3. For $n = 1, 2, 3, 4, 5$ and $m = 1, 2, \ldots, n$, check that the Delannoy number $D(m, n)$ can be calculated using the formula $D(m, n) = \sum_{k=0}^{m} C_m^k \cdot C_{m+n-k}^m$. Prove this formula.
4. For $n = 1, 2, 3, 4, 5$, $m = 1, 2, \ldots, n$, and $k = 0, 1, 2$, find $D^k(m, n)$, the number of lattice paths from the point $(0, 0)$ to the point $(m, n+k)$ such that diagonal steps are forbidden after height n. Check that $D^k(m, n) = \sum_{j=0}^{n} C_n^j \cdot C_{m+n-j+k}^{n+k}$.

5. For $n = 1, 2, 3, 4, 5$, $m = 1, 2, \ldots, n$, and $k = 0, 1, 2$, find $D_k(m, n) = \sum_{j=0}^{n} C_n^j \cdot C_{m+n-j-k}^{n-k}$. Give a lattice path representation of the number $D_k(m, n)$.
6. For $n = 1, 2, 3, 4, 5, 6, 7, 8$, check that for central Delannoy numbers $D(n)$ the following recurrent relation holds: $nD(n) = 3(2n-1)D(n-1) - (n-1)D(n-2)$. Prove this recurrent relation.

5.7. Binomial Coefficients on Integer Lattice

Path-related construction of binomial coefficients

5.7.1. Define the *combination number* $C(m, n)$ as the number of lattice paths from the point $(0, 0)$ to the point (m, n), with steps in $S = \{(1, 0), (0, 1)\}$. Let us call such paths *combination paths*.

If $m = n$, we obtain the *central combination number* $C(n) = C(n, n)$ that counts the combination paths from $(0, 0)$ to (n, n), i.e., in the square $n \times n$ grid.

Denoting a right (east) step $(1, 0)$ by r and an up (north) step $(1, 0)$ by u, we represent each combination path from $(0, 0)$ to (m, n) (i.e., in the rectangular $m \times n$ grid) as a word, consisting of the symbols r and u.

It is easy to see that there exists exactly one combination path $rrr \cdots rrr$ from $(0, 0)$ to $(m, 0)$, i.e., $C(m, 0) = 1$, for any positive integer m. Similarly, there exists exactly one combination path $uuu \cdots uuu$ from $(0, 0)$ to $(0, n)$, i.e., $C(0, n) = 1$, for any positive integer n. Moreover, it is natural to assume that there exists exactly one combination path from the point $(0, 0)$ to the same point $(0, 0)$. So, $C(0, 0) = 1$.

It is easy to check that there are exactly two combination paths, ru and ur, from $(0, 0)$ to $(1, 1)$; exactly three combination paths, rru, rur, and urr, from $(0, 0)$ to $(2, 1)$; exactly three combination paths, uur, uru, and ruu, from $(0, 0)$ to $(1, 2)$; and exactly six combination paths, $rruu$, $ruru$, $urru$, $uurr$, $urur$, and $ruur$, from $(0, 0)$ to $(2, 2)$.

Starting with $C(0,0) = 1$, we can construct the following table.

```
1
1  6
1  5  15
1  4  10  20
1  3   6  10  15
1  2   3   4   5  6
1  1   1   1   1  1  1
```

It turns out that we obtain a Pascal's triangle, "starting" with the point $(0,0)$, corresponding to $C(0,0) = 1 = \binom{0}{0}$.

5.7.2. However, there exists another representation of the elements of Pascal's triangle as the number of certain lattice paths.

Define the *2-combination number* $C_{(m,n)}$ as the number of lattice paths from the point $(0,0)$ to the point (m,n), with steps in $S = \{(1,0),(1,1)\}$. Let us call such paths *2-combination paths*.

Denoting a right (east) step $(1,0)$ by r and a diagonal (northeast) step $(1,1)$ by d, we represent each 2-combination path from $(0,0)$ to (m,n) (i.e., in the rectangular $m \times n$ grid) as a word, consisting of the symbols r and d.

It is easy to see that there exists exactly one 2-combination path $rrr \cdots rrr$ from $(0,0)$ to $(m,0)$, i.e., $C_{(m,0)} = 1$, for any positive integer m. Similarly, there exists exactly one combination path $ddd \cdots ddd$ from $(0,0)$ to (n,n), i.e., $C_{(n,n)} = 1$, for any positive integer n. Moreover, it is natural to assume that there exists exactly one lattice path from the point $(0,0)$ to the same point $(0,0)$. So, $C_{(0,0)} = 1$. Note that it is impossible to reach the point (m,n) with $m < n$ using any 2-combinations path; so, by construction, $C_{(m,0)} = 0$ and $C_{(m,n)} = 0$ for any positive integers $m < n$.

It is easy to check that there are exactly two 2-combination paths, dr and rd, from $(0,0)$ to $(1,1)$; exactly three 2-combination's paths, rrd, rdr, and drr, from $(0,0)$ to $(3,1)$; exactly three 2-combination's paths, ddr, rdd, and drd, from $(0,0)$ to $(3,2)$; exactly four 2-combination's paths, $rrrdr$, $rrdr$, $rdrr$, and $drrr$, from $(0,0)$ to $(4,1)$, etc.

Starting with $C_{(0,0)} = 1$, we can construct the following table.

```
                  1
               1  6
            1  5  15
         1  4  10 20
      1  3  6  10 15
   1  2  3  4  5  6
1  1  1  1  1  1  1
```

We again obtain a Pascal's triangle, "starting" with the point $(0, 0)$, corresponding to $C(0, 0) = 1 = \binom{0}{0}$.

5.7.3. There exists an equivalent representation of the combination number $C(m, n)$ as the number of certain lattice paths.

- The combination number $C(m, n)$ is equal to the number of lattice paths from the point $(0, 0)$ to the point $(m + n, m - n)$, with steps in $S = \{(1, 1), (1, -1)\}$.

□ In fact, these two approaches are equivalent. We should just replace any right step $(1, 0)$ with the right-up step $(1, 1)$ and replace any up step $(0, 1)$ with the right-down step $(1, -1)$. □

If $m = n$, a combination path in the square $n \times n$ grid corresponds to a lattice path from $(0, 0)$ to $(2n, 0)$.

- The central combination number $C(n) = C(n, n)$ is equal to the number of lattice paths from the point $(0, 0)$ to the point $(2n, 0)$, with steps in $S = \{(1, 1), (1, -1)\}$.

Thus, instead of the paths ru and ur from $(0, 0)$ to $(1, 1)$, we obtain the paths ↗↘ and ↘↗ from $(0, 0)$ to $(2, 0)$. Instead of the paths rru, rur, and urr from $(0, 0)$ to $(2, 1)$, we obtain the paths ↗↗↘, ↗↘↗, and ↘↗↗ from $(0, 0)$ to $(3, 1)$. Instead of the paths uur, uru, and ruu from $(0, 0)$ to $(1, 2)$, we obtain the paths ↘↘↗, ↘↗↘, and ↗↘↘ from $(0, 0)$ to $(3, -1)$.

5.7.4. There also exists an equivalent representation of the 2-combination number $C_{(m,n)}$ as the number of certain lattice paths.

- The 2-combination number $C_{(m,n)}$ is equal to the number of lattice paths from the point $(0, 0)$ to the point $(m + n, m - n)$, with steps in $S = \{(1, 1), (2, 0)\}$.

☐ In fact, these two approaches are equivalent. We should just replace any right step $(1,0)$ with the right-up step $(1,1)$ and replace any diagonal step $(1,1)$ with the double right step $(2,0)$. ☐

If $m = n$, the only "diagonal" 2-combination path in the square $n \times n$ grid corresponds to the only "horizontal" lattice path from $(0,0)$ to $(2n,0)$. If $n = 0$, the only "horizontal" 2-combination path in the square $n \times n$ grid corresponds to the only "diagonal" lattice path from $(0,0)$ to (n,n).

Thus, instead of the paths rd and dr from $(0,0)$ to $(2,1)$, we obtain the paths ↗→→ and →→↗ from $(0,0)$ to $(3,1)$. Instead of the paths rrd, rdr, and drr from $(0,0)$ to $(3,1)$, we obtain the paths ↗↗→→, ↗→→↗, and →→↗↗ from $(0,0)$ to $(4,2)$. Instead of the path dd from $(0,0)$ to $(2,2)$, we obtain the path →→→→ from $(0,0)$ to $(4,0)$. Instead of the path rr from $(0,0)$ to $(2,0)$, we obtain the path ↗↗ from $(0,0)$ to $(2,2)$.

Recurrent relation for binomial coefficients

5.7.5. Using the representation of combination numbers as the number of lattice paths from the point $(0,0)$ to the point (m,n), with steps in $S = \{(1,1),(1,-1)\}$, we can obtain a recurrent relation for combination numbers and prove that, indeed, any combination number can be interpreted as a binomial coefficient.

- *Combination numbers can be calculated using the following recurrent formula:*

$$C(m+1, n+1) = C(m+1, n) + C(m, n+1), \ C(0,0) = 1,$$
$$C(m, 0) = C(0, n) = 0 \ \text{for} \ m, n \geq 1.$$

☐ In fact, moving along a combination path from $(0,0)$ to $(m+1, n+1)$, we can reach the last step to the point $(m+1, n+1)$ from any of two possible positions: $(m+1, n)$ or $(m, n+1)$. Thus, the number $C(m+1, n+1)$ of combination paths from $(0,0)$ to $(m+1, n+1)$ is equal to the sum of the number $C(m+1, n)$ of combination paths leading to the point $(m+1, n)$ and the number

$C(m, n + 1)$ of combination paths leading to the point $(m, n + 1)$. The initial conditions of this recurrent structure were discussed earlier: $C(0,0) = C(0,n) = C(m,0) = 1$, $m, n \in \mathbb{N}$. □

For example, consider the algorithm of the proof for the number $C(4,3)$.

For any path counted by $C(4,3)$, we should do the last step to the point $(4,3)$ from the point $(4,2)$ or from the point $(3,3)$. As there are exactly $C(4,2) = 15$ combination paths from $(0,0)$ to $(4,2)$ and exactly $C(3,3) = 20$ combination paths from $(0,0)$ to $(3,3)$, we obtain that $C(4,3) = C(4,2) + C(3,3) = 15 + 20 = 35$.

Now, we can prove that any combination number is a binomial coefficient:

- The combination number $C(m,n)$ is equal to the number of m-combinations from an $(m+n)$-set or, equivalently, to the binomial coefficient $\binom{m+n}{m}$:

$$C(m,n) = C^m_{m+n} = \binom{m+n}{m}.$$

□ In fact, it is well known that the numbers C^m_{m+n} satisfy the following recurrent relation:

$$C^m_{m+n} = C^m_{m+n-1} + C^{m-1}_{m+n-1}.$$

This relation coincides with the above-obtained recurrent relation for the numbers $C(m,n)$: $C(m,n) = C(m, n-1) + C(m-1, n)$. As $C^0_0 = C^m_m = C^0_n = 1$, the initial conditions for the numbers C^m_{m+n} coincide with the initial conditions $C(0,0) = C(m,0) = C(0,n) = 1$ for the numbers $C(m,n)$. So, it holds that

$$C(m,n) = C^m_{m+n}, \quad m, n \geq 0.$$

The second statement is obvious. □

5.7.6. Using the representation of 2-combination numbers as the number of lattice paths from the point $(0,0)$ to the point (m,n), with steps in $S = \{(1,0), (1,1)\}$, we can obtain a recurrent relation for 2-combination numbers and prove that, indeed, any 2-combination number can be interpreted as a binomial coefficient.

- 2-combination numbers can be calculated using the following recurrent formula:

$$C_{(m+1,n+1)} = C_{(m,n+1)} + C_{(m,n)}, \quad C_{(0,0)} = C_{(m,0)} = C_{(n,n)} = 1$$

for $m, n \geq 1$.

☐ In fact, moving along a 2-combination's path from $(0,0)$ to $(m+1, n+1)$, we can reach the last step to the point $(m+1, n+1)$ from any of two possible positions: $(m, n+1)$ or (m, n). Thus, the number $C_{(m+1,n+1)}$ of 2-combination paths from $(0,0)$ to $(m+1, n+1)$ is equal to the sum of the number $C_{(m,n+1)}$ of 2-combination paths leading to the point $(m, n+1)$ and the number $C_{(m,n)}$ of 2-combination paths leading to the point (m, n). The initial conditions of this recurrent structure were discussed earlier: $C_{(0,0)} = 1$, and $C_{(m,0)} = C_{(n,n)} = 1$ for $m, n \in \mathbb{N}$. Moreover, $C_{(m,n)} = 0$ for $m < n$. ☐

For example, consider the algorithm of the proof for the number $C_{(4,3)}$.

For any path counted by $C_{(4,3)}$, we should do the last step to the point $(4, 3)$ either from the point $(3, 3)$ or from the point $(3, 2)$. As there are exactly $C_{(3,3)} = 1$ 2-combination paths from $(0, 0)$ to $(3, 3)$ and exactly $C_{(3,2)} = 3$ 2-combination paths from $(0, 0)$ to $(3, 2)$, we obtain that $C_{(4,3)} = C_{(3,3)} + C_{(3,2)} = 1 + 3 = 4$.

Now, we can prove that any 2-combination number is a binomial coefficient.

- The 2-combination number $C_{(m,n)}$ is equal to the number of n-combinations from an m-set or, equivalently, to the binomial coefficient $\binom{m}{n}$:

$$C_{(m,n)} = C_m^n = \binom{m}{n}.$$

☐ In fact, it is well known that the numbers C_m^n satisfy the following recurrent relation:

$$C_m^n = C_{m-1}^n + C_{m-1}^{n-1}.$$

This relation coincides with the obtained recurrent relation for the numbers $C_{(m,n)}$: $C_{(m,n)} = C_{(m-1,n)} + C_{(m-1,n-1)}$. As $C_0^0 = C_m^0 = C_n^n = 1$, the initial conditions for the numbers

C_m^n coincide with the initial conditions $C_{(0,0)} = C_{(m,0)} = C_{(n,n)} = 1$ for the numbers $C_{(m,n)}$. So, it holds that

$$C_{(m,n)} = C_m^n, m \geq n \geq 0.$$

The second statement is obvious. □

Properties of binomial coefficients and lattice paths

5.7.7. Classical combinatorics is based on a large set of combinatorial methods, and the lattice path representations of the numbers C_n^k of k-combinations of elements from an m-set are used relatively rarely.

However, there are many combinatorial identities that can be obtained using the lattice path representations of the numbers C_n^m.

For example, using the lattice path representation, it is easy to prove that:

- the sum of elements of the nth row of Pascal's triangle is 2^n:

$$C_n^0 + C_n^1 + C_n^2 + \cdots + C_n^n = 2^n.$$

□ In terms of the lattice paths, the sum $C_n^0 + C_n^1 + C_n^2 + \cdots + C_n^n$ is the number of all combination paths, leading to points $(0, n)$, $(1, n-1), (2, n-2), \ldots, (n-2, 2), (n-1, 1), (n, 0)$ of the diagonal $y = n - x$. Any such path contains exactly n steps, which can be chosen from the set $\{\rightarrow, \uparrow\}$. In other words, we have exactly two possibilities for each step; altogether, there are 2^n possibilities. □

Similarly, using the lattice path representation, we can prove that

$$C_{m+n}^k = C_n^0 C_m^k + C_n^1 C_m^{k-1} + C_n^2 C_m^{k-2} + \cdots + C_n^{k-2} C_m^2 + C_n^{n-1} C_m^1 + C_n^k C_m^0.$$

□ The number C_{m+n}^k is the number of paths from the point $(0,0)$ to the point $(m+n-k, k)$ of the diagonal $y = m+n-x$. Each such path passes through exactly one point $(n-t, t)$ of the diagonal $y = n - x$. Let us fix a point $(n-t, t)$ of the diagonal $y = n - x$. There are exactly C_n^t paths from $(0,0)$ to $(n-t, t)$. Moreover, there are exactly C_m^{k-t} paths from $(n-t, t)$ to $(m+n-k, k)$. So, the total number C_{m+n}^k of paths from the point $(0,0)$ to the point $(m+n-k, k)$ is equal to the sum $\sum_{t=0}^k C_n^t C_m^{k-t}$. □

Exercises

1. For $m = 1, 2, 3, 4, 5, 6$ and $n = 1, 2, \ldots, m$, construct all combination paths from $(0,0)$ to (m,n). Check that there are exactly C_{m+n}^m such paths.
2. For $m = 1, 2, 3, 4, 5$ and $n = 1, 2, \ldots, m$, construct all 2-combination paths from $(0,0)$ to (m,n). Check that there are exactly C_m^n such paths.
3. Using the lattice path representations of k-combination of elements of an n-set, prove that $C_n^k = C_n^{n-k}$.
4. Using the lattice path representations of k-combinations of elements of an n-set, prove that $C_n^k = C_{n-1}^{k-1} + C_{n-2}^{k-2} + \cdots + C_{k-1}^{k-1}$.
5. Using the lattice path representation of k-combinations of elements of an n-set, prove that C_n^k is the number of positive integer solutions of the equation $x_1 + x_2 + \cdots + x_k + x_{k+1} = n$.
6. Find a lattice path representation for the trinomial coefficients $P_{(n_1, n_2, n_3)}$ and for the polynomial coefficients $P_{(n_1, n_2, \ldots, n_k)}$.

References

[AbSt72], [Ande03], [Apos86], [Berg71], [BeSl95], [BiBa70], [Cofm75], [Coke03], [CoGu96], [Comt74], [Dave47], [Deza17], [Deza21], [Deza24], [DeMo10], [Dick05], [DZHKKP90], [FlSe09], [Gard89], [Gelf98], [GKP94], [Hara03], [HaPa77], [Knut97], [Knut76], [Kost82], [KoKo77], [Mazu10], [Ore48], [OsJe15], [Rior80], [Rose18], [Sier64], [SlPl95], [Sloa24], [Stan97], [Stra16], [Stru87], [Weis24], [Wiki24], [Went99].

Chapter 6
Zoo of Numbers

In this chapter, we collect some remarkable individual special numbers related to the Catalan numbers (see [Abra40], [Boro85], [CoGu96], [CoRo96], [DeDe12], [Deza17], [Deza18], [Deza21], [Deza23], [Deza24], [Dick05], [GKP94], [LeLi83], [Line86], [Litz63], [Mada79], [Malc86], [OLBC10], [Plat80], [Prim24], [RaTo57], [Ribe96], [Rybn82], [Rybn85], [Sier64], [SlPl95], [Sloa24], [Weis24], [Well86], [Wiki24], etc.).

- **1**: empty product; $0! = 1$; $\binom{0}{0} = 1$; $C_0 = C_1 = 1$; $M_0 = M_1 = 1$; $S_0 = 1$; $D(0) = D(0,0) = 1$; $N(0,0) = 1$; $S(0,0) = 1$; $s(0,0) = 1$; $B(0) = 1$.
- **2**: the only even prime; the value C_2; the value M_2; the value S_1; the value $B(2)$; the value $2!$; the first non-trivial Catalan number, Motzkin number, Schröder number, and Bell number.
- **3**: the first odd prime; the value $D(1,1)$, the first non-trivial Delannoy number; the value $N(3,2)$, the first non-trivial Narayana number; the value $S(3,2)$, the first non-trivial Stirling number of the second kind; the first Fermat prime, F_0; the first Mersenne prime, $\mathcal{M}_2$.
- **4**: the first composite element in Pascal's triangle; the value M_3.
- **5**: the value C_4; the value $B(4)$; the value $D(2,1)$; the only prime digit in which a perfect square can end; the only prime whose square is composed of only prime digits.

- **6:** the value S_3; the first perfect number; the first composite element in Narayana's triangle; the first composite element in the triangle of Stirling numbers of the second kind; the first composite element in the triangle of unsigned Stirling numbers of the first kind; the first composite factorial, 3!; the first non-trivial octagonal number; the only mean of a pair of twin primes which is triangular ($\frac{5+7}{2}$); the largest known number n such that there are n integers for which all pairwise sums are perfect squares.
- **7:** the second Mersenne prime, $\mathcal{M}_3$; the value $D(3,1)$; the only prime p such that $p+1$ is a perfect cube; the only prime equal to the difference between the product and the sum of the two previous primes; the biggest (besides 4 and 5) known solution to *Brocard's problem*: to find integers n such that $n!+1$ is a perfect square; the largest known prime that is not the sum of a triangular number, a square, and a cube, all of them positive.
- **8:** the first non-trivial hyperoctahedral number: the largest Fibonacci number of the form $p+1$ or $p-1$ for a prime p; the largest composite number such that all its proper divisors plus 1 are primes.
- **9:** the value $\mathcal{M}_4$; the only one-dimensional square that is a Delannoy number, $D(4,1)$.
- **10:** the first two-dimensional Narayana number; the first non-trivial four-dimensional hyperoctahedron number; the only known integer of the form $n!m!$, excluding $n = k! = m+1$ (10! = 7!6!).
- **11:** the first two-dimensional Schröder–Hipparchus number, $SH(3)$; the first two-dimensional Delannoy number, $D(5,1)$; the smallest prime repuint; the largest integer that cannot be expressed as a sum of at least two distinct primes; the largest number that is not expressible as a sum of two composite numbers.
- **14:** the value C_4; the first non-trivial seven-dimensional hypertetrahedron number.
- **15:** the smallest composite Bell number, $B(4)$; the smallest two-digit composite Delannoy number, $D(7,1)$; the smallest Mersenne composite, $\mathcal{M}_4$.
- **17:** the value $D(8,1)$; the only prime that is the average of two consecutive Fibonacci numbers ($\frac{13+21}{2}$); the only prime that is the

sum of four consecutive primes ($17 = 2+3+5+7$); the only prime of the form $p^q + q^p$, where p and q are primes: $17 = 2^3 + 3^2$; the only number n with n partitions of prime parts; the largest, if Goldbach's conjecture is true, integer that is not the sum of three distinct primes; the smallest prime whose sum of the digits is a cubic number; the only known prime that is equal to the sum of the digits of its cube: $17^3 = 4913$, and $4+9+1+3 = 17$.
- **21**: the value M_5; the first non-trivial octagonal number.
- **22**: the value S_4; the maximum number of regions into which five intersecting circles divide the plane.
- **23**: the smallest odd prime that is not a twin prime; the largest integer n such that no factor of a binomial coefficient $\binom{n}{k}$ is a perfect square; the biggest prime, besides $2, 3, 5, 7$, and 11, that is uniquely expressible as a sum of at most four squares.
- **24**: the first two-digit composite Stirling number of the first kind, $s(5,1)$; the sum of the twin primes 11 and 13; the only integer $n > 1$ such that $\sum_{i=1}^{n} i^2$ is a perfect square: $\sum_{i=1}^{24} i^2 = 70^2$.
- **25**: the first two-digit Delannoy number, $D(3,3)$, which is a square number; the first non-trivial Stirling number of the second kind, $S(5,3)$, which is a square number; the only perfect square of the form $k^3 - 2$.
- **26**: the largest integer n such that the segment $[m, m+100]$ contains n primes (it happens only for $m = 2$).
- **28**: the second perfect number; the only perfect number of the form $n^k + m^k$ with $k > 1$: $28 = 3^3 + 1^3$.
- **29**: the smallest prime equal to the sum of three consecutive squares: $29 = 2^2 + 3^2 + 4^2$; the smallest multi-digit prime which, on adding its reverse, gives a perfect square: $29 + 92 = 11^2$; the only *non-titanic prime* (i.e., with less than 1,000 decimal digits) of the form $p^p + 2$.
- **31**: the third Mersenne prime, $\mathcal{M}_5$; the smallest two-digit prime Stirling number of the second kind, $S(6,2)$; the smallest prime that can be represented as a sum of two triangular numbers in two different ways: $31 = 21+10 = 28+3$; the smallest prime that can be represented as a sum of two triangular numbers with prime indices; there are only 31 numbers that cannot be

expressed as a sum of distinct squares; $3 + 5 + 7 + 11 + \cdots + 89 = 31^2$, i.e., a sum of the first 31 odd primes is a square of this prime.

- **33**: the smallest odd repdigit that is not a prime number; the largest integer that is not a sum of distinct triangular numbers.
- **36**: the first non-trivial square Lah number, $L(4,2)$; the smallest perfect square expressible as a sum of four consecutive primes, which are also two pairs of prime twins: $36 = 5 + 7 + 11 + 13$; the smallest triangular number whose sum of divisors as well as the sum of its proper divisors are also triangular numbers.
- **41**: the smallest prime whose cube can be written as the sum of three cubes in two ways: $41^3 = 40^3 + 17^3 + 2^3 = 33^3 + 32^3 + 6^3$; the sum of two consecutive squares: $4^2 + 5^2$.
- **42**: the value C_5; the smallest number n such that n^2 is the mean of cubed twin primes, $42^2 = \frac{11^3 + 13^3}{2}$.
- **51**: the value M_6; the product of the distinct Fermat primes 3 and 17; a regular polygon with 51 sides is constructible with a compass and a straightedge.
- **52**: the value $B(5)$; a vertically symmetrical number; an untouchable number since it is never the sum of proper divisors of any number.
- **53**: the largest known integer that can be expressed as a sum of three non-negative triangular numbers in exactly one way.
- **65**: the only number that gives a square of a prime when adding its reverse to it as well as subtracting its reverse from it: $65 + 56 = 11^2$, $65 - 56 = 3^2$; the only number that is the difference $(3^4 - 2^4)$ between two biquadratic numbers with prime indices; $(65!)^2 + 1$ is a prime.
- **67**: the smallest multi-digit prime whose square, 4489, and cube, 300763, consist of different digits.
- **83**: the largest known prime that can be expressed as a sum of three positive triangular numbers in exactly one way; the only prime equal to a sum of squares of odd primes: $83 = 3^2 + 5^2 + 7^2$; the only prime of the form $p^4 + 2$, where p is a prime.
- **89**: the smallest positive integer whose square, 7921, and cube, 704969, are likewise primes upon reversal.

- **90:** the value S_4; the value $S(6,3)$; expressible as the sum of distinct non-zero squares in six ways, more than any smaller number.
- **100:** the smallest perfect square whose summation of the differences between itself and each of its digits, where each difference is raised to the power of the corresponding digit, is equal to a prime: $101 = (100-1)^1 + (100-0)^0 + (100-0)^0$ is a prime.
- **109:** the smallest number (coincidentally prime) that has more distinct digits than its square, 11881.
- **113:** the smallest prime that is a sum of three biquadratic numbers with prime indices: $113 = 2^4 + 2^4 + 3^4$.
- **121:** the only perfect square of the form $1+p+p^2+p^3+p^4$, $p \in P$: $121 = 1 + 3 + 3^3 + 3^3 + 3^4$; the only perfect square, besides 4, of the form $n^3 - 4$ ($121 = 5^3 - 4$).
- **127:** the value M_7; the fourth Mersenne prime, $\mathcal{M}_7$; the smallest three-digit prime Stirling number of the second kind, $S(8,2)$; the exponent for the 12th Mersenne prime, $\mathcal{M}_{127} = 2^{127} - 1$ (it is the largest prime ever discovered by hand calculations, as well as the largest known double Mersenne prime); the smallest prime that can be written as the sum of the first two or more odd primes ($127 = 3 + 5 + 7 + 11 + 13 + 17 + 19 + 23 + 29$).
- **132:** the value C_6; the 99th composite number; the number of irreducible trees with 15 vertices.
- **144:** the largest, besides 0 and 1, perfect square that is a Fibonacci number; the sum of a twin prime pair (71; 73).
- **149:** the only known prime in the concatenate square sequence.
- **173:** the largest known prime whose square, 29929, and cube, 5177717, consist of different digits.
- **203:** the value $B(6)$; 203 different triangles can be made from three rods with integer lengths of at most 12.
- **211:** the largest known prime that cannot be written as the sum of a prime and a positive triangular number.
- **239:** the largest integer, besides 23, that is not a sum of less than nine cubic numbers: $239 = 2 \cdot 4^3 + 4 \cdot 3^3 + 3 \cdot 1^3$.
- **257:** the fourth Fermat prime, F_3; the largest prime in a sequence of 15 primes of the form $2n + 17$, where n runs through the first 15 triangular numbers.

- **289:** the square of the sum of the first four primes: $289 = (2+3+5+7)^2$.
- **323:** the value M_8; a semiprime, $323 = 17 \cdot 19$; the sum of nine consecutive primes $(19+\cdots+53)$; the sum of 13 consecutive primes $(5 + \cdots + 47)$.
- **343:** the only cubic number (7^3), besides 1, such that the sum of its divisors is a perfect square: $1 + 7 + 7^2 + 7^3 = 20^2$.
- **367:** the largest number (in fact, a prime) whose square, 134689, has strictly increasing digits.
- **394:** the value S_5; a semiprime, $394 = 2 \cdot 197$.
- **400:** the only known square of the form $1 + k + k^2 + k^3$, where $k \in \mathbb{N}\setminus\{1\}$ (in fact, $k = 7$).
- **407:** the largest integer, besides $1, 153, 370$, and 371, that is the sum of the cubes of their decimal digits.
- **429:** the value C_7; $429 = 3 \cdot 11 \cdot 13$, a Sphenic number.
- **463:** the smallest multi-digit prime such that both the sum of the digits and the product of the digits of its square remain squares.
- **496:** the third perfect number; the perfect Stirling number of the second kind, $S(32, 31)$; the smallest triangular number such that the sum of the cubes of its digits is prime: $4^3 + 9^3 + 6^3 = 1009$.
- **541:** the 100th prime number, the 10th hexagonal star (i.e., Star of David) number.
- **576:** the only known perfect square that can be represented as a difference between a squared sum of consecutive primes and the sum of their squares: $576 = 24^2 = (2+3+5+7+11)^2 - (2^2+3^2+5^2+7^2+11^2)$; it is the only such case for all primes up to $2 \cdot 10^9$.
- **613:** a prime that presents a mathematical enigma in the story *Number of the End* by Jason Earls: *bring the first digit back to get 136, it is triangular; now, bring the first digit of that back to get 361, it is a square*; the square 375769 of 613 is the largest known perfect square that divides a number of the form $n! + 1$, which happens when $n = 229$, another prime.
- **631:** a prime that is the reverse concatenation of the first three triangular numbers.

- **691:** the only known prime that is a square, 169, when turned upside down and another square, 196, when reversed; moreover, 169 and 196 are the smallest consecutive squares using the same digits.
- **701:** the smallest prime whose square 491401 contains all square digits only; it is equal to $5^4 + 4^3 + 3^2 + 2^1 + 1^0$.
- **727:** the first prime whose square (528529) can be represented as the concatenation of two consecutive numbers.
- **773:** replacing each digit of the prime 773 with its square and cube, respectively, results in two new primes: 49499 and 34334327.
- **786:** the largest known number n such that the binomial coefficient $\binom{2n}{n}$ is not divisible by the square of an odd prime.
- **835:** the value M_9; a semiprime: $835 = 5 \cdot 167$.
- **877:** the value $B(7)$; a Chen prime, as $879 = 3 \cdot 293$ is a semiprime; a prime index prime: $877 = p_{151}$.
- **900:** the smallest perfect square that is a sum of different primes that uses all the 10 digits: $900 = 503 + 241 + 89 + 67 = 509 + 283 + 61 + 47$.
- **2047:** the first Mersenne composite with prime index, $\mathcal{M}_{11}$.
- **65537:** the biggest known Fermat prime, F_4.
- **4294967297:** the smallest Fermat composite (in fact, semiprime), F_5.
- 10^{100}: one *googol*; in decimal notation, it is written as the digit 1 followed by one hundred zeroes; a googol is approximately 70!; 1 googol $\approx 2^{332.19280949}$.
- **1000...0007** $= 10^{999} + 7$: the smallest *titanic prime*, i.e., a prime of at least 1,000 decimal digits.
- $2^{4253}-1$: the smallest Mersenne titanic prime, $\mathcal{M}_{4253}$; the 19th Mersenne prime; it has 1,281 digits.
- **1000...00033603** $= 10^{9999} + 33603$: the smallest *gigantic prime*, i.e., a prime number with at least 10,000 decimal digits; it has exactly 10,000 digits.
- $2^{44497}-1$: the smallest Mersenne gigantic prime, $\mathcal{M}_{44497}$; the 27th Mersenne prime; it has 13,395 digits.

- $191273 \cdot 2^{3321908}-1$: the smallest known *megaprime*, i.e., a prime with at least one million decimal digits; it has exactly 1,000,000 digits.
- $2^{6972593}-1$: the smallest Mersenne megaprime, $\mathcal{M}_{6972593}$; the 38th Mersenne prime; it has 2,098,960 digits.
- $2^{82589933}-1$: the biggest known Mersenne prime, $\mathcal{M}_{82589933}$; the 51st known Mersenne prime; it has 24,862,048 digits.
- $10^{10^{100}} = 10^{googol}$: one *googolplex*; written out in ordinary decimal notation, it is 1 followed by 10^{100} zeroes, that is, a unity followed by googol zeroes.

Chapter 7
Mini Dictionary

In this chapter, a mini dictionary, i.e., a list of all special numbers related to the Catalan numbers, is presented (see [Abra40], [CoGu96], [DeDe12], [Deza17], [Deza18], [Deza21], [Deza23], [Deza24], [Dick05], [Dorr65], [Goul85], [GGL95Û], [Hons91], [Line86], [Plat80], [Prim24], [SlPl95], [Stan15], [Weis24], [Wiki24], etc.).

- $S_d(n,k)$ — *associated* (in fact, *d-associated*) *Stirling numbers of the second kind*; the number of partitions of an n-set into k subsets such that each of these subsets contains at least d elements; $S_1(n,k) = S(n,k)$ are ordinary *Stirling numbers of the second kind*; $S_2(n,k)$ are *Ward numbers*.
- $|s_d(n,k)|$ — *associated* (in fact, *unsigned d-associated*) *Stirling numbers of the first kind*; the number of partitions of an n-set into k cycles such that each of these cycles contains at least d elements; $|s_1(n,k)| = |s(n,k)|$ are ordinary *unsigned Stirling numbers of the first kind*; $|s_2(n,k)|$ gives the number of derangements of n elements with k cycles.
- $BN(n,k) = \frac{k}{2n+k}\binom{2n+k}{n}$ — *ballot numbers*; $C_n = BN(n,1)$.
- $B(n) = \sum_{k=0}^{n-1}\binom{n-1}{k}B(k)$, with $B(0) = 1$ — *Bell numbers*: $B(n)$ is the number of partitions of an n-set into non-empty parts.
- $B_n = -\frac{1}{n+1}\sum_{k=1}^{n}\binom{n+1}{k+1}B_{n-k}$ with $B_0 = 0$ — *Bernoulli numbers*.

- $S_n = \sum_{k=0}^n \frac{\binom{n}{k}\binom{n+k}{k}}{k+1}$, $n \geq 0$ — *big Schröder numbers* (or *Schröder numbers, large Schröder numbers*).
- $\binom{n}{m}$, $n = 0, 1, 2, \dots, m = 0, 1, \dots, n$ — *binomial coefficients*; they form *Pascal's triangle* — a number triangle, the sides of which are formed by 1 and any internal entry is obtained by adding the two entries diagonally above.
- $C_n = \frac{1}{n+1}\binom{2n}{n}$, $n = 0, 1, 2, \dots$ — *Catalan numbers*.
- $CS_m(n) = \frac{mn^2 - mn + 2}{2}$, $n = 1, 2, 3, \dots$ — *centered m-gonal numbers*.
- $CO(n) = \frac{(2n-1)(2n^2-2n+3)}{3}$, $n = 1, 2, 3, \dots$ — *centered octahedral numbers*.
- $CS_4(n) = \frac{(2n+1)^2 + 1}{2}$, $n = 1, 2, 3, \dots$ — *centered square numbers*.
- $D(n) = D(n,n)$ — *central Delannoy numbers*: $D(n) = \sum_{k=0}^n 2^k \cdot (C_n^k)^2$.
- $\widetilde{B}(n)$ — *complementary Bell numbers* (or *Uppuluri-Carpenter numbers*): $\widetilde{B}(n) = S(n,0) - S(n,1) + S(n,2) - \cdots + (-1)^n S(n,n)$, where $S(n,k)$ are the Stirling numbers of the second kind.
- $D(n,k)$ — *Delannoy numbers*: $D(m,n) = D(m-1,n) + D(m-1,n-1) + D(m,n-1)$, $D(0,0) = 1$, $n, k = 0, 1, 2, \dots$; $D(n) = D(n,n)$ — *central Delannoy numbers*.
- $n!!$ — *double factorials* (or *semifactorials*): $n!! = n(n-2)(n-4)\cdots$.
- E_n — *Euler numbers*: $\frac{2}{e^x + e^{-x}} = \sum_{n=0}^\infty \frac{E_n}{n!} \cdot x^n$.
- A_n — *Euler zigzag numbers* (or *up-down numbers*): numbers of alternating up-down permutations of $\{1, 2, \dots, n\}$; the number Z_n of all alternating (up-down and down-up) permutations of $\{1, 2, \dots, n\}$ is $2A_n$, $n \geq 2$; A_{2k+1} — *tangent numbers* (or *zag numbers*); A_{2n} — *secant numbers* (or *zig numbers*).
- $E(n,k)$ — *Eulerian numbers of the first kind*: $E(n,k) = (n-k)E(n-1,k-1) + (k+1)E(n-1,k)$, $E(0,0) = 1$, $n = 0, 1, 2, \dots$, $k = 0, 1, 2, \dots, n$.
- $\left\langle\!\!\left\langle {n \atop k} \right\rangle\!\!\right\rangle$ — *Eulerian numbers of the second kind*: $\left\langle\!\!\left\langle {n \atop k} \right\rangle\!\!\right\rangle = (2n-k-1)\left\langle\!\!\left\langle {n-1 \atop k-1} \right\rangle\!\!\right\rangle + (k+1)\left\langle\!\!\left\langle {n-1 \atop k} \right\rangle\!\!\right\rangle$, $\left\langle\!\!\left\langle {0 \atop 0} \right\rangle\!\!\right\rangle = 1$, $n = 0, 1, 2, \dots$, $k = 0, 1, 2, \dots, n$.

- $n! = 1 \cdot 2 \cdot \ldots \cdot n$, $n = 1, 2, 3, \ldots$; $0! = 1$ — *factorial numbers* (or *factorials*).
- $n^{\underline{k}}$ — *falling factorials*: $n^{\underline{k}} = \frac{n!}{(n-k)!}$.
- $F_n = 2^{2^n} + 1$, $n = 0, 1, 2, \ldots$ — *Fermat numbers*.
- $u_{n+2} = u_{n+1} + u_n$, $u_1 = u_2 = 1$ — *Fibonacci numbers*.
- $OB(n)$ — *Fubini numbers* (or *ordered Bell numbers*): $OB(n) = \sum_{k=0}^{n} k! S(n,k)$, where $S(n,k)$ are the Stirling numbers of the second kind.
- $A_m(p, r) = \frac{r}{mp+r}\binom{mp+r}{m}$ — *Fuss–Catalan numbers* (or *two-parameter Fuss–Catalan numbers*, *Raney numbers*); one-parameter Fuss–Catalan numbers are considered the numbers $A_m(p,1) = \frac{1}{mp+1}\binom{mp+1}{m}$.
- $D^k(m,n) = \sum_{j=0}^{n} C_n^j \cdot C_{m+n-j+k}^{n+k}$, $D_k(m,n) = \sum_{j=0}^{n} C_n^j \cdot C_{m+n-j-k}^{n-k}$ — *generalized Delannoy numbers*; $D^0(m,n) = D_0(m,n) = D(m,n)$.
- $|s(n,k)|_\alpha$, $\alpha \in \mathbb{N}$ — *generalized Stirling numbers of the first kind* (or *generalized α-factorial coefficients*): $|s(n,k)|_\alpha = (\alpha n + 1 - 2\alpha)|s(n-1,k)|_\alpha + |s(n-1,k-1)|_\alpha + \delta_{n,0}\delta_{k,0}$.
- $S(n, m, k)$, $n, m, k \in \mathbb{N}$ — *generalized Stirling numbers of the second kind*: $S(n, m, k)$ is that divided by $k!$ number of ways to place n different elements into m different boxes such that none of the k fixed boxes are empty; $S(n, k, k) = S(n, k)$.
- $p \in P$, $p > 10^{9999}$ — *gigantic primes*: prime numbers with at least 10,000 decimal digits.
- H_n — *harmonic numbers*: $H_n = \sum_{k=1}^{n} \frac{1}{k}$, $n = 1, 2, 3, \ldots$; $H_n^{(m)}$ — *generalized harmonic numbers*: $H_n^{(m)} = \sum_{k=1}^{n} \frac{1}{k^m}$, $m \in \mathbb{R}$, $n = 1, 2, 3, \ldots$.
- $SH_0 = 1$, $SH_n = \sum_{k=0}^{n} \frac{\binom{n}{k}\binom{n+k}{k}}{2k+2}$, $n \geq 1$ — *Hipparchus numbers* (or *Schröder–Hipparchus numbers*, *super-Catalan numbers*, *little Schröder numbers*).
- $hf(n) = 1^1 \cdot 2^2 \cdot \ldots \cdot n^n$, $n \in \mathbb{N}$ — *hyperfactorials*.
- $HO(n) = HO^4(n) = \frac{n^2(n^2+2)}{3}$ — *hyperoctahedral numbers* (or *four-dimensional hyperoctahedron numbers*, *hexadecahoron numbers*,

4-cross-polytope numbers, 4-orthoplex numbers, 16-cell numbers, β^4-numbers, $\{3,3,4\}$ numbers).

- $S_3^4(n) = \frac{n(n+1)(n+2)(n+3)}{24}$ — hypertetrahedral numbers (or triangulo-triangular numbers, pentatope numbers).
- $T(n)$, $n \in \mathbb{N}$ — involution numbers (or telephone numbers): $T(n)$ is the number of involutions in the symmetric group S_n of all permutations of a given n-set.
- $C_{n,k,d} = \frac{d}{(k-1)n+d}\binom{kn+d-1}{n}$ — (k,d)-Catalan numbers; $C_{n,k,1} = C_{n,k} = d_n^k$ — k-Catalan numbers (or Fuss numbers); $C_n = C_{n,2,1}$ — Catalan numbers.
- $C^k(n) = n^k$, $n = 1, 2, 3, \ldots$ — k-dimensional hypercube numbers, $k \in \mathbb{N}$, $k > 1$.
- $HO^k(n) = \sum_{j=0}^{k-1}(-1)^j\binom{k-1}{j}2^{k-j-1}S_3^{k-j}(n)$ — k-dimensional hyperoctahedron numbers, $k \in \mathbb{N}$, $k > 1$.
- $S_3^k(n) = \frac{n(n+1)(n+2)\cdots(n+(k-1))}{k!}$ — k-dimensional hypertetrahedron numbers (or k-dimensional simplicial numbers).
- $L(n,k)$ — Lah numbers: $L(n,k)$ is the number of partitions of an n-set into k totally ordered parts; $L(n,k) = C_{n-1}^{k-1}\frac{n!}{k!}$, $n = 0, 1, 2, \ldots$, $k = 0, 1, 2, \ldots, n$.
- $S_n = \sum_{k=0}^{n}\frac{\binom{n}{k}\binom{n+k}{k}}{k+1}$, $n \geq 0$ — large Schröder numbers (or Schröder numbers, big Schröder numbers).
- $SH_0 = 1$, $SH_n = \sum_{k=0}^{n}\frac{\binom{n}{k}\binom{n+k}{k}}{2k+2}$, $n \geq 1$ — little Schröder numbers (or Schröder–Hipparchus numbers, super-Catalan numbers, Hipparchus numbers).
- $L_{m,n} = \frac{2m+1}{m+n+1}\binom{2n}{m+n}$, $n \geq m \geq 0$ — Lobb numbers.
- $L_n = L_{n-1} + L_{n-2}$, $L_0 = 2$, $L_1 = 1$ — Lucas numbers.
- $[x^n]\prod_{m=1}^{n}\left(\sum_{i=0}^{m-1}x^i\right)$ — Mahonian numbers: numbers of permutations of n elements with k inversions.
- $p \in P$, $p > 10^{999999}$ — megaprimes: prime numbers with at least one million decimal digits.
- $\mathcal{M}_n = 2^n - 1$, $n = 1, 2, 3, \ldots$ — Mersenne numbers.
- $S_m(n) = \frac{n((m-2)n-m+4)}{2}$, $n = 1, 2, 3, \ldots$ — m-gonal numbers, $m \in \mathbb{N}$.
- $M_n = M_{n-1} + \sum_{i=0}^{n-2}M_iM_{n-2-i}$, $M_0 = M_1 = 1$ — Motzkin numbers.

- $n!_\alpha$, $\alpha \in \mathbb{N}$ — *multifactorials* (or *α-factorials*): $n!_\alpha = n(n-\alpha)(n-2\alpha)\cdots$; $n!_1 = n!$; $n!_2 = n!!$.
- $\binom{n}{k_1, k_2, ..., k_m}$ — *multinomial coeffficients*: $\binom{n}{k_1, k_2, ..., k_m} = \frac{n!}{k_1!k_2!...k_m!}$, $k_1, k_2, \ldots, k_m$ — non-negative integers, such that $k_1 + k_2 + \cdots + k_m = n$; $\binom{n}{k,l}$ — *trinomial coefficients*.
- $N(n,k)$ — *Narayana numbers*: $N(n,k) = \frac{1}{n} \cdot C_n^k \cdot C_n^{k-1}$, $n = 0,1,2,\ldots$, $k = 0,1,2,\ldots,n$.
- $cn(n)$, $n = 0,1,2,\ldots$ — *Narayana's cow numbers*; $cn(0) = cn(1) = cn(2) = 1$ and $cn(n) = cn(n-1) + cn(n-3)$, $n \geq 3$.
- $N_m(n,k)$ — *Narayana numbers of order m*: $N_m(n,k) = \frac{m+1}{n+1} \cdot C_{n+1}^{k+1} \cdot C_{n-m-1}^k$, $n = 0,1,2,\ldots$, $k = 0,1,2,\ldots,n$; $N(m,k) = N_0(m,k)$.
- $OB(n) = \sum_{k=0}^n k!S(n,k)$ — *ordered Bell numbers* (or *Fubini numbers*).
- $OS(n,k) = k!S(n,k)$, $n = 0,1,2,\ldots$, $k = 0,1,2,\ldots,n$ — *ordered Stirling numbers of the second kind*.
- $S_3^4(n) = \frac{n(n+1)(n+2)(n+3)}{24}$ — *pentatope numbers* (or *hypertetrahedral numbers, triangulo-triangular numbers*).
- $2^{n-1}(2^n - 1)$, $n \in \mathbb{N}$, where $2^n - 1 \in P$ — *perfect numbers* (in fact, *even perfect numbers*): positive integers equal to the sum of their proper divisors.
- $n\#$, $n \in \mathbb{N}$ — *primorials*: $n\#$ is the product of prime numbers less than or equal to n.
- $O(n) = \frac{n(2n^2+1)}{3}$, $n = 1,2,3,\ldots$ — *octahedral numbers*.
- $A_m(p,r) = \frac{r}{mp+r}\binom{mp+r}{m}$ — *Raney numbers* (or *Fuss–Catalan numbers, two-parameter Fuss–Catalan numbers*).
- $S^d(n,k)$ — *reduced* (in fact, *d-reduced*) *Stirling numbers of the second kind*: $S^d(n,k)$ is the number of partitions of the set $\{1,2,\ldots,n\}$ into k non-empty subsets so that the distance between any two elements in each subset is not less than d; $S^1(n,k) = S(n,k)$.
- $R(n) = (n-1)(2R(n-1) + 3R(n-2))/(n+1)$, $R(0) = 1$, $R(1) = 0$ — *Riordan numbers* (or *Motzkin sums, Motzkin summands, ring numbers*): $M_n = R(n) + R(n+1)$.
- $n^{\overline{k}}$ — *rising factorials*: $n^{\overline{k}} = \frac{(n+k-1)!}{(n-1)!}$.
- $R_n = 2^n \cdot 1 \cdot 3 \cdot 5 \cdots (2n-1) = 2^n(2n-1)!!$, $n \in \mathbb{N}$ — *Rodriguez numbers*.

- $S_n = \sum_{k=0}^{n} \frac{\binom{n}{k}\binom{n+k}{k}}{k+1}$, $n \geq 0$ — *Schröder numbers* (or *large Schröder numbers*, *big Schröder numbers*).
- $SH_0 = 1$, $SH_n = \sum_{k=0}^{n} \frac{\binom{n}{k}\binom{n+k}{k}}{2k+2}$, $n \geq 1$ — *Schröder–Hipparchus numbers* (or *super-Catalan numbers*, *little Schröder numbers*, *Hipparchus numbers*).
- $n!!$ — *semifactorials* (or *double factorials*): $n!! = n(n-2)(n-4)\cdots$.
- $S_4(n) = n^2$, $n = 1, 2, 3, \ldots$ — *square numbers*.
- $|s(n,k)|$ — *unsigned Stirling numbers of the first kind*: $|s(n,k)|$ is the number of partitions of an n-set into k cycles; $|s(n,k)| = |s(n-1,k-1)| + (n-1)|s(n-1,k)|$, $|s(0,0)| = 1$, $n = 0, 1, 2, \ldots$, $k = 0, 1, 2, \ldots, n$.
- $s(n,k)$ — *Stirling numbers of the first kind*: $s(n,k) = (-1)^{n+k}|s(n,k)|$, $n = 0, 1, 2, \ldots$, $k = 0, 1, 2, \ldots, n$.
- $S(n,k) = \frac{1}{k!}\sum_{i=0}^{k}(-1)^i\binom{k}{i}(k-i)^n$, $n = 1, 2, 3, \ldots$, $k = 1, 2, 3, \ldots, n$ — *Stirling numbers of the second kind*: $S(n,k)$ is the number of partitions of an n-set into k parts.
- $!n = [\frac{n!}{e}]$, $n \in \mathbb{N}$ — *subfactorials* (or *derangement numbers*, *de Montmort numbers*).
- $C^S(n,m) = \frac{(2m)!(2n)!}{(m+n)!m!n!}$, $m, n \in \mathbb{Z}$, $m \geq 0$, $n \geq 0$ — *Super Catalan numbers*.
- $SH_0 = 1$, $SH_n = \sum_{k=0}^{n} \frac{\binom{n}{k}\binom{n+k}{k}}{2k+2}$, $n \geq 1$ — *super-Catalan numbers* (or *Schröder–Hipparchus numbers*, *little Schröder numbers*, *Hipparchus numbers*).
- $sf(n) = 1! \cdot 2! \cdots n!$, $n \in \mathbb{N}$ — *superfactorials*.
- $T(n)$, $n \in \mathbb{N}$ — *telephone numbers* (or *involution numbers*): the number of ways n people can be connected by person-to-person telephone calls.
- $S_3^3(n) = \frac{n(n+1)(n+2)}{6}$, $n = 1, 2, 3, \ldots$ — *tetrahedral numbers*.
- $p \in P$, $p > 10^{999}$ — *titanic primes*: prime numbers with at least 1,000 decimal digits.
- $\binom{n}{k,l}$ — *trinomial coefficients*: $\binom{n}{k,l} = \frac{n!}{(n-k-l)!k!l!}$, $0 \leq l + k \leq n$.
- $S_3(n) = \frac{n(n+1)}{2}$, $n = 1, 2, 3, \ldots$ — *triangular numbers*.
- $t_{n+3} = t_{n+2} + t_{n+1} + t_n$, $t_1 = 0$, $t_2 = t_3 = 1$ — *tribonacci numbers*.
- $S_3^4(n) = \frac{n(n+1)(n+2)(n+3)}{24}$ — *triangulo-triangular numbers* (or *pentatope numbers*, *hypertetrahedral numbers*).

- $\widetilde{B}(n)$ — *Uppuluri–Carpenter numbers* (or *complementary Bell numbers*): $\widetilde{B}(n) = S(n,0) - S(n,1) + S(n,2) - \cdots + (-1)^n S(n,n)$, where $S(n,k)$ are the Stirling numbers of the second kind.
- $|s_2(n,k)|$ — *Ward numbers* (or *2-associated Stirling numbers of the second kind*): $|s_2(n,k)|$ is the number of partitions of an n-set into k subsets such that each of these subsets contains at least two elements.

Chapter 8
Exercises

In this chapter, we present some interesting problems concerning the Catalan numbers as well as their generalizations and relations, and we provide (sketches of) their solutions.

Problems Connected with Catalan Numbers

1. The sequence of *Rodriguez numbers* is a sequence whose nth term is defined as $R_n = 2^n \cdot 1 \cdot 3 \cdot 5 \cdots (2n-1)$, $n \in \mathbb{N}$. Find the first five members of this sequence; check that $R_n = 2(2n-1)R_{n-1}$; prove that $C_n = \frac{R_n}{(n+1)!}$.

2. Prove that $C_n = \frac{4^{n+1} \cdot \Gamma(n+\frac{1}{2})}{\sqrt{\pi}\Gamma(n+2)}$, where $\Gamma(z) = \int_0^\infty t^{z-1}e^{-t}dt$, is the *Euler gamma function*.

3. Using the result of the previous problem, prove that $C_n \sim \frac{4^n}{\sqrt{\pi}n^{1.5}}$, i.e., $\lim_{n\to\infty} \frac{C_n}{\frac{4^{n-1}}{\sqrt{\pi}n^{1.5}}} = 1$.

4. Prove that $C_n \sim \frac{4^n}{\sqrt{\pi}}(x^{-3/2} - \frac{9}{8}x^{-5/2} + \frac{145}{128}x^{-7/2} + \cdots)$.

5. Prove that $C_n = \frac{1}{2\pi}\int_0^4 x^n \sqrt{\frac{4-x}{x}}\,dx = \frac{2}{\pi}4^n \int_{-1}^1 t^{2n}\sqrt{1-t^2}\,dt$.

6. Prove that $\sum_{n=0}^\infty \frac{C_n}{4^n} = 2$; $\sum_{n=1}^\infty \frac{C_n}{4^n} = 1$; $\sum_{n=0}^\infty \frac{1}{C_n} = 2 + \frac{4\sqrt{3}\pi}{27}$.

7. Prove that
$$\sum_{\substack{i_1+\cdots+i_m=n \\ i_1,\ldots,i_m \geq 0}} C_{i_1} \cdots C_{i_m}$$
$$= \begin{cases} \dfrac{m(n+1)(n+2)\cdots(n+m/2-1)}{2(n+m/2+2)(n+m/2+3)\cdots(n+m)} \\ \quad \times C_{n+m/2}, \quad m \text{ even,} \\[2mm] \dfrac{m(n+1)(n+2)\cdots(n+(m-1)/2)}{(n+(m+3)/2)(n+(m+3)/2+1)\cdots(n+m)} \\ \quad \times C_{n+(m-1)/2}, \quad m \text{ odd.} \end{cases}$$

8. Prove that, starting with $C_0 = 1$, we can obtain the sequence of Catalan numbers by the recurrent relation $C_n = \binom{2n}{n} - \sum_{k=0}^{n-1} C_k \binom{2n-2k-1}{n-k}$.

9. Prove that, starting with $C_0 = 1$, we can obtain the sequence of Catalan numbers by the recurrent relation $(n+1)C_n = 4^n - \frac{1}{2}\sum_{k=0}^{n-1} 4^{n-k} C_k$.

10. Check for small n and prove the following equalities, involving sums over Catalan numbers:

$$C_n = \sum_{k=0}^{n-1} C_k 2^{n-2k-1} \binom{n-1}{2k}; \quad C_n = \frac{1}{n}\sum_{k=0}^{n-1} C_{n-k+1}\binom{2k+1}{k+1};$$

$$C_n = \sum_{k=0}^{n} (-1)^k 2^{n-k} \binom{n}{k}\binom{k}{\lfloor k/2 \rfloor};$$

$$C_n = \sum_{k=0}^{\lfloor n/2 \rfloor} \left(\frac{n-2k+1}{n-k+1}\binom{n}{n-k}\right)^2.$$

11. Prove that the *exponential generating function* for the sequence of Catalan numbers has the form $e^{2x}(I_0(2x) - I_1(2x))$, where $I_n(x)$ is a *modified Bessel function of the first kind*:

$$e^{2x}(I_0(2x) - I_1(2x))$$
$$= \sum_{n=0}^{\infty} C_n \frac{x^n}{n!} = 1 + x + x^2 + \frac{5}{6}x^3 + \frac{7}{12}x^4 + \frac{7}{20}x^5 + \cdots.$$

12. Prove that $\sum_{n=0}^{\infty} \frac{C_n x^{2n}}{(2n)!} = \frac{I_1(2x)}{x}$, where $I_n(x)$ is a *modified Bessel function of the first kind*.
13. For $n = 1, 2, 3, 4, 5$, construct the $n \times n$ *Henkel matrix* (or *Catalan–Henkel matrix*), in which the position (i, j), $1 \leq i, j \leq n$, is occupied by the Catalan number C_{i+j-2}. Verify that the determinant of this matrix is equal to unity regardless of its dimension. Obtain a similar result for a *shifted Henkel matrix*, in which the position (i, j), $1 \leq i, j \leq n$, is occupied by the number C_{i+j-1}.
14. Prove the *Jonah formula* $\binom{n}{k-1} = \sum_{i=1}^{k} C_k \binom{n-2i}{k-i}$.
15. Let $p > 1$ be a positive integer and $q \leq p - 1$. Define $d_{q0}^p = 1$ and d_{qk}^p as the number of *p-good paths* from $(1, q - 1)$ to $(k, (p-1)k - 1)$, $k \geq 0$. Prove the *generalized Jonah formula* $\binom{n-q}{k-1} = \sum_{i=1}^{k} d_{qi}^p \binom{n-pi}{k-i}$. Prove the closed formula $d_{qk}^p = \frac{p-q}{pk-q} \binom{pk-q}{k-1}$. Prove the following recurrent relation:

$$d_{qk}^p = \sum_{i,j} d_{p-r,i}^p d_{q+r,j}^p, \quad i, j, r \geq 1, k \geq 1, q < p - r,$$

and $i + j = k + 1$.

16. Let p be an odd prime, m be a positive integer not divisible by p, and C_k be a Catalan number. Prove that $C_{p-1} \equiv -1 \pmod{p}$;

$$\sum_{k=1}^{p-1} \frac{C_k}{m^k} \equiv \frac{m-4}{2} \cdot \left(1 - \frac{m(m-4)}{p}\right) \pmod{p}.$$

17. Define *k-power* as a construction composed of the numbers $k+1, n, \ldots, 3, 2$ (in this specified order) only using the operation of exponentiation and the corresponding arrangement of brackets. Prove that the number of $(n + 1)$-powers is the nth Catalan number C_n, $n \geq 0$. Find several methods of proof.
18. Define a *mountain with n ascents and n descents* as a broken line composed of n segments of the same length inclined at an angle of $45°$ to the positive direction of the OX-axis and of n segments of the same length inclined at an angle of $135°$ to the positive direction of the OX-axis, such that the resulting "start" and "end" are located on the same level. Prove that the number of mountains with n ascents and n descents is equal to the nth Catalan number C_n, $n \geq 0$. Give several proofs.

19. Define a k-board as an infinite table, composed of k rows and satisfying the following conditions: the first and last rows of the table consist of unities; for any table's part of the form

$$\begin{array}{ccc} & b & \\ a & & c \\ & d & \end{array}$$

the equality $ac+1 = bd$ holds. Prove that the number of different diagonals in an $(n+1)$-board is equal to the nth Catalan number C_n, $n \geq 0$.

20. Prove that the Catalan numbers list the numbers of random walks on a half-line: if the object is on a straight line at the zero position and can move either left or right par one position in one step, never moving to the left of zero, then the number of $2n$-step ways from the zero position back to the zero position is equal to the nth Catalan number C_n, $n \geq 0$.

21. Prove that C_n is the number of ways to tile a stairstep shape of height n with n rectangles.

22. Prove that C_n is the number of standard *Young tableaux* whose diagram is a $2 \times n$ rectangle.

23. Prove that C_n is the number of length n sequences that start with 1 and can increase by either 0 or 1 or decrease by any number (to at least 1).

24. Prove that C_n is the number of possible *parse trees* for a sentence (assuming binary branching) in natural language processing.

Problems Connected with Number Triangles

1. Construct the first 10 rows of the *factorial triangle*: a number triangle in which the rightmost element of the nth row, $n \geq 0$, is equal to 1, the leftmost element of the nth row of the triangle is $n!$, and any internal element of the nth row is obtained as the sum of the element positioned in the previous row on the left and the element increased by n times positioned in the previous row on the right. Check that the elements of the factorial triangle $e(n, m)$, where n is the row's number and m is the

number of the element in the row, satisfy the following recurrent relation: $e(n+1, m) = e(n, m-1) + (n+1)e(n, m)$. Find the initial conditions.
2. Check that the zeroth, first, second, and third rows of the factorial triangle represent the coefficients of the expansion of the polynomials $1, 1+x, (1+x)(2+x), (1+x)(2+x)(3+x)$, respectively. Prove that the nth row of the triangle gives the coefficients of the expansion of the polynomial $(1+x)(2+x)\cdots(n+x)$.
3. Prove that the sum of the elements of the nth row of the factorial triangle is equal to $(n+1)!$: $\sum_{m=0}^{n} e(n, m) = (n+1)!$.
4. Construct the first 10 rows of *Lukas's triangle*, each element L_n^k of which is determined using Pascal's rule, $L_n^k = L_{n-1}^k + L_{n-1}^{k-1}$, under the initial conditions $L_0^0 = 2$, $L_n^0 = 1$, and $L_n^n = 2$ for all positive integers n.
5. Prove that Lukas's triangle consists of the coefficients of the expansion of the polynomial $(x+2y)(x+y)^n$.
6. Make sure that the sums of the elements of the ascending diagonals of Lukas's triangle give the sequence of *Lukas numbers* that are defined recursively: $L_n = L_{n-1} + L_{n-2}$ and $L_0 = 2, L_1 = 1$.
7. Prove that there is the following relationship between elements L_n^k of Lukas's triangle and binomial coefficients: $L_n^k = C_n^k + C_{n-1}^{k-1}$: the nth row of Lukas's triangle can be obtained by adding the nth and the $(n-1)$th rows of Pascal's triangle.
8. The figure shows the first few rows of the *tribonacci triangle*.

```
        1
       1 1
      1 3 1
     1 5 5 1
    1 7 13 7 1
   1 9 25 25 9 1
```

Formulate the recurrent rule for the construction of the triangle. Check that the sums of the elements positioned on the ascending diagonals form the sequence of tribonacci numbers 1, 1, 2, 4, 7, 13, 24, 44, 81, 149, ... (sequence A000073 in the OEIS).
9. Construct the first 10 rows of the number triangle with the following recurrent rule: $T(0,0) = 1$, $T(n, j) = T(n-1, j) + T(n-2, j) + T(n-3, j)$ (every interior number can be obtained by

adding the three previous numbers on its vertical) and $T(n,n) = T(n,0)$. Find in this triangle the tribonacci numbers. Prove the corresponding statement.

10. Construct the first 10 rows of *Losanitsch's triangle*, named in honor of a Serbian chemist, S. Losanitsch. It is a number triangle whose right- and left-hand sides consist of unities, and each internal element is equal to the sum of the two numbers that are directly above it, except for the elements located at an odd position k in an even row with the number $2t$. For their construction, one should subtract from the sum of the two numbers standing directly above this element, the element located at the position with the number $\frac{k-1}{2}$ in the $(t-1)$th row of Pascal's triangle.

11. Check that Losanitsch's triangle is symmetrical with respect to the principal diagonal. Check that the first right (left) descending diagonal of Losanitsch's triangle consists of the consecutive positive integers $1, 2, 3, \ldots$, each of which is repeated twice.

12. Prove that the second right (left) descending diagonal of Losanitsch's triangle is formed from the alternating *square numbers* n^2 and *pronic numbers* $n(n+1)$.

13. Prove that the sum of the elements of the nth row of Losanitsch's triangle is $2^{n-2} + 2^{\lfloor n/2 \rfloor - 1}$.

14. Check that the sum of elements of the ascending diagonal of Losanitsch's triangle is either $\frac{u_{2n-1} + u_{n+1}}{2}$ or $\frac{u_{2n} + u_n}{2}$, where u_n is the nth Fibonacci number.

15. The *Bell triangle* (or *Aitken's array*, *Peirce triangle*) is a number triangle obtained by beginning the first row with the number one and beginning subsequent rows with the last number of the previous row. Rows are filled out by adding the number in the previous column to the number above it. Construct the first 10 rows of this triangle. Find the Bell numbers in this triangle. Give a combinatorial interpretation for the elements of the constructed triangle.

16. *Clark's triangle* is a number triangle created by setting the vertex equal to 0, filling one diagonal with unities, the other diagonal with multiples of an integer α, and filling in the remaining entries by summing the elements on either side from one row

above. Construct the first five rows of Clark's triangle with $\alpha = 1, 2, 3, 4, 5, 6$. Obtain the recurrent relation for the entries $T^\alpha_{m,n}$, $n = 0, 1, 2, \ldots$, $m = 0, 1, 2, \ldots, n$, of the triangle. Prove that $T^\alpha_{m,n} = \alpha\binom{m}{n+1} + \binom{m-1}{n-1}$.

17. The *Leibniz harmonic triangle* is a triangular arrangement of unit fractions in which the sides consist of the reciprocals of the row's numbers, and each internal element is the sum of two elements below it. Construct the first five rows of the triangle. Prove that for the elements $L_{(n,k)}$, $n = 1, 2, 3, \ldots$, $k = 1, 2, 3, \ldots, n$, of the triangle, the following recurrent relation holds: $L_{(n,1)} = \frac{1}{n}$, and $L_{(n,k)} = L_{(n-1,k-1)} - L_{(n,k-1)}$. Prove that $L_{(n,k)} = \frac{1}{n\binom{n-1}{k-1}} = \frac{1}{k\binom{n}{k}}$.

18. Define a *generalized Pascal's triangle* (more exactly, an (α, β, L, R)-*Pascal's triangle*) as a number triangle constructed by the following row: if $GT(n,k)$ is the kth element of the nth row of the triangle, then $GT(n,k) = \alpha GT(n-1, k-1) + \beta GT(n-1, k)$, while $GT(0, n) = L(n)$ and $GT(n, 0) = R(n)$. Prove that the classical Pascal's triangle can be obtained for $\alpha = \beta = 1$, $L(n) = R(n) = 1$. Give several examples of generalized Pascal's triangles. Is it possible to obtain a closed formula for $GT(n,k)$?

19. The *Bernoulli triangle* is a number triangle composed of the partial sums of binomial coefficients: $BT_{nk} = \sum_{i=0}^{k} \binom{n}{i}$. Construct the first five rows of this triangle. Prove that $BT_{nn} = 2^n$; $BT_{n,n-1} = 2^n - 1$.

20. A *Magog triangle* of order n is a number triangle of order n with entries from 1 to n, such that entries are non-decreasing across its rows and columns and all entries in column k, $k = 1, 2, \ldots, n$, are less than or equal to k. Construct several Magog triangles. Which properties of Magog triangles can we prove?

21. A *monotone triangle* (or *strict Gelfand pattern*, *Gog triangle*) of order n is a number triangle with n numbers along each side, with the base containing entries between 1 and n, such that there is a strict increase across rows and a weak increase diagonally up or down toward the right. Construct monotone triangles of order n, $n = 1, 2, 3, 4, 5$. Which properties of monotone triangles can we prove?

Other Related Problems

1. Prove that $\binom{x}{m}\binom{x}{n} = \sum_{k=0}^{m} \binom{m+n-k}{k,m-k,n-k}\binom{x}{m+n-k}$.
2. For integers $j, k,$ and n, $0 \le j \le k \le n$, prove the identity $\sum_{m=0}^{n} \binom{m}{j}\binom{n-m}{k-j} = \binom{n+1}{k+1}$.
3. Check that the proven "club rule" $\sum_{m=0}^{n} \binom{m}{k} = \binom{n+1}{k+1}$ is a special case of the identity from the previous problem.
4. Prove the identity $\sum_{j=k}^{n}(n+1-j)\binom{j-1}{k-1} = \binom{n+1}{k+1}$ as another special case of the same problem.
5. For non-negative integers $n \ge q$, prove the identity $\sum_{k=q}^{n}\binom{n}{k}\binom{k}{q} = 2^{n-q}\binom{n}{q}$, using a combinatorial approach.
6. Prove that $\sum_{n=0}^{p}(-1)^{p+n}\binom{p}{n}n^p = p!$. Check that, for an exponent greater than p, there is a recurrent relation $\sum_{n=0}^{p}(-1)^{p+n}\binom{p}{n}n^{p+a} = p!P_{2a}(p)$, where the polynomial $P_{2a}(p)$ is defined by another recurrence: $P_{2a+2}(p) = \sum_{x=1}^{p} xP_{2a}(x); P_0(p) = 1$.
7. Prove that, for a fixed positive integer n, the sequence C_n^k, $k = 0, 1, 2, \ldots, n$ is *unimodal*, that is, it has a single maximum, which is achieved for an even n at one point, $k_n = \lfloor \frac{n}{2} \rfloor = \frac{n}{2}$, and for an odd n at two consecutive points, $k_n = \lfloor \frac{n}{2} \rfloor = \frac{n-1}{2}$ and $k_n + 1 = \lfloor \frac{n}{2} \rfloor + 1 = \frac{n+1}{2}$.
8. Prove that, for any non-negative integer k, it holds that $\binom{\frac{1}{2}}{k} = \binom{2k}{k}\frac{(-1)^{k+1}}{2^{2k}(2k-1)}$.
9. Prove that, as $k \to \infty$, it holds that $\binom{z}{k} \sim \frac{(-1)^k}{\Gamma(-z)k^{z+1}}$; $\binom{z+k}{k} \sim \frac{e^{z(H_k-\gamma)}}{\Gamma(z+1)}$. Here, H_k is the kth harmonic number, $\Gamma(z)$ is the Euler gamma function, and $\gamma = 0.5772215\ldots$ is the Euler–Mascheroni constant.
10. Check that the sequence of central binomial coefficients $\binom{2n}{n} = \frac{(2n)!}{(n!)^2}$, $n \ge 0$, starts with the elements 1, 2, 6, 20, 70, 252, 924, 3432, 12870, 48620, ... (sequence A000984 in the OEIS). For the central binomial coefficient $\binom{2n}{n}$, obtain the following estimations:
    (a) $\frac{4^n}{2n+1} \le \binom{2n}{n} \le 4^n$, $n \ge 1$;
    (b) $\frac{4^n}{\sqrt{4n}} \le \binom{2n}{n} \le \frac{4^n}{\sqrt{3n+1}}$, $n \ge 1$;
    (c) $\binom{2n}{n} = \frac{4^n}{\sqrt{\pi n}}\left(1 - \frac{c_n}{n}\right)$, where $\frac{1}{9} < c_n < \frac{1}{8}$, and $n \ge 1$.

11. Prove that the number of factors of 2 in the central binomial coefficient $\binom{2n}{n}$ is equal to the number of unities in the binary representation of n. Prove that the only odd central binomial coefficient is 1.
12. Prove that the generating function for the sequence of central binomial coefficients is $\frac{1}{\sqrt{1-4x}}$.
13. Prove that, as $n \to \infty$, it holds that $\binom{2n}{n} \sim \frac{4^n}{\sqrt{\pi n}}$.
14. The *central polynomial coefficient* is defined as any polynomial coefficient $\binom{n}{k_1, k_2, \ldots, k_m}$, for which it holds that $\lfloor \frac{n}{m} \rfloor \leq k_i \leq \lceil \frac{n}{m} \rceil$, $\sum_{i=1}^m k_i = n$. Find the central polynomial coefficients for $m = 3, 4, 5$ at small n. Check that, for a fixed n, they are all equal among themselves and are greater than all other polynomial coefficients corresponding to this n. Prove this statement.
15. Check that the nth horizontal section of Pascal's pyramid is a triangle consisting of the coefficients of the expansion $(x+y+z)^n = \sum_{k_1+k_2+k_3=n} \binom{n}{k_1,k_2,k_3} x^{k_1} y^{k_2} z^{k_3}$ placed in the following way:

$\binom{n}{n,0,0}$ $\binom{n}{n-1,1,0}$ $\cdots$ $\binom{n}{1,n-1,0}$ $\binom{n}{0,n,0}$
$\binom{n}{n-1,0,1}$ $\binom{n}{n-2,1,1}$ $\cdots$ $\binom{n}{0,n-1,1}$
$\vdots$
$\binom{n}{1,0,n-1}$ $\binom{n}{0,1,n-1}$
$\binom{n}{0,0,n}$

16. *Pascal's s-simplex* is a generalization of Pascal's triangle and Pascal's pyramid for an arbitrary dimension s. Pascal's triangle corresponds to the case $s = 2$, while Pascal's pyramid corresponds to the case $s = 3$. Characterize Pascal's s-simplex for $s = 1$ and for $s = 4$. Describe the construction's scheme and the recurrent relation for Pascal's s-simplex, $s \in \mathbb{N}$. What is the combinatorial meaning of the elements of Pascal's s-simplex? What properties of this object can we highlight?
17. Prove that, for a fixed positive integer $n \geq 2$, the sequence $D(n-k, k)$, $k = 0, 1, 2, \ldots, n$, has a single maximum, which is achieved for an even n at one point, $k_n = \lfloor \frac{n}{2} \rfloor = \frac{n}{2}$, and for an odd n at two consecutive points, $k_n = \lfloor \frac{n}{2} \rfloor = \frac{n-1}{2}$, and $k_n + 1 = \lfloor \frac{n}{2} \rfloor + 1 = \frac{n+1}{2}$.
18. Prove that $N_n(z) = \sum_0^n \frac{1}{n+1} \binom{n+1}{k} \binom{2n-k}{n} (z-1)^k$, where $N_n(z)$ is the nth Narayana polynomial.

19. For integers $n \geq 0$, let d_n denote the number of underdiagonal paths from $(0,0)$ to (n,n) in an $n \times n$ grid, with the step set $S = \{(k,0) \mid k \in \mathbb{N}^+\} \cup \{(0,k) \mid k \in \mathbb{N}^+\}$. Prove that $d_n = N_n(4)$, where $N_n(z)$ is the nth Narayana polynomial.

20. Prove that $N_n(z) = (z-1)^{n+1} \int_0^{\frac{z}{z-1}} P_n(2x-1)\,dx$, $n \geq 1$, where $N_n(z)$ is the nth Narayana polynomial and $P_n(x)$ is the nth-degree *Legendre polynomial*.

21. An *Aztec diamond* of order n is a set of squares on the integer lattice whose centers have coordinates (x, y) and satisfy the inequality $|x| + |y| \leq n$. Prove that the determinant of the $(2n-1) \times (2n-1)$ Hankel matrix of the Schröder numbers, that is, the square matrix whose (i,j)th entry is S_{i+j-1}, is the number of domino tilings of the *order n Aztec diamond*, which is $2^{\frac{n(n+1)}{2}}$.

22. Let us define the *Wedderburn–Etherington numbers* a_n by the following recurrent relation:
$$a_{2n-1} = \sum_{i=1}^{n-1} a_i a_{2n-i-1},$$
$$a_{2n} = \frac{a_n(a_n+1)}{2} + \sum_{i=1}^{n-1} a_i a_{2n-i}, \quad a_1 = 1, a_0 = 0.$$
Find several first elements of this sequence. Prove that a_n gives the number of unordered rooted trees with n leaves in which all nodes, including the root, have either zero or exactly two children (so-called *Otter trees*).

23. Define the nth *telephone number* (or *involution number*) $T(n)$, $n = 0, 1, 2, \ldots$, as the number of ways n people can be connected by person-to-person telephone calls. Check that this number can be defined as the number of permutations on n elements that are involutions. Check that the first values of the sequence $T(n)$, $n = 0, 1, 2, \ldots$, are $1, 1, 2, 4, 10, 26, 76, 232, 764, 2620, 9496, \ldots$ (sequence A000085 in the OEIS). Prove that the telephone numbers satisfy the recurrence relation $T(n) = T(n-1) + (n-1)T(n-2)$.

24. Prove that $T(n) = \sum_{k=0}^{\lfloor n/2 \rfloor} \binom{n}{2k}(2k-1)!! = \sum_{k=0}^{\lfloor n/2 \rfloor} \frac{n!}{2^k(n-2k)!k!}$.

Solutions to Problems Connected with Catalan Numbers

1. The sequence starts with the elements $R_1 = 2$, $R_2 = 12$, $R_3 = 120$, $R_4 = 1680$, and $R_5 = 30240$ (see the sequence A001813 in the OEIS). In fact, it holds that $R_n = 2^n \cdot (2n-1)!!$ and $R_n = (4n-2)!!!!$.

2. It is easy to show ([Kara83] and [Deza24]) that the function $\Gamma(z)$ satisfies the functional equation $\Gamma(z+1) = z \cdot \Gamma(z)$ with $\Gamma(1) = 1$. Then,

$$\Gamma(2) = 1 \cdot \Gamma(1) = 1, \quad \Gamma(3) = 2 \cdot \Gamma(2) = 2 \cdot 1 = 2, \ldots, \Gamma(n+1) = n!$$

for any positive integer n; as $\Gamma\left(\frac{1}{2}\right) = \sqrt{\pi}$ ([Kara83] and [Deza24]),

$$\Gamma\left(\frac{3}{2}\right) = \Gamma\left(\frac{1}{2}+1\right) = \frac{1}{2} \cdot \Gamma\left(\frac{1}{2}\right) = \frac{1}{2}\sqrt{\pi},$$

$$\Gamma\left(\frac{5}{2}\right) = \Gamma\left(\frac{3}{2}+1\right) = \frac{3}{2} \cdot \Gamma\left(\frac{3}{2}\right) = \frac{1 \cdot 3}{2^2}\sqrt{\pi}, \ldots, \Gamma\left(n+\frac{1}{2}\right)$$

$$= \frac{1 \cdot 3 \cdots (2n-1)}{2^n}\sqrt{\pi}, \quad n \in \mathbb{N}.$$

Therefore, it holds that

$$\frac{4^n \cdot \Gamma(n+\frac{1}{2})}{\sqrt{\pi} \cdot \Gamma(n+2)} = \frac{4^n \cdot 1 \cdot 3 \cdots (2n-1)\sqrt{\pi}}{2^n \sqrt{\pi} \cdot (n+1)!}$$

$$= \frac{2^n \cdot (2n-1)!!}{(n+1)!} = C_n.$$

3. For any real $z \neq 0$, we have the *Stirling formula* ([Kara83] and [Deza24]):

$$\Gamma(z) = e^{-z} z^{z-\frac{1}{2}} \sqrt{2\pi} \left(1 + O\left(\frac{1}{|z|}\right)\right).$$

Then, it holds that

$$\Gamma(n+1) = n! = e^{-(n+1)}(n+1)^{n+\frac{1}{2}}\sqrt{2\pi}\left(1 + O\left(\frac{1}{n+1}\right)\right).$$

Noting that
$$\lim_{n\to\infty}\left(\frac{n+1}{n}\right)^n = \lim_{n\to\infty}\left(1+\frac{1}{n}\right)^n = e, \quad \lim_{n\to\infty}\left(\frac{n+1}{n}\right)^{\frac{1}{2}} = 1,$$
we obtain the formula $\Gamma(n+1) = n! \sim e^{-n} \cdot n^n \cdot \sqrt{2\pi n}$. Similarly,
$$\Gamma\left(n-\frac{1}{2}\right) = e^{-(n-\frac{1}{2})}\left(n-\frac{1}{2}\right)^{n-1}\sqrt{2\pi}\left(1 + O\left(\frac{1}{n-1}\right)\right).$$
As
$$\lim_{n\to\infty}\left(\frac{n-\frac{1}{2}}{n}\right)^{n-1} = \lim_{n\to\infty}\left(1+\frac{-0,5}{n}\right)^n \left(\frac{n}{n-\frac{1}{2}}\right) = e^{-0,5},$$
we get the formula
$$\Gamma\left(n-\frac{1}{2}\right) \sim e^{-n} \cdot n^{n-1} \cdot \sqrt{2\pi}.$$
Then, we have that
$$C_{n-1} \sim \frac{4^{n-1} \cdot e^{-n} \cdot n^{n-1} \cdot \sqrt{2\pi}}{\sqrt{\pi} \cdot e^{-n} \cdot n^n \cdot \sqrt{2\pi n}} \sim \frac{4^{n-1}}{n^{1,5}\sqrt{\pi}}.$$
Going from $n-1$ to n and noting that $n^{1,5} \sim (n+1)^{1,5}$, we get the result.
4. See [Vars91] and [GKP94].
5. See, for example, [FeBa17].
6. This fact immediately follows from the previous problem. This result has a simple probabilistic interpretation. Consider a random walk on the integer line, starting at 0. Let -1 be a "trap" state, such that if the walker arrives at -1, it will remain there. The walker can arrive at the trap state at times $1, 3, 5, 7, \ldots$, and the number of ways the walker can arrive at the trap state at time $2k+1$ is C_k. Since the considered random walk is recurrent, the probability that the walker eventually arrives at -1 is $\sum_{n=0}^{\infty} \frac{C_n}{2^{2n+1}} = 1$. See, for example, [DeMh20] and [ChDe20].
7. This equality is called the *Catalan m-fold convolution*. See [BoRe14].

8. For small n, it holds that $1 = C_0$, $\binom{2}{1} - \binom{1}{1} = 2 - 1 = C_1$. $\binom{4}{2} - \binom{3}{2} - \binom{1}{1} = 6 - 3 - 1 = 2$.
9. For $c_n = \frac{C_n}{4^n}$, we now have a simple recurrent relation: $c_n = \frac{1}{n+1} - \frac{1}{2(n+1)} \sum_{k=0}^{n-1} c_k$. This implies $\sum_{k=0}^{\infty} \frac{C_k}{4^k} = \sum_{k=0}^{\infty} c_k = 2$.
10. Use a combinatorial approach. Here, $\lfloor x \rfloor$ is the floor function.
11. See the sequences A144186 and A144187 in the OEIS.
12. See [Weis24].
13. In fact, for $n = 1$, the result is trivial: the determinant of the 1×1 matrix (1) is equal to 1. For $n = 2$, we have the matrix $\begin{pmatrix} 1 & 1 \\ 1 & 2 \end{pmatrix}$ with determinant 1. For $n = 3$, we get the matrix $\begin{pmatrix} 1 & 1 & 2 \\ 1 & 2 & 5 \\ 2 & 5 & 14 \end{pmatrix}$; its determinant is also equal to 1. Moreover, if the indexing is shifted so that the (i,j)th entry is filled with the Catalan number C_{i+j-1}, then the determinant is still 1, regardless of the value of n. For $n = 1$, the result is trivial. For $n = 2$, we have the matrix $\begin{pmatrix} 1 & 2 \\ 2 & 5 \end{pmatrix}$ with determinant 1. For $n = 3$, we get the matrix $\begin{pmatrix} 1 & 2 & 5 \\ 2 & 5 & 14 \\ 5 & 14 & 42 \end{pmatrix}$ with determinant 1.

Taken together, these two conditions uniquely define the sequence of Catalan numbers.

Another feature unique to the Catalan–Hankel matrix is that the $n \times n$ submatrix starting at 2 has determinant $n+1$ (see, for example, [MaWo00]):

$$\det \begin{bmatrix} 2 \end{bmatrix} = 2; \quad \det \begin{bmatrix} 2 & 5 \\ 5 & 14 \end{bmatrix} = 3; \quad \det \begin{bmatrix} 2 & 5 & 14 \\ 5 & 14 & 42 \\ 14 & 42 & 132 \end{bmatrix} = 4.$$

14. In fact, $\binom{n}{0} = C_1 \binom{n-2}{0} (1 = 1)$ and $\binom{n}{1} = C_1 \binom{n-2}{1} + C_2 \binom{n-4}{0}$ ($n = (n-2) + 2$).
15. See the previous problem and [HiPe91]. For the closed formula, compare it with the following closed formula for C_k: $C_k = \frac{1}{k} \binom{2k}{k-1}$. For the recurrence, compare it with the classical recurrent relation for C_k: $C_k = \sum_i C_i C_{k-i-1}$.

16. For all $k = 1, 2, \ldots, p-1$, we have $C_{\frac{p-1}{2}}^k \equiv C_{\frac{-1}{2}}^k = \frac{C_{2k}^k}{(-4)^k}$ (mod p). Then, it holds that $C_{p-1} = \frac{1}{2p-1} \prod_{k=1}^{p-1} \frac{p+k}{k} \equiv -1$ (mod p). Therefore, it holds that $\sum_{k=1}^{p-1} \frac{C_k}{m^k} \equiv \sum_{0<k<p-1} C_{\frac{p-1}{2}}^k \cdot \frac{1}{k+1} \cdot \left(-\frac{4}{n}\right)^k + \frac{C_{p-1}}{m^{p-1}} \equiv -\frac{m-4}{2} \cdot \left(\frac{m(m-4)}{p}\right) + \frac{m}{2} - 2$ (mod p).

17. For $n = 0$, the situation is trivial: there is only one such construction: 2. For $n = 1$, we also have exactly one $(n+1)$-power (i.e., 2-power): $3^2 = 9$; for $n = 2$, we have two $(n+1)$-powers (in fact, 3-powers): $(4^3)^2 = 4^6$ and $4^{(3^2)} = 4^9$; for $n = 4$ we have five 4-powers: $((5^4)^3)^2 = 5^{24}$, $5^{(4^{(3^2)})} = 5^{262144}$, $(5^{(4^3)})^2 = 5^{128}$, $(5^4)^{(3^2)} = 5^{36}$, and $5^{((4^3)^2)} = 5^{4096}$.

Consider a proof based on the enumeration of the set of plane binary trees. In fact, let us build a one-to-one correspondence between the set of $(n+1)$-powers and the set of plane binary trees with n internal vertices according to the following law: put the power a^b to match the planted plane binary tree with one internal vertex, drawing one "branch" of the tree to a and the other "branch" of the tree to b; if the base a (exponent b), in turn, has the form c^d, continue the branch corresponding to the base a (to the exponent b), drawing from its upper vertex the edges to c and to d; if the base a (the exponent b) is a single number, the corresponding branch will be the last branch of the constructed tree, that is, its upper vertex is a hanging vertex (leaf); use similar considerations for all subpowers of a given $(n+1)$-power.

Thus, each $(n+1)$-power corresponds to a plane binary tree with n internal vertices. For example, the power $(5^{(4^3)})^2$ corresponds to the following tree:

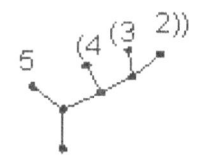

Conversely, for each planted binary tree with n internal vertices, we can restore the corresponding $(n+1)$-power. In fact, by marking all leaves (excluding the lowest vertex) of a given tree by the numbers $n+1, n, \ldots, 3, 2$ in clockwise order (from left to right) and marking all internal vertices (excluding the lowest one) by the opening brackets "(", we restore the corresponding $(n+1)$-power by moving along the tree from the "ground" in the clockwise direction and gathering existing marks; the closed brackets are now uniquely restored.

In the following figure, we can see all constructions corresponding to the case $n = 3$. They give us the powers $5^{(4^{(3^2)})}$, $5^{((4^3)^2)}$, $((5^4)^3)^2$, $(5^{(4^3)})^2$, $(5^4)^{(3^2)}$, correspondingly.

Thus, we obtain a one-to-one correspondence that matches the set of all $(n+1)$-powers with the set of all plane binary trees with n internal vertices. Since the number of plane binary trees with n internal vertices is C_n, the number of $(n+1)$-powers is also C_n.

18. Let us build a one-to-one correspondence between the set of all mountains with n ascents and n descents and the set of plane rooted trees with n edges according to the following law: fix the root vertex and draw from it, one by one, as many edges as "increasing" segments form the left slope of the first "peak" of the mountain; then, "go back" along the constructed edges, making the number of steps equal to the number of "descending" segments formed by the right slope of the first "peak" of the mountain; afterward, draw from the corresponding vertex, one by one, as many edges as "increasing" segments form the left slope of the second "peak" of the mountain, etc.

Conversely, for each tree, you can easily restore the corresponding mountain: the first "peak" obtained by the up-down movement along the first (left) "branch" of the tree, the second "peak" obtained by the up-down movement along the second "branch" of the tree, etc.

In the following, all possibilities for $n = 3$ are presented:

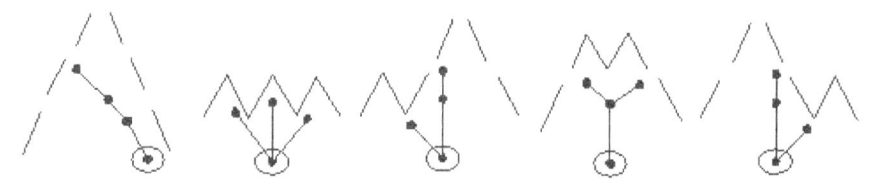

19. For $n = 0$, the $(n+1)$-board (i.e., the 1-board) has the following form:
$$1\ 1\ 1\ 1\ 1\ 1\ \ldots$$
For $n = 1$, the $(n+1)$-board looks like this:
$$1\ 1\ 1\ 1\ 1\ 1\ 1\ \ldots$$
$$1\ 1\ 1\ 1\ 1\ 1\ 1\ \ldots$$
For $n = 2$, we have
$$1\ 1\ 1\ 1\ 1\ 1\ 1\ \ldots$$
$$1\ 2\ 1\ 2\ 1\ 2\ 1\ \ldots$$
$$1\ 1\ 1\ 1\ 1\ 1\ 1\ \ldots$$
For $n = 3$, we have
$$1\ 1\ 1\ 1\ 1\ 1\ \ldots$$
$$1\ 2\ 2\ 1\ 3\ 1\ \ldots$$
$$1\ 3\ 1\ 2\ 2\ 1\ \ldots$$
$$1\ 1\ 1\ 1\ 1\ 1\ \ldots$$

It is easy to see that the number of different diagonals for the constructed boards is equal to 1 (we have the only diagonal: 1), 1 (we have the only diagonal: 11), 2 (we have two diagonals: 111 and 121), and 5 (we have five diagonals: 11111, 1231, 1211, 1121, and 1321).

It is easy to check that the number of different diagonals of an $(n+1)$-board is equal to the number of Euler's triangulations of a convex $(n+2)$-gon. In fact, there exists a one-to-one correspondence between the set of diagonals of an $(n+1)$-board and the set of partitions of a convex $(n+2)$-gon into triangles constructed according to the following law: fix one of the vertices of the $(n+2)$-gon, and assign to it the number 0; two neighboring vertices of the $(n+2)$-gon will obtain the number 1; for each triangle of a given partition, the last, that is, the third, vertex will obtain the number equal to the sum of the numbers of the other two vertices of the triangle. Then, each right-to-left numbering of n vertices of the $(n+2)$-gon from the "first" vertex, numbered by 1, will correspond to one of the diagonals of the $(n+1)$-board. In the following figure, we can see all possibilities for $n = 3$:

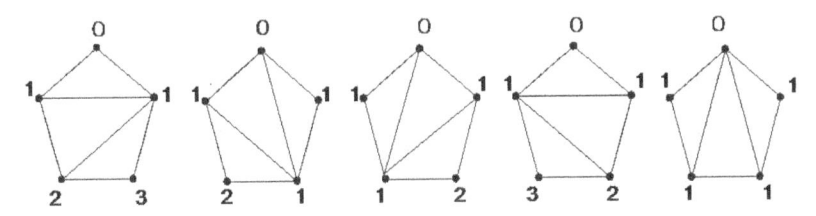

Conversely, for each diagonal of the $(n+1)$-board, we can restore the corresponding partition of a convex $(n+2)$-gon into triangles. For example, the sequence 1, 1, 3, 2, 1 specifies the following splitting of a hexagon into triangles. (Fix a vertex 0; mark the remaining vertices clockwise by numbers 1, 1, 3, 2, 1; draw the diagonal that gives a triangle with a sum of vertex numbers equal to 2 (diagonals a and b); draw a diagonal that gives a triangle with the sum of the vertex numbers equal to 3 (diagonal c).)

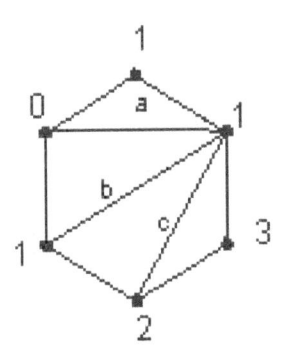

20. $1 =$ step forward; $-1 =$ step back. So, we get a bijection with the set of all correct $(1, -1)$-sequences. Note that this problem is a special case of the *ballot problem* (see [Gess92]).
21. Cutting across the anti-diagonal and looking at only the edges gives plane binary trees.

438   *Catalan Numbers*

22. In other words, it is the number of ways the numbers $1, 2, \ldots, 2n$ can be arranged in a $2 \times n$ rectangle so that each row and each column is increasing. As such, the formula can be derived as a special case of the *hook-length formula* ([Youn00]). The following are all five possibilities for $n = 3$:

$$123\ 124\ 125\ 134\ 135$$
$$456\ 356\ 346\ 256\ 246$$

However, there exists a simple bijection: for any correct sequence of brackets with n pairs of brackets, consider a natural numbering of the brackets with the numbers $1, 2, 3, \ldots, 2n$. If a number i corresponds to an opening bracket, put it in the first row of the table; otherwise, put it in the second row.

23. For $n = 4$, these are 1234, 1233, 1232, 1231, 1223, 1222, 1221, 1212, 1211, 1123, 1122, 1121, 1112, 1111.

24. A parse tree is an ordered rooted tree that represents the syntactic structure of a string according to some context-free grammar. For details, see, for example, [ChWi07]. See also encoding general trees as binary trees.

Solutions to Problems Connected with Number Triangles

1. The following figure shows the first six rows of the triangle.

$$
\begin{array}{ccccccccccc}
 & & & & & 1 & & & & & \\
 & & & & 1 & & 1 & & & & \\
 & & & 2 & & 3 & & 1 & & & \\
 & & 6 & & 11 & & 6 & & 1 & & \\
 & 24 & & 50 & & 35 & & 10 & & 1 & \\
120 & & 274 & & 225 & & 85 & & 15 & & 1
\end{array}
$$

2. Let $(1+x)(2+x)(3+x)\cdots(n+x) = E_n^0 + E_n^1 x + E_n^2 x^2 + \cdots + E_n^n x^n$. As $E_n^0 = n!$, $E_n^n = 1$, it holds that

$$(1 + x)(2 + x)(3 + x) \cdots (n + x)((n + 1) + x)$$
$$= (E_n^0 + E_n^1 x + E_n^2 x^2 + \cdots + E_n^n x^n)((n + 1) + x)$$

$$= (n+1)E_n^0 + (n+1)E_n^1 x + (n+1)E_n^2 x^2 + \cdots$$
$$+ (n+1)E_n^n x^n + E_n^0 x + E_n^1 x^2 + E_n^2 x^3 + \cdots + E_n^n x^{n+1}$$
$$= E_n^0(n+1) + (E_n^0 + (n+1)E_n^1)x$$
$$+ (E_n^1 + (n+1)E_n^2)x^2 + \cdots + E_n^n x^{n+1}$$
$$= E_{n+1}^0 + E_{n+1}^1 x + E_{n+1}^2 x^2 + \cdots + E_{n+1}^{n+1} x^{n+1}.$$

So, $E_{n+1}^0 = (n+1)!$, $E_{n+1}^{n+1} = 1$, and $E_n^{k-1} + (n+1)E_n^k = E_{n+1}^k$, $1 \leq k \leq n-1$. As we obtain the same recurrent relation with the same initial conditions, it holds that $E_n^k = e(n,k)$.

3. For $n=0$, it holds that $e(0,0) = (0+1)! = 1$. For $n=1$, it holds that $e(1,0) + e(1,1) = (1+1)! = 2$. Let the statement be true for $n=k$: $e(k,0) + e(k,1) + \cdots + e(k,k) = (k+1)!$. Then, for $n=k+1$, it holds that $e(k+1,0)+e(k+1,1)+\cdots+e(k+1,k+1) = (k+2)!$ In fact,

$$e(k+1,0) + e(k+1,1) + \cdots + e(k+1,k+1)$$
$$= (k+1)! + e(k,0) + (k+1)e(k,1) + e(k,1)$$
$$+ (k+1)e(k,2) + \cdots + e(k,k-1) + (k+1)e(k,k)$$
$$= (k+1)! + (e(k,0) + e(k,1) + \cdots + e(k,k-1)$$
$$+ e(k,k)) + (k+1)(e(k,1) + e(k,2) + \cdots + e(k,k-1))$$
$$+ ke(k,k) = 2(k+1)! + (k+1)((k+1)! - e(k,0)$$
$$- e(k,k)) + ke(k,k) = 2(k+1)! + (k+1)((k+1)!$$
$$- k! - 1) + k = 2(k+1)! + k(k+1)! = (k+1)!(k+2)$$
$$= (k+2)!.$$

4. The first five rows of Lukas's triangle are given as follows:
$$2$$
$$1\ 2$$
$$1\ 3\ 2$$
$$1\ 4\ 5\ 2$$
$$1\ 5\ 9\ 7\ 1$$
$$1\ 6\ 14\ 16\ 9\ 2$$

5. It can be proven by direct computation.

6. Similarly to how the original Pascal's triangle gives the Fibonacci numbers as the sums of its diagonals, this triangle gives the Lucas numbers $2, 1, 3, 4, 7, 11, 18, 29, 47, 76, \ldots$ (sequence A000032 in the OEIS). The proof is similar.
7. It follows from the construction. As it is true for small values of the parameters, then

$$L_n^k = L_{n-1}^k + L_{n-1}^{k-1} = \left(C_{n-1}^k + C_{n-2}^{k-1}\right) + \left(C_{n-1}^{k-1} + C_{n-2}^{k-2}\right) = C_n^k + C_n^{k-1}.$$

8. It follows from the construction of the triangle ($TT_0^0 = 1$, $TT_n^0 = TT_n^n = 1$, and $TT_n^k = T_{n-1}^k + TT_{n-1}^{k-1} + T_{n-2}^{k-1}$) and the recurrent definition of the tribonacci numbers $t_{n+3} = t_{n+2} + t_{n+1} + t_n$ with initial conditions $t_0 = t_1 = 1, t_2 = 2$.
9. The triangle, often also called the *tribonacci triangle*, starts with the rows

$$\begin{array}{c} 1 \\ 1\ 1 \\ 2\ 1\ 2 \\ 4\ 2\ 2\ 4 \\ 7\ 4\ 4\ 4\ 7 \\ 13\ 7\ 8\ 8\ 7\ 13 \end{array}$$

(sequence A082793 in the OEIS). By the construction, the tribonacci numbers form its sides.
10. The first few rows of the triangle are as follows (sequence A034851 in the OEIS):

$$\begin{array}{c} 1 \\ 1\quad 1 \\ 1\quad 1\quad 1 \\ 1\quad 2\quad 2\quad 1 \\ 1\quad 2\quad 4\quad 2\quad 1 \\ 1\quad 3\quad 6\quad 6\quad 3\quad 1 \\ 1\quad 3\quad 9\quad 10\quad 9\quad 3\quad 1 \\ 1\quad 4\quad 12\quad 19\quad 19\quad 12\quad 4\quad 1 \end{array}$$

11. It follows from the construction.
12. The second pair of diagonals contain the "quarter-squares" (sequence A002620 in the OEIS), or the square numbers and pronic numbers interleaved.

The next pair of diagonals contains the *alkane numbers* $l(6, n)$ (sequence A005993 in the OEIS). And the next pair of diagonals

contains the alkane numbers $l(7,n)$ (sequence A005994 in the OEIS), while the next pair has the alkane numbers $l(8,n)$ (sequence A005995 in the OEIS), then the alkane numbers $l(9,n)$ (sequence A018210 in the OEIS), then $l(10,n)$ (sequence A018211 in the OEIS), then $l(11,n)$ (sequence A018212 in the OEIS), then $l(12,n)$ (sequence A018213 in the OEIS), etc.
13. The first few elements are $1, 2, 3, 6, 10, 20, 36, 72, 136, 272, \ldots$ (sequence A005418 in the OEIS).
14. Use the recurrent definition of the Fibonacci numbers.
15. The first rows of the Bell triangle are

$$\begin{array}{ccccc} & & 1 & & \\ & 1 & & 2 & \\ 2 & & 3 & & 5 \\ 5 & 7 & & 10 & 15 \\ 15 & 20 & 27 & 37 & 52 \end{array}$$

(sequence A011971 in the OEIS). Numbering the rows of the resulting triangle, starting with the zeroth row, and the elements of a given row from a zero element, we get for the numbers t_n^k, $n = 0, 1, 2, 3, \ldots$, $k = 0, 1, 2, \ldots, n$, the relations immediately following from the above algorithm:

$$t_0^0 = 1, \quad t_n^0 = t_{n-1}^{n-1}, \quad n \geq 1, \quad t_n^k = t_n^{k-1} + t_{n-1}^{k-1}, \quad 1 \leq k \leq n.$$

It turns out that the number t_n^k is the number of partitions of the set $\{1, 2, \ldots, n+1\}$ such that the element $n+1$ does not fall into the same class with any of the elements $k+1, k+2, \ldots, n$.
16. The following illustration shows Clark's triangle for $n = 6$ (sequence A090850 in the OEIS):

$$\begin{array}{cccccc} & & & 0 & & \\ & & 6 & & 1 & \\ & 12 & & 7 & & 1 \\ 18 & & 19 & & 8 & 1 \\ 24 & 37 & & 27 & 9 & 1 \end{array}$$

In this case, $T_{m2}^6 = (m-1)^3$ and $T_{m3}^6 = \left(\frac{(m-1)(m-2)}{2}\right)^2 = (S_3(m-2))^2$.

17. The first eight rows of the triangle are

$$
\begin{array}{c}
1 \\
\frac{1}{2} \quad \frac{1}{2} \\
\frac{1}{3} \quad \frac{1}{6} \quad \frac{1}{3} \\
\frac{1}{4} \quad \frac{1}{12} \quad \frac{1}{12} \quad \frac{1}{4} \\
\frac{1}{5} \quad \frac{1}{20} \quad \frac{1}{30} \quad \frac{1}{20} \quad \frac{1}{5} \\
\frac{1}{6} \quad \frac{1}{30} \quad \frac{1}{60} \quad \frac{1}{60} \quad \frac{1}{30} \quad \frac{1}{6} \\
\frac{1}{7} \quad \frac{1}{42} \quad \frac{1}{105} \quad \frac{1}{140} \quad \frac{1}{105} \quad \frac{1}{42} \quad \frac{1}{7} \\
\frac{1}{8} \quad \frac{1}{56} \quad \frac{1}{168} \quad \frac{1}{280} \quad \frac{1}{280} \quad \frac{1}{168} \quad \frac{1}{56} \quad \frac{1}{8} \\
\vdots \quad \vdots \quad \vdots
\end{array}
$$

(see sequence A003506 in the OEIS).

18. For example, the conditions $GT(n,k) = 2GT(n-1, k-1) + T(n-1, k)$ with $GT(n, 0) = n, GT(n, n) = n^2$, give the following triangle (sequence A228576 in the OEIS):

$$
\begin{array}{c}
0 \\
1 \quad 1 \\
2 \quad 3 \quad 4 \\
3 \quad 7 \quad 10 \quad 9 \\
4 \quad 13 \quad 24 \quad 29 \quad 19
\end{array}
$$

Clark's triangle gives another example. See also *Riordan arrays*, [Spru94].

19. The first few rows of the triangle are presented as follows:

$$
\begin{array}{c}
1 \\
1 \quad 2 \\
1 \quad 3 \quad 4 \\
1 \quad 4 \quad 7 \quad 8 \\
1 \quad 5 \quad 11 \quad 15 \quad 16
\end{array}
$$

The properties are obvious. So, the triangle contains all Mersenne numbers.

20. Magog triangles are in one-to-one correspondence with totally symmetric self-complementary plane partitions (see [Weis24]). An example is

$$
\begin{array}{l}
1 \\
1\ 1 \\
1\ 1\ 1 \\
1\ 1\ 1\ 3 \\
1\ 2\ 3\ 4\ 5
\end{array}
$$

21. A monotone triangle of order 5 is presented as follows:

$$
\begin{array}{ccccccccc}
 & & & & 4 & & & & \\
 & & & 2 & & 5 & & & \\
 & & 1 & & 4 & & 5 & & \\
 & 1 & & 3 & & 4 & & 5 & \\
1 & & 2 & & 3 & & 4 & & 5
\end{array}
$$

There is a bijection between monotone triangles of order n and alternating sign matrices of order n (see [Weis24]).

Solutions to Other Related Problems

1. Find the connection with a distribution of $m + n - k$ signs for a pair of combinatorial objects.
2. Consider the coefficient at x^{n+1} in the decomposition $x \left(\frac{x^j}{(1-x)^{j+1}} \right) \left(\frac{x^{k-j}}{(1-x)^{k-j+1}} \right) = \frac{x^{k+1}}{(1-x)^{k+2}}$, and use the representation $\frac{x^l}{(1-x)^{l+1}} = \sum_{p=0}^{\infty} \binom{p}{l} x^p$.
3. Take $j = k$.
4. Take $j = k - 1$.
5. The left-hand side of the identity is the number of ways to select some subset of the set $\{1, 2, \ldots, n\}$ which has at least q elements and "mark" in this subset q elements. The right-hand side gives the same number of ways since there are exactly $\binom{n}{q}$ ways to choose a set of q "marked" elements and possibly add to them some additional elements, which can be done in 2^{n-q} ways.
6. Prove the formula $f_a(p+1) = \sum_{x=0}^{a} (p+1)^{x+1} f_{a-x}(p)$, where $f_a(p) = P_{2a}(p)$.
7. See [Jablo01]. See also a similar proof for the sequence $S(n, k)$ in [Deza24].
8. The statement can be checked by direct computation.
9. Use the decomposition $(-1)^k \binom{z}{k} = \binom{-z+k-1}{k} = \frac{1}{\Gamma(-z)} \frac{1}{(k+1)^{z+1}} \prod_{j=k+1} \frac{(1+\frac{1}{j})^{-z-1}}{1 - \frac{z+1}{j}}$.
10. (a) These simple bounds follow from the relation $4^n = (1+1)^{2n} = \sum_{k=0}^{2n} \binom{2n}{k}$.

11. Use *Legendre's formula* (see Chapter 4): $\nu_p(n!) = \sum_{i=1}^{\infty} \lfloor \frac{n}{p^i} \rfloor$.
12. In fact, $\frac{1}{\sqrt{1-4x}} = \sum_{n=0}^{\infty} \binom{2n}{n} x^n = 1 + 2x + 6x^2 + 20x^3 + 70x^4 + 252x^5 + \cdots$. This decomposition can be proved using the binomial series and the equality $\binom{2n}{n} = (-1)^n 4^n \binom{-1/2}{n}$, where $\binom{-1/2}{n}$ is a generalized binomial coefficient.
13. It follows from the Stirling formula.
 More precisely, $\binom{2n}{n} = \frac{4^n}{\sqrt{\pi n}} \left(1 - \frac{1}{8n} + \frac{1}{128n^2} + \frac{5}{1024n^3} + O(n^{-4})\right)$.
14. Compare with a similar property of the binomial coefficients. See, for example, [Bond93].
15. It follows from the construction of the pyramid. See, for example, [Bond93].
16. Recall that the numbers in Pascal's pyramid can be found in the trinomial distribution. In general, Pascal's s-simplex is a multi-dimensional arrangement of the coefficients of the decomposition $(x_1 + x_2 + \cdots + x_s)^n$, $n = 0, 1, 2, \ldots$.
17. See [Jablo01] for the proof of a similar property of the binomial coefficients. See also a similar proof for the sequence $S(n,k)$ in [Deza24].
18. See, for example, [ChRe14].
19. See, for example, [Coke03].
20. In fact, it holds that $P_n(x) = 2^{-n} \sum_{k=0}^{\lfloor n/2 \rfloor} (-1)^k \binom{n-k}{k} \binom{2n-2k}{n-k} x^{n-2k}$.
21. That is, $\begin{vmatrix} S_1 & S_2 & \cdots & S_n \\ S_2 & S_3 & \cdots & S_{n+1} \\ \vdots & \vdots & \ddots & \vdots \\ S_n & S_{n+1} & \cdots & S_{2n-1} \end{vmatrix} = 2^{\frac{n(n+1)}{2}}$.

 For example, $|2| = 2 = 2^1$; $\begin{vmatrix} 2 & 6 \\ 6 & 22 \end{vmatrix} = 8 = 2^3$; $\begin{vmatrix} 2 & 6 & 22 \\ 6 & 22 & 90 \\ 22 & 90 & 394 \end{vmatrix} = 64 = 2^6$. See [EuFu05].
22. The Wedderburn–Etherington numbers are an integer sequence named after Ivor Malcolm Haddon Etherington (1908–1994) and Joseph Wedderburn (1882–1948). The first few numbers in the sequence are $0, 1, 1, 1, 2, 3, 6, 11, 23, 46, \ldots$ (sequence A001190 in the OEIS).

In terms of the interpretation of these numbers as counting rooted binary trees with n leaves, the summation in the recurrence counts the different ways of partitioning these leaves into two subsets and of forming a subtree with each subset as its leaf. The formula for even values of n is slightly more complicated than the formula for odd values in order to avoid double counting trees with the same number of leaves in both subtrees.

23. One way to explain this recurrence is to partition the $T(n)$ connection patterns of the n subscribers to a telephone system into the patterns in which the first person is not calling anyone else and the patterns in which the first person is making a call. There are $T(n-1)$ connection patterns in which the first person is disconnected, explaining the first term of the recurrence. If the first person is connected to someone, there are $n-1$ choices for that person and $T(n-2)$ patterns of connection for the remaining $n-2$ people, explaining the second term of the recurrence.

24. In each term of the first sum, k gives the number of matched pairs, the binomial coefficient $\binom{n}{2k}$ counts the number of ways of choosing the $2k$ elements to be matched, and the double factorial $(2k-1)!! = \frac{(2k)!}{2^k k!}$ is the product of the odd integers up to its argument and counts the number of ways of completely matching the $2k$ selected elements.

It follows from the summation formula and Stirling's approximation that, asymptotically, $T(n) \sim \left(\frac{n}{e}\right)^{n/2} \frac{e^{\sqrt{n}}}{(4e)^{1/4}}$. Moreover, the exponential generating function of the telephone numbers is $e^{\left(\frac{x^2}{2}+x\right)}$ (see [Deza24]).

Bibliography

[Abra40]   Abramov, A.N. *Amazing Numbers*. M.-L.: Childish Publishing House, 1940.
[Abra74]   Abramovitch, V. Bernoulli numbers. *Kvant*, 1974, 6, 10–14.
[AbSt72]   Abramowitz, M. and Stegun, I.A. *Handbook of Mathematical Functions with Formulas, Graphs, and Mathematical Tables*. New York: Dover Publications, 1972.
[ABP06]   Ackerman, E., Barequet, G. and Pinter, R.Y. On the number of rectangulations of a planar point set. *Journal of Combinatorial Theory, Series A*, 2006, 113, 1072–1091.
[Adam97]   Adamchik, V. On Stirling numbers and Euler sums. *Journal of Computational and Applied Mathematics*, 1997, 79, 119–130.
[Aign82]   Aigner, M. *Combinatorial Theory*. Moscow: Mir, 1982.
[Aign98]   Aigner, M. Motzkin numbers. *European Journal of Combinatorics*, 1998, 19, 663–675.
[Alte71]   Alter, R. Some remarks and results on Catalan numbers. *Proceedings of the 2nd Louisiana Conference on Combinatorics, Graph Theory and Computing*, 1971, 109–132.
[AlKu73]   Alter, R. and Kubota, K.K. Prime and prime power divisibility of Catalan numbers. *Journal of Combinatorial Theory*, 1973, A 15, 243–256.
[Ande03]   Anderson, D.A. *Discrete Mathematics and Combinatorics*. Moscow: Williams, 2003.
[Andr79]   André, D. Developpements de séc x et de tang x. *Comptes rendus de l'Académie des sciences*, 1879, 88, 965–967.

[Andr98]   Andrews, G.E. *The Theory of Partitions.* Cambridge: Cambridge University Press, 1998.
[Apos86]   Apostol, T.M. *Introduction to Analytic Number Theory.* New-York: Springer-Verlag, 1986.
[BaCo87]   Ball, W.W.R. and Coxeter, H.S.M. *Mathematical Recreations and Essays*, 13th edn. New York: Dover, 1987.
[BPS91]   Barcucci, E., Pinzani, R. and Sprugnoli, R. The Motzkin family. *Pure Mathematics and Applications Series A*, 1991, 2, 249–279.
[Bell34]   Bell, E.T. Exponential numbers. *The American Mathematical Monthly*, 1934, 41, 411–419.
[Bell38]   Bell, E.T. The iterated exponential integers. *Annals of Mathematics*, 1938, 39, 539–557.
[Berg71]   Berge, C. *Principles of Combinatorics.* Cambridge: Academic Press, 1971.
[Bern99]   Bernhart, F.R. Catalan, Motzkin, and Riordan numbers. *Discrete Mathematics*, 1999, 204(1–3), 73–112.
[BeSl95]   Bernstein, M. and Sloane, N.J.A. Some canonical sequences of integers. *Linear Algebra and Its Applications*, 1995, arXiv:math/0205301. doi:10.1016/0024-3795(94)00245-9.
[Bhar00]   Bhargava, M. The factorial function and generalizations. *The American Mathematical Monthly*, 2000, 107(9), 783–799.
[BiBa70]   Birkhoff, G. and Bartee, T.C. *Modern Applied Algebra.* McGraw Hill Text, 1970.
[Bond93]   Bondarenko, B.A. *Generalized Pascal's Triangles and Pyramids, Their Fractals, Graphs, and Applications.* Fibonacci Association, 1993.
[Boro85]   Boro, V. *Live Numbers.* Moscow: Mir, 1985.
[BoRe14]   Bowman, D. and Regev, A. Counting symmetry: Classes of dissections of a convex regular polygon. *Advances in Applied Mathematics*, 2014, 5, 35–55.
[BoBa03]   Borwein, J. and Bailey, D. *Mathematics by Experiment: Plausible Reasoning in the 21-st Century.* Wellesley, MA: A K Peters, 2003.
[Bron01]   Bronstein, E.M. Generating functions. *Sorosovsky Educational Journal*, 2001, 7(2), 10.
[Brua97]   Brualdi, R.A. *Introductory Combinatorics*, 4th edn. New York: Elsevier, 1997.
[Buch09]   Buchstab, A.A. *Number Theory.* Moscow: Nauka, 2009.

Bibliography 449

[Burn16]   Burns, R. Structure and asymptotics for Motzkin numbers modulo small primes using automata, 2016. arXiv:1612.08146.
[Call09]   Callan, D. A combinatorial survey of identities for the double factorial, 2009. arXiv:0906.1317.
[Camp84] Campbell, D. The computation of Catalan numbers. *Mathematics Magazine*, 1984, 57, 195–208.
[Cata44]   Catalan, E. Note Extraite d'une Lettre Adressée à l'Éditeur. *Journal für die Reine und Angewandte Mathematik*, 1844, 27, 192.
[Cata74]   Catalan, E. Question 1135. *Nouvelles Annales de Mathématiques: Journal des Candidats aux Écoles Polytechnic et Normale, Series 2*, 1874, 13, 207.
[ChRe14]   Chen, R.X.F. and Reidys, C.M. Narayana polynomials and some generalizations, 2014, arXiv:1411.2530.
[CWZ20]   Chen, X., Wang, Y. and Zheng, S.-N. Analytic properties of combinatorial triangles related to Motzkin numbers. *Discrete Mathematics*, 2020, 343(12), 112–133.
[ChDe20]   Chebotarev, P. and Deza, E. Hitting time quasi-metric and its forest representation. *Optimization Letters*, 2020, 14, 291–307.
[ChWi07]   Chiswell, I. and Wilfrid, H. *Mathematical Logic*. Oxford: Oxford University Press, 2007.
[CYY20]   Choi, H., Yeh, Y.-N. and Yoo, S. Catalan-like number sequences and Hausdorff moment sequences. *Discrete Mathematics*, 2020, 343(5), arXiv:1809.07523.
[ChMo75]   Chorneyko, I.Z. and Mohanty, S.G. On the enumeration of certain sets of planted trees. *Journal of Combinatorial Theory, Series B*, 1975, 18, 209–221.
[Chu87]   Chu, W. A new combinatorial interpretation for generalized Catalan numbers. *Discrete Mathematics*, 1987, 65, 91–94.
[ClKi08]   Claesson, A. and Kitaev, S. Classification of bijections between 321- and 132-avoiding permutations. *Séminaire Lotharingien de Combinatoire*, 2008, 60. Article B60d. elibm.org/ft/10009522003.
[Cofm75]   Cofman, A. *Introduction to Applied Combinatorics*. Moscow: Nauka, 1975.
[Coke03]   Coker, C. Enumerating a class of lattice paths. *Discrete Mathematics*, 2003, 271(1–3), 13–28.
[Coke04]   Coker, C. A family of Eigensequences. *Discrete Mathematics*, 2004, 282 (1–3), 249–250.

[Comt74]   Comtet, L. *Advanced Combinatorics: The Art of Finite and Infinite Expansions*. Netherlands: Reidel, 1974.

[CoGu96]   Conway, J.H. and Guy, R.K. *The Book of Numbers*. New York: Springer-Verlag, 1996.

[CoRo96]   Courant, R. and Robbins, H. *What Is Mathematics? An Elementary Approach to Ideas and Methods*, 2nd edn. Oxford: Oxford University Press, 1996.

[Dave47]   Davenport, H. The geometry of numbers. *Mathematical Gazette*, 1947, 31, pp. 206–210.

[Dave99]   Davenport, H. *The Higher Arithmetic: An Introduction to the Theory of Numbers*, 8th edn. Cambridge: Cambridge University Press, 2008.

[Dede63]   Dedekind, R. *Essays on the Theory of Numbers*. Cambridge: Cambridge University Press, 1963.

[DeZa80]   Dershowitz, N. and Zaks, S. Enumeration of ordered trees. *Discrete Mathematics*, 1980, 31, 9–28.

[DeMh20]   Deza, E. and Mhanna, B. On special properties of some quasi-metrics. *Chebyshevskii sbornik*, 2020, 21(1), 145–164.

[DeMo10]   Deza, E. and Model, D. *Elements of Discrete Mathematics*. Moscow: URSS, 2010.

[DeDe12]   Deza, E. and Deza, M.M. *Figurate Numbers*. World Scientific, 2012.

[DeKo13]   Deza, E. and Kotova, L. *Collection of Problems on Number Theory*. Moscow: URSS, 2013.

[Deza17]   Deza, E. *Special Positive Integer Numbers*. Moscow: URSS, 2017.

[Deza18]   Deza, E. *Special Combinatorial Numbers*. Moscow: URSS, 2018.

[Deza21]   Deza, E. *Mersenne Numbers and Fermat Numbers*. World Scientific, 2021.

[Deza23]   Deza, E. *Prefect and Amicable Numbers*. World Scientific, 2023.

[Deza24]   Deza, E. *Stirling Numbers*. World Scientific, 2024.

[Dick05]   Dickson, L.E. *History of the Theory of Numbers*. New York: Dover, 2005.

[DoRe69]   Dobson, A.J. and Rennie, B.C. On Stirling numbers of the second kind. *Journal of Combinatorial Theory*, 1968, 7, 212–214.

[DZHKKP90]   Dokin, V.N., Zhukov, V.D., Kolokolnikova, N.A., Kuzmin, O.V., and Platonov, M.L. *Combinatorial Numbers and Polynomials in Models of Discrete Distributions*. Irkutsk: Irkutsk University, 1990.

[DoSh77]　　Donaghey, R. and Shapiro, L.W. Motzkin numbers. *Journal of Combinatorial Theory, Series A*, 1977, 23(3), 291–301.

[Dorr65]　　Dörrie, H. *100 Great Problems of Elementary Mathematics: Their History and Solutions*. New York: Dover, 1965.

[Dutk91]　　Dutka, J. The early history of the factorial function. *Archive for History of Exact Sciences*, 1991, 43(3), 225–249.

[EdGr16]　　Edwards, S. and Griffiths, W. A combinatorial identity related to cross polytope numbers. *The Fibonacci Quarterly*, 2016, 154, 253–258.

[EgGa88]　　Eggleton, R.B. and Guy, R.K. Catalan strikes again! How likely is a function to be convex? *Mathematics Magazine*, 1988, 61, 211–219.

[Erus00]　　Erusalemskii, J.M. *Discrete Mathematics*. Moscow: University Book, 2000.

[EuFu05]　　Eu, S.-P. and Fu, T.-S. A simple proof of the Aztec diamond theorem. *Electronic Journal of Combinatorics*, 2005, 12, 1077–8926.

[Eule48]　　Euler, L. *Introductio in analysin infinitorum*, 1, 2. Marc Michel Bousquet, 1748.

[FeBa17]　　Feng, Q. and Bai-Ni, G. Integral representations of the Catalan numbers and their applications. *Mathematics*, 2017, 5(3), 40.

[Ficht01]　　Fichtenholtz, G.M. *Course of Differential and Integral Calculus* (in three volumes). Moscow: Phtismatlit, 2001.

[FlSe09]　　Flajolet, P. and Sedgewick, R. *Analytic Combinatorics*. Cambridge: Cambridge University Press, 2009.

[Ford61]　　Forder, H.G. Some problems in combinatorics. *The Mathematical Gazette*, 1961, 45, 199–201.

[Gard76]　　Gardner, M. Catalan numbers: An integer sequence that materializes in unexpected places. *Scientific American*, 1976, 234, 120–125.

[Gard61]　　Gardner, M. *Second Book of Mathematical Puzzles and Diversions*. New York: Freeman, 1961.

[Gard76]　　Gardner, M. Catalan numbers: An integer sequence that materializes in unexpected places. *Scientific American*, 1976, 234, 120–125.

[Gard88]　　Gardner, M. *Time Travel and Other Mathematical Bewilderments*. New York: Freeman, 1988.

[Gard89]　　Gardner, M. *Penrose Tiles to Trapdoor Ciphers*. New York: Freeman, 1989.

452   Catalan Numbers

[GaSa92]   Gavrilov, G.P. and Sapozhenko, A.V. *Tasks and Exercises on the Course of Discrete Mathematics*. Moscow: Nauka, 1992.
[Gelf98]   Gelfand, I.M. *Lectures on Linear Algebra*. Courier Dover Publications, 1998.
[Gelf59]   Gelfond, A.O. *Calculus of Finite Differences*. Moscow: GIFML, 1959.
[Gess92]   Gessel, I.M. Super ballot numbers. *Journal of Symbolic Computation*, 1992, 14, 179–194.
[GeXi05]   Gessel, I.M. and Xin, G. A combinatorial interpretation of the numbers $6(2n)!/n!(n+2)!$. *Journal of Integer Sequences*, 2005, 8, Article 05.2.3, arXiv:math/0401300.
[Glai77]   Glaisher, J.W.L. On the product $1^1 \cdot 2^2 \cdot 3^3 \cdot \ldots \cdot n^n$. *Messenger of Mathematics*, 1877, 7, 43–47.
[GKP94]   Graham, R.L., Knuth, D.E. and Patashnik, O. *Concrete Mathematics: A Foundation for Computer Science*, 2nd edn. Reading, MA: Addison-Wesley, 1994.
[Goul85]   Gould, H.W. *Bell and Catalan Numbers: Research Bibliography of Two Special Number Sequences*, 6th edn. Morgantown, WV: Math Monongliae, 1985.
[GoJa83]   Goulden, I.P. and Jackson, D.M. *Combinatorial Enumeration*. New York: Wiley, 1983.
[GGL95Û]   Graham, R., Grötschel, M. and Lovász, L. *Handbook of Combinatorics*. Elsevier Science B.V., 1995.
[GPP01]   Guibert, O., Pergola, E. and Pinzani, R. Vexillary involutions are enumerated by Motzkin numbers. *Annals of Combinatorics*, 2001, 5(2).
[GuJa90]   Gulden, J. and Jacson, D. *Enumerating Combinatorics*. Moscow: Nauka, 1990.
[Guy58]   Guy, R.K. Dissecting a polygon into triangles. *Bulletin of the Malaysian Mathematical Sciences Society*, 1958, 5, 57–60.
[Hagg04]   Haggarty, R. *Discrete Mathematics for Programmers*. Moscow: Technosphera, 2004.
[Hall70]   Hall, M. *Combinatorics*. Moscow: Mir, 1970.
[Hara03]   Harari, F. *Graph Theory*. Moscow: Mir, 2003.
[HaPa77]   Harari, F. and Palmer, E. *Graph Enumeration*. Moscow: Mir, 1977.
[HaWr79]   Hardy, G.H. and Wright, E.M. *An Introduction to the Theory of Numbers*, 5th edn. Oxford: Clarendon Press, 1979.
[HiPe91]   Hilton, P. and Pedersen, J. Catalan numbers, their generalization, and their uses. *The Mathematical Intelligencer*, 1991, 13, 64–75.

[Hogg69]	Hoggatt Jr., V.E. *The Fibonacci and Lucas Numbers.* Boston, MA: Houghton Mifflin, 1969.
[Hons73]	Honsberger, R. *Mathematical Gems I.* Washington, DC: Mathematical Association of America, 1973.
[Hons85]	Honsberger, R. *Mathematical Gems III.* Washington, DC: Math. Assoc. Amer., 1985.
[Hons91]	Honsberger, R. *More Mathematical Morsels.* Washington: Mathematical Association of America, 1991.
[Jablo01]	Jablonskii, S.V. *Introduction to Discrete Mathematics.* Moscow: Higher School, 2001.
[IrRo90]	Ireland, K. and Rosen, M. *A Classical Introduction to Modern Number Theory.* New York: Springer-Verlag, 1990.
[Kara83]	Karatsuba, A.A. *Basic Analytic Number Theory.* Moscow: Nauka, 1983.
[Klar70]	Klarner, D.A. Correspondences between plane trees and binary sequences. *Journal of Combinatorial Theory*, 1970, 9, 401–411.
[Knot77]	Knott, G.D. A numbering system for binary trees. *Communications of the ACM*, 1977, 20(2), 113–115.
[Knut76]	Knuth, D.E. *Mathematics and Computer Science. Coping with Finiteness.* Moscow: Nauka, 1976.
[Knut97]	Knuth, D.E. *The Art of Computer Programming, Vol. 1: Fundamental Algorithms,* 3rd edn. Reading, MA: Addison-Wesley, 1997.
[Kord95]	Kordemskii, B.A. *Great Lives in Mathematics.* Moscow: Prosvestcheniye, 1995.
[KoKo77]	Korn, G. and Korn, T. *Handbook of Mathematics (for Scientists and Engineers).* Moscow: Nauka, 1977.
[KoSa06]	Koshy, T. and Salmassi, M. Parity and primality of Catalan numbers. *The College Mathematics Journal*, 2006, 37(1), 52–53.
[Kost82]	Kostrikin, A. *Introduction to Algebra.* Berlin: Springer Verlag, 1982.
[KLS01]	Krízek, M., Luca, F. and Somer, L. *17 Lectures on Fermat Numbers: From Number Theory to Geometry.* CMS Books in Mathematics, Vol. 9. New York: Springer-Verlag, 2001.
[KuAD88]	Kuznetsov, O.P. and Adelson-Velsky, G.B. *Discrete Mathematics for Engineers.* Moscow: Energoatomizdat, 1988.
[Lah54]	Lah, I. A new kind of numbers and its application in the actuarial mathematics. *Boletim do Instituto dos Actuários Portugueses*, 1954, 9, 7–15.

454 *Catalan Numbers*

[Land94]   Lando, S.K. *Lectures on Combinatorics*. Moscow: MCCME, 1994.
[Land02]   Lando, S.K. *Lectures on Generating Functions*. Moscow: MCCME, 2002.
[Lege79]   Legendre, A.-M. *Théorie des Nombres*, 4th edn. Paris: A. Blanchard, 1979.
[LiNi96]   Lidl, R. and Niederreiter, H. *Finite Fields*. Cambridge: Cambridge University Press, 1996.
[Line86]   Lines, M.E. *A Number for Your Though*. Bristol: Adam Hilger, 1986.
[LiWi92]   Lint, J.H. and Wilson, R.M. *Course in Combinatorics*. Cambridge: Cambridge University Press, 1992.
[LeLi83]   Le Lionnais, F. *Les Nombres Remarquables*. Paris: Hermann, 1983.
[Lips88]   Lipsky, V. *Combinatorics for Programmers*. Moscow: Mir, 1988.
[Litz63]   Litzman, V. *Fun and Entertaining about Numbers and Figures*. Moscow: Fizmatgiz, 1963.
[Mada79]   Madachy, J.S. *Madachy's Mathematical Recreations*. New York: Dover, 1979.
[Malc86]   Malcolm, E. *A Number for Your Thoughts: Facts and Speculations about Number from Euclid to the Latest Computers*. CRC Press, 1986.
[Male11]   Malenfant, J. Finite, closed-form expressions for the partition function and for Euler, Bernoulli, and Stirling numbers, 2011, arXiv:1103.1585v6.
[MaSc15]   Mansour, T. and Schork, M. *Commutation Relations, Normal Ordering, and Stirling Numbers*. CRC Press, 2015.
[MaSu08]   Mansour, T. and Sun, Y. Identities involving Narayana polynomials and Catalan numbers, 2008. arXiv:0805.
[Mark0]   Markushevich, A.I. *Return Sequences*. Leningrad: Technical and Theoretical Literature, 1950.
[Mazu10]   Mazur, D.R. *Combinatorics. A Guided Tour*. The Mathematical Association of America, 2010.
[MSY96]   Millar, J., Sloane, N.J.A. and Young, N.E. A new operation on sequences: The Boustrouphedon transform. *Journal of Combinatorial Theory, Series A*, 1996, 76(1), 44–54.
[MSC96]   Mitrinovic, D.S., Sandor, J. and Crstici, B. *Handbook on Number Theory*. Kluwer Academic Publishers, 1996.
[Motz48]   Motzkin, T.S. Relations between hypersurface cross ratios, and a combinatorial formula for partitions of a polygon, for

permanent preponderance, and for non-associative products. *Bulletin of the American Mathematical Society*, 1948, 54, 352–360.

[MaWo00] Mays, M.E. and Wojciechowski, J. A determinant property of Catalan numbers. *Discrete Mathematics*, 2000, 211, 125–133.

[NeSz22] Németh, L. and Szalay, L. Properties of Motzkin triangle and t-generalized Motzkin sequences. *Aequationes Mathematicae*, 2022, https://doi.org/10.1007/s00010-021-00864-0.

[OLBC10] Olver, F.W.J., Lozier, D.M., Boisvert, R.F. and Clark, C.W. *Handbook of Mathematical Functions*. Cambridge: Cambridge University Press, 2010.

[Ore48] Ore, O. *Number Theory and Its History*. Dover Publications, Inc., 1948.

[Ore80] Ore, O. *Graph Theory*. Moscow: Nauka, 1980.

[OsJe15] Oste, R. and Van der Jeugt, J. Motzkin paths, Motzkin polynomials and recurrence relations. *The Electronic Journal of Combinatorics*, 2015, 22(2), 2–8.

[Otte48] Otter, R. The number of trees. *Annals of Mathematics, Second Series*, 1948, 49(3), 583–599.

[Pasc54] Pascal, B. *Traité du triangle arithmétique, avec quelques autres petits traitez sur la mesme matière*. Paris, 1654.

[PeSk03] Pemmaraju, S. and Skiena, S. *Computational Discrete Mathematics: Combinatorics and Graph Theory with Mathematica*. Cambridge: Cambridge University Press, 2003.

[PWZ96] Petkovšek, M., Wilf, H.S. and Zeilberger, D. $A = B$. London: CRC Press, 1996.

[Plat80] Platonov, M.L. *Combinatorial Numbers*. Irkutsk: Irkutsk University, 1980.

[Plou92] Plouffe, S. 1031 *Generating Functions and Conjectures*. Montréal: Université du Québec à Montréal, 1992.

[Poly56] Pólya, G. On picture-writing. *The American Mathematical Monthly*, 1956, 63, 689–697.

[Prim24] The Prime Pages (prime number research, records and resources). https://primes.utm.edu/ (10.03.2024).

[Pudw24] Pudwell, L. Catalan numbers and permutations. Accepted to appear in *Mathematics Magazine*.

[RaTo57] Rademacher, H. and Toeplitz, O. *The Enjoyment of Mathematics: Selections from Mathematics for the Amateur*. Princeton: Princeton University Press, 1957.

[Ribe96] Ribenboim, P. *The New Book of Prime Numbers Records*. New York: Springer-Verlag, 1996.

[Rior79] Riordan, J. *Combinatorial Identities.* New York: Wiley, 1979.

[Rior80] Riordan, J. *An Introduction to Combinatorial Analysis.* New York: Wiley, 1980.

[RoTe09] Roberts, F.S. and Tesman, B. *Applied Combinatorics.* CRC Press, 2009.

[Roge78] Rogers, D.G. Pascal triangles, Catalan numbers and renewal arrays. *Discrete Mathematics,* 1978, 22, 301–310.

[Rose18] Rosen, K.H. (ed.) *Handbook of Discrete and Combinatorial Mathematics.* CRC Press, 2018.

[RoYa13] Rowland, E. and Yassawi, R. Automatic congruences for diagonals of rational functions, 2013–2014, arXiv:1310.8635.

[Rybn82] Rybnikov, K.A. *Combinatorial Analysis. Tasks and Exercises.* Moscow: Nauka, 1982.

[Rybn85] Rybnikov, K.A. *Introduction to Combinatorial Analysis.* Moscow: Moscow State University, 1985.

[Robb55] Robbins, H. A remark on Stirling's formula. *The American Mathematical Monthly,* 1955, 62(1), 26–29.

[Schw97] Schwermer, J. Motzkin, Theodor Samuel. *Neue Deutsche Biographie,* 1997, 231–249.

[Sand78] Sands, A.D. On generalized Catalan numbers. *Discrete Mathematics,* 1978, 21, 218–221.

[Sier64] Sierpiński, W. *Elementary Theory of Numbers.* Warszawa, 1964.

[Sing78] Singmaster, D. *An Elementary Evaluation of the Catalan Numbers.* Amer. Math. Monthly, 1978, 85, 366–368.

[SlPl95] Sloane, N.J.A. and Plouffe, S. *The Encyclopedia of Integer Sequences.* San Diego: Academic Press, 1995.

[Sloa24] Sloane, N.J.A., *et al.* On-line encyclopedia of integer sequences. http://oeis.org/ (31.03.2024).

[Schm10] Schmidt, M.D. Generalized j-factorial functions, polynomials, and applications. *Journal of Integer Sequences,* 2010, 3, Article 10.6.7, 1–54.

[Smit84] Smith, D.E. *A Source Book in Mathematics.* New York: Dover, 1984.

[Spru94] Sprugnoli, R. Riordan arrays and combinatorial sums. *Discrete Mathematics,* 1994, 132 (1–3), 267–290.

[Stan97] Stanley, R.P. *Enumerative Combinatorics,* Vols. 1, 2. Cambridge: Cambridge University Press, 1997.

[Stan10] Stanley, R.P. A survey of alternating permutations. *Combinatorics and Graphs.* Contemporary Mathematics,

	Vol. 531. Providence, RI: American Mathematical Society, 2010.
[Stan13]	Stanley, R.P. *Catalan Addendum to Enumerative Combinatorics*, Vol. 2, 2013. https://math.mit.edu/~rstan/ec/catadd.pdf.
[Stan15]	Stanley, R.P. *Catalan Numbers*. Cambridge: Cambridge University Press, 2015.
[StWa78]	Stein, P.R. and Waterman, M.S. On some new sequences generalizing the Catalan and Motzkin numbers. *Discrete Mathematics*, 1978, 26, 261–272.
[Stra16]	Strang, G. *Linear Algebra and Its Applications*, 5th edn. Wellesley-Cambridge Press, 2016.
[Stru87]	Struik, D.J. *A Concise History of Mathematics*. Dover Publications, 1987.
[Sula01]	Sulanken, R.A. Bijective recurrences for Motzkin paths. *Advances in Applied Mathematics*, 2001, 27, 627–640.
[Sun14]	Sun, Z.-W. Congruences involving generalized central trinomial coefficients. *Science China Mathematics*, 2014, 57, 1375–1400.
[Temm93]	Temme, N.M. Asymptotic estimates of Stirling numbers. *Studies in Applied Mathematics*, 1993, 89(3), 31–40.
[Tsan10]	Tsang, C. *Fermat Numbers*. Washington: University of Washington, 2010.
[Uspe76]	Uspensky, V.A. *Pascal's Triangle: Certain Applications of Mechanics to Mathematics*. Moscow: Mir, 1976.
[Vars91]	Vardi, I. *Computational Recreations in Mathematica*. Redwood City, CA: Addison-Wesley, 1991.
[Vile14]	Vilenkin, N.I. *Combinatorics*. Moscow: Academic Press, 2014.
[VVV06]	Vilenkin, N.Ia., Vilenkin, A.N. and Vilenkin, P.A. *Combinatorics*. Moscow: MCCME, 2006.
[Voro61]	Vorob'ev, N.N. *Fibonacci Numbers*. New York: Blaisdell, 1961.
[VoKu84]	Voronin, S. and Kulagin, A. Method of generating functions. *Quant*, 1984, 5, 12–19.
[WaZh15]	Wang, Y. and Zhang, Z.-H. Combinatorics of generalized Motzkin numbers. *Journal of Integer Sequences*, 2015, 18(2), Article 15.2.4, 1–12.
[Weis24]	Weisstein, E.W. MathWorld — A Wolfram Web Resource. http://mathworld.wolfram.com/ (31.03.2024).
[Well86]	Wells, D. *The Penguin Dictionary of Curious and Interesting Numbers*. Middlesex: Penguin Books, 1986.

[Went99] Wentzel, E.S. *Probability Theory*. Moscow: Higher School, 1999.
[West95] West, J. Generating trees and the Catalan and Schröder numbers. *Discrete Mathematics*, 1995, 146(1–3), 247–262.
[Wiki24] Wikipedia, the free Encyclopedia. http://en.wikipedia.org (31.03.2024).
[Youn00] Young, A. On quantitative substitutional analysis. *Proceedings of the London Mathematical Society, Series 1*, 1900, 33(1), 97–145.
[ZhQi17] Zhao, J.-L. and Qi, F. Two explicit formulas for the generalized Motzkin numbers. *Journal of Inequalities and Applications*, 2017. Article 44, 1–24.

Index

A

associated Stirling numbers of the first kind, 413
associated Stirling numbers of the second kind, 413

B

ballot numbers, 413
Bell number, 21, 29, 157, 203–204, 338, 413
Bernoulli numbers, 413
big Schröder number, 250, 414
binary tree, 19, 149
binomial coefficients, 7, 414
binomial theorem, 72

C

Catalan numbers, 133, 414
Catalan's problem, 139
centered m-gonal numbers, 414
centered octahedral numbers, 279, 414
centered square numbers, 279, 414
central Delannoy numbers, 286, 414
complementary Bell numbers, 346, 414

D

Delannoy number, 276, 414
derangements, 39
double factorial, 356, 414

E

Euler numbers, 207, 414
Euler zigzag numbers, 37, 203, 206, 414
Eulerian number, 35
Eulerian numbers of the first kind, 414
Eulerian numbers of the second kind, 414

F

factorial, 348, 415
factorial numbers, 348, 415
falling factorial, 329, 355, 415
Fermat numbers, 115, 415
Fibonacci numbers, 40, 415
Fubini number, 21, 347, 415
Fuss numbers, 416
Fuss–Catalan numbers, 210, 415

G

generalized binomial coefficient, 239, 356

generalized Delannoy numbers, 296, 415
generalized harmonic number, 334, 415
generalized Motzkin numbers, 248
generalized Pascal's triangles, 119
generalized Stirling numbers of the first kind, 415
generalized Stirling numbers of the second kind, 415
gigantic prime, 411, 415

H

harmonic numbers, 334–335, 415
Hipparchus numbers, 270, 415
hyperfactorial, 359, 415
hyperoctahedral numbers, 292, 415
hyperoctahedron numbers, 291
hypertetrahedral numbers, 109, 416

I

involution numbers, 357, 416

K

(k,d)-Catalan numbers, 416
k-Catalan numbers, 416
k-dimensional hypercube numbers, 416
k-dimensional hyperoctahedron numbers, 416

L

Lah numbers, 416
large Schröder number, 250, 416
little Schröder number, 269–270, 416
Lobb number, 208, 416
Lucas numbers, 41, 416

M

m-gonal numbers, 416
Mahonian number, 36, 416
megaprime, 412, 416
Mersenne numbers, 112, 203, 335, 416

Motzkin numbers, 215, 416
multifactorials, 358, 417
multinomial coefficients, 9, 417

N

Narayana numbers, 298, 417
Narayana numbers of order m, 417
Narayana's cow number, 323, 417

O

octahedral numbers, 292, 417
ordered Bell number, 21, 347, 417
ordered Stirling numbers of the second kind, 417

P

Pascal's pyramid, 126
Pascal's triangle, 61, 199, 335, 414
Pentatope numbers, 109, 417
perfect numbers, 113, 417
plane binary tree, 150
plane rooted quasi-tree, 228
plane rooted trees, 155
planted binary quasi-tree, 226
primorial, 359, 417

Q

quasi-tree, 226

R

Raney numbers, 210, 415, 417
reduced Stirling numbers of the second kind, 417
Riordan numbers, 417
rising factorial, 355, 417
Rodriguez numbers, 417, 421
rooted tree, 18, 149

S

Schröder numbers, 250, 418
Schröder–Hipparchus numbers, 270, 418
secant numbers, 414
second kind, 203

semifactorial, 356, 418
square numbers, 418
Stirling number, 29, 417
Stirling number of the first kind, 104, 203, 327, 413, 418
Stirling number of the second kind, 102, 301, 322, 326, 413, 418
subfactorials, 418
super Catalan numbers, 209, 418
superfactorials, 418

T

tangent numbers, 414
telephone numbers, 357, 418
tetrahedral numbers, 108, 418
titanic prime, 411, 418
tree, 17, 149
triangular numbers, 66, 108, 211, 322, 335, 418
triangulo-triangular numbers, 109, 418
tribonacci number, 280–281, 418
trinomial coefficient, 127, 417–418

U

unsigned Stirling number of the first kind, 33, 35, 327, 418
up-down numbers, 37, 206, 414
Uppuluri–Carpenter numbers, 346, 419

W

Ward numbers, 419

Z

zag numbers, 414
zig numbers, 414